OBJECT-ORIENTED SYSTEMS ANALYSIS AND DESIGN

PRENTICE HALL SERIES
IN INFORMATION MANAGEMENT

William King, *Series Editor*

OBJECT-ORIENTED SYSTEMS ANALYSIS AND DESIGN

Ronald J. Norman

San Diego State University

Prentice Hall · Upper Saddle River, NJ 07458

Library of Congress Cataloging-in-Publication Data

Norman, Ronald J.
    Object-oriented systems analysis and design / Ronald J. Norman.
      p.   cm. — (Prentice Hall series in information management)
    Includes bibliographical references and index.
    ISBN 0-13-122946-X
    1. Object-oriented programming (Computer science)   2. System
analysis. 3. System design. I. Title. II. Series.
    (William R. King, series editor)
    QA76.64.N67   1996
    005.1'2—dc20   95-36248
                                              CIP

Editor-in-Chief: *Rich Wohl*
Managing Editor: *Nicholas Radhuber*
Project Manager: *Susan Rifkin*
Manufacturing Buyer: *Paul Smolenski*
Editorial Assistant: *Jane Avery*
Cover Designer: *Karen Salzbach*
Interior Design: *Lorraine Castellano*

©1996 by Prentice-Hall, Inc.
A Simon & Schuster Company
Upper Saddle River, New Jersey 07458

Printed in the United States of America

10 9 8 7 6 5 4

ISBN: 0-13-122946-X

PRENTICE-HALL INTERNATIONAL (UK) LIMITED, *London*
PRENTICE-HALL OF AUSTRALIA PTY. LIMITED, *Sydney*
PRENTICE-HALL CANADA INC., *Toronto*
PRENTICE-HALL HISPANOAMERICANA, S.A., *Mexico*
PRENTICE-HALL OF INDIA PRIVATE LIMITED, *New Delhi*
PRENTICE-HALL OF JAPAN, INC., *Tokyo*
SIMON & SCHUSTER ASIA PTE. LTD., *Singapore*
EDITORA PRENTICE-HALL DO BRASIL, LTDA., *Rio de Janeiro*

To God, who has blessed me beyond belief.
To Caralie, my wife and best friend.
To Merilee, my ambitious and loving daughter.
To Wendy, my easygoing and loving daughter.
To Julie, my fun-loving and loving daughter.
To Dr. John C. Maxwell, my mentor and friend.
I love you all very much.
Thank you for allowing me to stretch to my potential.

Contents in Brief

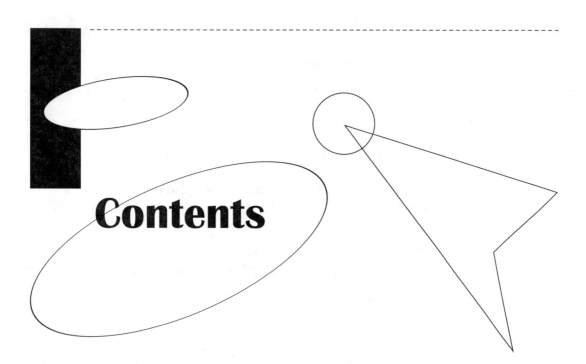

Contents

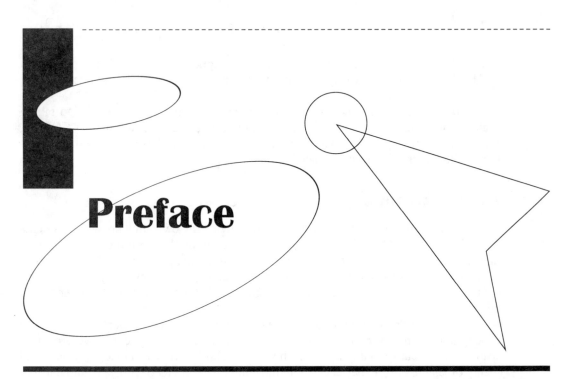

Preface

Information systems. They are everywhere. But you can't touch them, see them, or hold them in the same sense that you can touch a statue, see a sunset, or hold a hamburger in your hand. For the most part, information systems are invisible except for the human-computer interface portion such as a keyboard, joystick, monitor, mouse, or printer.

Business and government rely heavily on information systems to support their missions and goals. In fact it is almost impossible to conduct substantial business these days without "computers." Many people use the word "computers" as a surrogate for the information system on which they are relying. Keeping track of names and addresses, paying bills, playing games, creating term papers, and doing income tax returns are just a few of the information systems for which people use their computer.

The purpose of this book is to be an introductory resource for systems analysis, design, and implementation of information systems. Systems analysis, design, and implementation are, in the broadest sense, the processes used by professional men and women to create or maintain information systems.

THE INTENDED AUDIENCE FOR THIS BOOK

Object-Oriented Systems Analysis and Design is intended to be used in a single-term course or a two-term sequence of courses (e.g., semester or quarter) in information systems development. Four-year universities and colleges, community or junior colleges, and trade-technical/vocational schools that have one or more courses devoted to information systems development or software engineering are perhaps the best candidates for its use. Master's programs that also have one or more courses in information systems development or software engineering would also benefit from this book although the writing style is admittedly geared primarily to undergraduate students.

I recommend that students have basic computer literacy skills prior to taking such a course. In addition, experience with a programming language such as COBOL, Visual (anything), or C/C++ would contribute to a better understanding and appreciation of the information systems development process.

WHY I WROTE THIS BOOK

My sole goal for writing this book is to provide my own students with a systems analysis, design, and implementation resource textbook that would effectively communicate the essence of these activities in the most meaningful way possible. The anonymous feedback that I have received on earlier drafts of this book from my students has encouraged me to pursue its formal publication so that other students may benefit from it as well.

The current generation of students taking computer-related courses has come to expect more than just theory in the classroom because many students have grown up with computers at school and in their homes. In almost every instance, they have had plenty of "hands- on" experiences with the computer and video games. Computer theory without practical hands-on application is a little like peanut butter without jelly or graham crackers without milk—some things are just made for each other.

I have used many good systems analysis and design textbooks over the years, none of which were perfect. This one isn't either. However, in this book, I have tried to address some of the perceived issues that my students consistently bring up as shortcomings in other textbooks. Chief among their complaints are books that:

- are intended to be introductory yet assume a high degree of related background knowledge.
- are too conceptual or theoretical.
- minimize concepts and theories and substitute anecdotal experiences.
- use examples that students have a difficult time relating to.
- claim to be presenting the best systems analysis and design methodology with the implication that everybody should be using it.

A FEW WORDS ABOUT INFORMATION SYSTEMS JARGON

Jargon is a way of life. Understanding of and use of jargon is equated with knowledge of the subject area. Misuse of jargon can distinguish the fake from the real. Doctors, attorneys, accountants, engineers, mechanics, and surfers all have their jargon. Even young people have their own jargon.

Information systems and computer technology have jargon also. Lots of it! The accepted name of the field itself has changed over the years. A few of the accepted (at least at one time) names for this field are: data processing, management information systems, business information systems, information systems, software engineering, systems engineering, and information systems engineering. For purposes of this book these terms are synonymous. Information systems is the preferred one for use throughout this book. Is this the best term? Who knows! My goal is to be consistent throughout the book.

Systems analysis and design is known by a few other terms also, such as software engineering, systems engineering, information systems development, and information systems engineering. You probably noticed that some of these terms were included as accepted terms for this field of study. Not surprising for a field of study as young (about 50 years) and as diverse as this one. I have selected systems analysis and design as the preferred term for use in this book. On rare occasion, due to sentence structure, I may use one of the other terms.

The titles assigned to the people who do systems analysis and design are also diverse. Titles such as systems analyst, analyst/programmer, software engineer, and systems engineer are quite common. This book will use systems analyst as the preferred term.

Finally, there is the ongoing discussion of the most appropriate term used to classify the people who use information systems, such as knowledge worker, user, and customer. Knowledge worker appears to have a very small following these days. Perhaps the oldest term is the word "user." This term conjures up all kinds of thoughts. So, the term "customer" has become popular in recent years. This term has a positive connotation that is good. However, students tend to think of themselves as customers—customers of banks, restaurants, stores, super markets, and even the university. They, in turn, may even wait on customers in their role as employees of a bank, restaurant, store, and so on. Because of this potential confusion, I have chosen the term "user" for this book. When used, the term can refer to a single person or a group of people.

HOW TO USE THIS BOOK

Object-Oriented Systems Analysis and Design is divided into three parts. My personal experience with several other systems analysis and design books suggests that instructors can readily omit sections of chapters or entire chapters and modules to suit their own teaching style and content preferences. Rearranging of chapter and module order to meet instructional goals is also an option. Instructors can do the same with this book.

Part I, Systems Analysis and Conceptual Design, is made up of seven chapters. Chapter 1 presents an introduction to systems analysis and design that introduces the reader to the systems development process, most often referred to as the systems development life cycle. Foundational information related to the systems development process is covered along with additional foundational information related to systems analysts, the people who perform the systems development process. Chapter 2 introduces the reader to feasibility analysis and the requirements determination activities within systems analysis. Chapter 3 presents an object-oriented methodology and model in its entirety. This shows the reader the "big picture" before moving on to necessary details. Chapters 4-7 present the details of the object-oriented methodology and model.

Part II, Physical Design and Implementation, consists of six chapters. Chapter 8 presents an overview of the design portion of systems analysis and design. Chapter 9 focuses on output design. Chapters 10 and 11 address input design and file and database design. Chapter 12 presents software construction and testing concepts followed by presentation of information systems implementation from both a technical and organizational behavior perspective in Chapter 13.

Part III, Miscellaneous Systems Analysis and Design Topics, presents eight supplementary and complementary topics that either enable systems analysis and design or are additional aspects of systems analysis and design. For example, Module A, Information Systems Planning, is an additional aspect of systems analysis and design, while Module G, Communication and Electronic Meetings, presents these topics which contribute in an enabling way to systems analysis and design. Each of the other six modules—Prototyping, CASE, Software Process Improvement, The Systems Development Challenge, Project Management, and Business Process Reengineering—presents topics that are important to the systems analysis and design process. These eight modules are intended to be used at the instructor's discretion and at the point in time during the course that best fits his or her course goals.

SUPPLEMENTS

As you, the instructor, knows one of the significant differences between textbooks and trade books is the supplemental materials that accompany the former. *Object-Oriented Systems Analysis and Design* has four supplements including:

- An instructor's guide (with sample object models)
- A test item file
- Classroom presentation materials
- Object Modeling software
- Electronic communication with the author

The instructor's guide contains suggestions for presenting each of the chapters and modules in the book along with the answers to the end of chapter questions. The test item file contains objective (e.g., true and false/multiple-choice) and subjective (e.g., short answers required) questions and answers. The classroom materials include overhead transparencies of most figures and other important chapter concepts that may not have figures in the book. In addition, a full-color electronic version of the classroom materials in Microsoft PowerPoint (IBM) format is available by contacting the publisher.

Another supplement is object-oriented modeling software. Object International, Inc. (OI) graciously makes its object-oriented modeling whiteboard software package called Playground™ available to both students and instructors. Playground™ is an easy-to-use tool for building object models. It uses the same modeling notation presented in this book.

The "classroom and personal study only" version is available at no cost to a student or an instructor. Playground™ requires the Microsoft Windows operating system. Due to constant upgrades from both Microsoft and OI, specific system requirements are not listed here. Specific requirements are available at the time of downloading the Playground™ software directly from OI's world-wide web home page which is http://www.oi.com/business/object/. The file that is downloaded is a self-extracting file and at the time of this book's publication took between 1 and 2 megabytes of disk space to download.

Again, each student is allowed his or her own personal copy of Playground™. The instructor may choose to download Playground™ and then make copies available to students in a manner he or she chooses. The only difference between the "classroom and personal study only" version of Playground™ and the registered (fee) version is the "classroom and personal study only" watermark that appears at the bottom of the display screen and on printed output.

The final supplement is one that is not generally found in textbooks—electronic addresses to communicate directly with the author. Recognizing that change is a constant, these addresses are correct as of the publication date for this book; however there is no guarantee that these addresses will be correct in the future. If you, the instructor who adopts this text or is considering its adoption, send me electronic communication, I will do my best to reply back to you as soon as possible. I welcome your comments and questions, feedback and suggestions, as well as corrections for a future revision of the book.

My World Wide Web Home Page Universal Resource Locator (URL) address is: http://rohan.sdsu.edu/dept/cbaweb/IDS/RNorman.html (case sensitive)

My electronic mail address is: ronald.norman@sdsu.edu (not case sensitive)

San Diego State University's (SDSU) Home Page URL is: http://www.sdsu.edu/

SDSU's College of Business Administration (CBA) Home Page URL is: http://rohan.sdsu.edu/dept/cbaweb/

SDSU's CBA's Information and Decision Systems Department Home page URL is: http://rohan.sdsu.edu/dept/cbaweb/IDS/index.html (case sensitive)

ACKNOWLEDGMENTS

I am indebted to many people for their contributions in the development of this book and would like to acknowledge them here.

I am grateful to many students over several semesters and years in my undergraduate and graduate systems analysis and design courses who used preliminary versions of this book. I appreciate not only their helpful suggestions but also their patience and support of this work. Two students in particular, Ron Dashwood and Vicky Smith, combined to create the first basic object-oriented primer for my students to use in 1992.

I am grateful to the following reviewers of preliminary versions or portions of this book. My only regret is that I could not include all of their suggestions for a variety of reasons:

Terrence K. Babbitt, Former CIO Foodmaker Corporation; John Borton, University of Southern Colorado; Peter Coad, Object International, Inc.; Gail F. Corbitt, California State University, Chico; Bill Hardgrave, University of Arkansas; Joseph Morrell, Metro State College of Denver; Stevan Mrdalj, Eastern Michigan University; Richard G. Ramirez, Iowa State University; Ramesh Subramanian, University of Alaska, Anchorage; Thaddeus Usowicz, San Francisco State University; and David C. Yen, Miami of Ohio University.

I am grateful for the help of Germaine Kang, a former graduate student, who assisted with some of the object-oriented database material in the file and database design chapter of the book. I am also grateful for the help of Jin-Jue Tsai, a former graduate student, for his diligent and persistent search for object technology research materials and resources.

Thanks to Elliot Chikofsky, Northeastern University and DMR Group, Inc., for his friendship and long-term support and assistance with my CASE and software development research. Many professional opportunities became possible because of his involvement.

Thanks also to Dr. Stephen Robbins for mentoring me during this writing process. His many years of experience as an author have been a source of inspiration to me when the going got tough.

Thanks to Jeff Whitten and Lonnie Bentley, both of Purdue University, for their work in systems analysis and design over the years. It is a privilege to know and work with them.

I would like to thank Dr. Gail Corbitt, Associate Professor at California State University, Chico, for her support and contributions during the years we worked together at San Diego State University. She crystallized a significant part of the organizational behavior portion of the implementation chapter for me.

My thanks also to the Prentice Hall staff—Rich Wohl, P.J. Boardman, Susan Rifkin, Paul Becker, and others—that assisted with the publication of this book.

I apologize to any contributing person I may have overlooked at this time.

Finally, I hope that you will derive value from this book. Any shortcomings, errors, omissions, or inconsistencies attributed to the book are the sole responsibility of me. Best wishes for your career.

Ronald J. Norman
1995

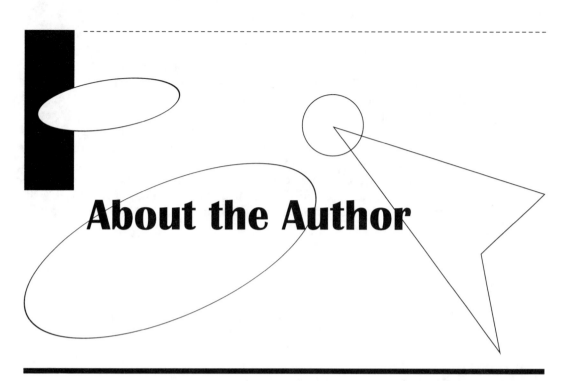

About the Author

Ronald J. Norman is currently a Professor of Information and Decision Systems at San Diego State University. He teaches executive, graduate, and undergraduate level courses in information systems management, software engineering/systems analysis and design, and information as an organizational resource. He received a Ph.D. in Management Information Systems and Organizational Behavior from the University of Arizona in 1987. He is also a Certified Computer Professional (CCP) and has over 25 years of industry MIS experience including software development, consulting, and management.

Dr. Norman's research interests are object-oriented systems development, organizational behavior related to the introduction of technology, and CASE technology. He has served as facilities chairman for the International Conference on Information Systems and was the program committee chairman for CASE '88, the Second International Workshop on Computer-Aided Software Engineering. He was the General Chairman for CASE '90, co-sponsored by the IEEE Computer Society, a mini-track coordinator on IPSE/Software Development Environments, and a mini-track coordinator on CASE technology for the Hawaiian International Conference on Systems Sciences (HICSS-24) in 1991 and repeated the CASE mini-track at HICSS-25 in January 1992. He was guest editor for the March 1992 issue of IEEE Software focusing on integrated CASE and was also guest editor for the April 1992 issue of the Communications of the ACM on a similar topic. He served as program co-chair for CASE '95 in Toronto, Canada and was on the program committee for the European CASE conference in 1995 and again in 1996. He is also co-editor of a special CASE theme issue of the Automated Software Engineering Journal to appear in 1996.

Dr. Norman has published or co-published over two dozen articles which have appeared in journals including the Communications of the ACM, IEEE Software, International Journal of Software Engineering and Knowledge Engineering, CIO JOURNAL, Information & Software Technology, Journal of Computer Information systems, and Journal of Systems Management.

   Dr. Norman's industry work/consulting engagements include several international, national, and local organizations, such as Avco Finance, Unisys Corporation, Tymshare, Inc., NCR Corporation, BASF, US Navy, American Management Systems, Performance Research Corporation, Science Applications Inc, University of California Medical Center, San Diego Police Department, Solar Turbines International, General Dynamics Corporation, National Liberty Corporation, and INJOY.

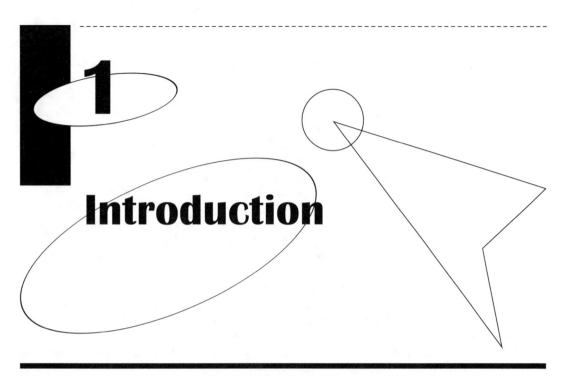

Introduction

CHAPTER OBJECTIVES (YOU SHOULD BE ABLE TO)

1. Define a system, information system, and automated information system.
2. Define the basic components and the basic characteristics of an automated information system.
3. Define systems analysis and design and discuss why it is a difficult human endeavor.
4. Describe the skills and activities of a systems analyst.
5. Describe a general model of the analysis, design, and implementation process.
6. Discuss systems analysis and design as a career.
7. Discuss what a systems analyst does.
8. Discuss systems analysis and design projects and where they come from.
9. Discuss the need for creating information systems requirements specifications.
10. Define and describe the information systems life cycle.
11. Define and describe the information systems development life cycle.
12. Discuss the principles used to guide systems analysis and design.

A computer salesperson, a computer hardware engineer, and a computer software engineer were traveling the freeway together one afternoon to make a sales presentation. Suddenly the right front tire of the automobile they were riding in burst. After the driver calmly pulled the vehicle over to the side of the freeway, each got out of the car and looked at the problem tire. The salesperson said, "No problem, we should just call the auto club and have them take care of it for us." The hardware engineer said, "We don't need to call them. Let's just put the spare tire on ourselves." Finally, the software engineer said, "Let's all get back in the car and hope that the problem goes away all by itself."

Welcome to the world of systems analysis and design, a discipline where we sometimes wish that "problems would just go away all by themselves." The fact is, they rarely do. Systems analysis and design is an exciting, challenging, and ever evolving discipline. Its challenge is the development of quality information systems that meet users' requirements and have a minimum of problems.

I have been involved with systems analysis and design since the late 1960s and am glad that you are studying it during this very exciting time of computer technology. I hope that you, too, have captured or soon will capture the excitement and challenge that this discipline has to offer men and women today. There are many different ways that the information in this book can be useful to you as you pursue your professional goals. So, let's get going.

The purpose of this chapter is to give you a broad overview of systems analysis and design as a discipline. Although the book's primary goal is to introduce you to an object-oriented methodology for analyzing, designing, and implementing information systems, this chapter is, by design, an overview of the discipline, void of much of the technical strategies, methodologies, and details used when analyzing, designing, and implementing information systems. As such, this chapter could be considered an introductory chapter for almost any systems analysis and design book. My hope is that the remainder of the book will bring the information in this chapter to life for you by introducing you to a specific and practical object-oriented methodology for systems analysis, design, and implementation.

SYSTEMS ANALYSIS AND DESIGN HAS MANY OTHER NAMES

As with many disciplines, systems analysis and design has many terms that in a general sense refer to the same or similar topic. The actual differences are so subtle that their debate is of little value in an introductory book such as this one. For example, systems analysis and design is also known as information systems engineering, software engineering, systems engineering, software development, and systems development, to name a few. For purposes of this book, these terms will be considered synonymous. Professional men and women working in the field of systems analysis and design often have their own personal preference for the use of these terms. So, no matter which term I chose to emphasize in this book, I could not satisfy everyone. For example, the term *information systems engineering* could refer to the entire systems development discipline, while *systems analysis, systems design,* and *systems implementation* could refer to the three major partitions of information systems engineering.

Bottom line for this book? **When the term *systems analysis and design* is used throughout the book, it covers the entire systems development process from planning all the way through implementation, maintenance, and evolution.** At various points throughout the book, I may refer to one of the other previously mentioned synonyms for systems analysis and design. This is not done to confuse you, but merely because in some situations the use of one synonym fits better in a given sentence structure than does another synonym. Just remember that each is a synonym for systems analysis and design in this book. At other times I may refer to the individual term—*analysis, design,* or *implementation*—which is intended to refer to a smaller portion of systems analysis and design.

There is no doubt that systems analysis and design is about developing software, but it is much more than that. While most computer programming courses focus primarily on learning the syntax of the language and then using that language to develop software that has zero defects, systems analysis and design takes a much broader perspective and focuses on:

1. **Systems planning**—performing planning and initial feasibility activities to determine which information systems projects take priority over others.
2. **Systems analysis**—understanding and documenting the requirements of a specific problem domain. A **problem domain** refers to the business problem or function being planned, analyzed, designed, and ultimately implemented as an automated information system.
3. **Systems design**—designing an appropriate solution for the problem domain based on the documented requirements from systems analysis; a variation on this is an approach in which commercially available systems are evaluated for suitability to meet the requirements and one is selected.
4. **Systems implementation**—constructing, testing, and installing the information system and having the users use the information system.
5. **Systems evolution**—maintaining and enhancing an information system so that it continues to meet the needs of the business.

Think back to a computer programming course that you have taken. In that course you probably had the opportunity to create and test one or more computer programs. Your instructor probably gave you a few pages describing a particular problem domain and what your completed software was expected to accomplish. Your main task was to create the software to do it. To put this example in the context of systems analysis and design, the planning and systems analysis portion of your assignment was completed by your instructor. He or she gave you the results of the planning and analysis in the form of a few pages describing the problem domain and your program (software) requirements. The systems design was your responsibility to complete. You did so by thinking about the requirements, designing your solution, and finally coding and testing the program to generate results in keeping with the instructor-provided requirements. Implementation was probably ignored in this project because it was strictly a laboratory learning assignment for you and not a real project for a business. Therefore, systems analysis and design as a concept consisting of planning, systems analysis, systems design, and systems implementation is a process that includes all activities performed to produce an automated information system. The details of these activities will be described later, but first let's discuss systems, information systems, and automated information systems.

WHAT IS A SYSTEM?

A **system** is a set of interrelated components, working together for a common purpose. There are two types of systems: natural and fabricated. Natural systems include the human body, the solar system, and the earth's ecological systems. Fabricated systems are created by people to satisfy some purpose that we have. Philosophically speaking, fabricated systems should serve you, not the other way around. For example, an automobile can be thought of as a fabricated system, so can a bicycle, a telephone, and any other manufactured item. Our federal, state, and local governments are fabricated systems as well as public and private schools and churches and synagogues. Even businesses such as AT&T, IBM, General Motors, the U.S. government's Internal Revenue Service (IRS), San Diego Chargers, and your local supermarket, frozen yogurt, pizza, and video stores can be thought of as systems. The term *business* will be used throughout this book as a generic term meaning a for-profit or nonprofit company, governmental agency, organization, or any other similar entity.

Fabricated systems are all around us. Unless we hike deep into the wilderness without our cellular phones and other electronic gadgets, we can hardly avoid them. We create, expand, split, praise, criticize, and "beat" systems. Some of the best fabricated systems are the ones in which we are hardly aware that they are working for us. Televisions, VCRs, CD players, and stereos tend to be examples of ones that are taken for granted, mostly because they perform well. It's only when some part of the system fails or you want to use some lesser-known feature of the system that extra attention is paid to these systems.

As Figure 1.1a shows, a generic **systems model** consists of six components—inputs, processes, outputs, controls, feedback, and boundary. Using predetermined **controls,** a system accepts **inputs** at its **boundary**, **processes** them into **outputs**, and provides a **feedback** mechanism for taking any necessary corrective action. Your bicycle, having been designed to only allow the pedals to turn in one direction to give it forward motion (control), accepts the forward motion of the pedals (input) primarily from your legs and feet, and propels (process) the bicycle forward (output). By observing various road conditions (e.g., hills, potholes, traffic signals, stop signs, water, oil slicks, and so on), which translate into feedback mechanisms on your part, you control the bicycle's speed and balance to deal with them.

The last component, **boundary**, is the perimeter or border of the system. The boundary can also be thought of as the **scope** of the system, in other words, what elements, features, options, and so on will be included in the system. By definition then, everything else is excluded from the system. A bicycle or any other manufactured item is designed to include certain parts, all others being outside the boundary of the bicycle or other item.

As Figure 1.1b shows, virtually all systems are part of a larger system called a **suprasystem** from the systems vantage point. An airplane, boat, automobile, and bicycle are all systems in their own right, and each belongs to the larger suprasystem called a transportation system. Likewise, virtually all systems can be decomposed into smaller systems called **subsystems** from the vantage point of the system in question.

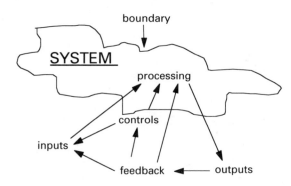

a) Systems Model with Six Components

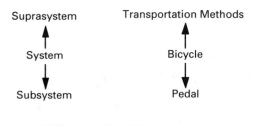

b) Systems Model Hierarchy

Figure 1.1 Systems

So, an airplane system consists of a wing subsystem, wheel subsystem, fuselage sub-system, electrical wiring subsystem, engine subsystem, fuel subsystem, and so on. Is a personal laser printer a suprasystem, system, or subsystem? Arguably, it could be any of these depending on your perspective. You could look at the printer and say, "the printer is a system, its component parts are its subsystems, and it is part of my per-sonal computer suprasystem," or you could say "the printer is a subsystem in my per-sonal computer system," or finally you could say, "the printer is a suprasystem, its paper, trays, toner cartridge, power cord, and so on are its systems."

WHAT IS AN INFORMATION SYSTEM?

An **information system** is a type of fabricated system that is used by one or more per-sons to help accomplish a task or assignment. Information systems come in all shapes and sizes and are limited only by human imagination. For example, you may have an address book that lists the names, addresses, and telephone numbers of your friends and relatives. When you want to call one of them, you refer to your book to get the phone number. If you desire to send your rich aunt a get well card (of course you are sincere about this!), you look up her address in your book when addressing the enve-lope. On a much larger scale, the major airline companies have massive information systems to handle passenger reservations and baggage.

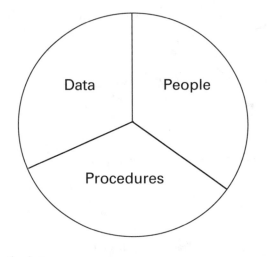

Figure 1.2   An Information System

Information systems are established to support policies and/or procedures. This can be done in virtually limitless ways; for example, you may have a mental policy stating that you will maintain an address book. You also have a procedure, probably in your head, of how to maintain the address book so that the names, addresses, and phone numbers are current. Your best friend may also have this same policy but with a different procedure for keeping it current.

In addition to having the six general system components, an information system has three additional components—people, procedures, and data—shown in Figure 1.2. **People** interface in some way with a system. Sometimes we provide the input, sometimes we do the processing, sometimes we do the output, sometimes we do the controlling, and sometimes we provide the feedback. The way we are supposed to interface with the information system is often documented in the form of **procedures**, and our interaction usually results in providing **data** to the system. When one of your relatives changes his or her address, you have a procedure (undocumented no doubt) for providing new data to replace the obsolete data in your address book. Using a voting information system as another example, when we vote, we interact as people, according to the established voting procedures, and we provide our voting data to the voting system.

As the amount of time to maintain an information system increases or the number of people needed to maintain it increases, it becomes economically justifiable to invest in an automated information system to accomplish the task. For example, many people who write personal checks have purchased personal money manager software to help them with the task of paying their bills by check and balancing their checkbook. Most people can do these tasks manually, but a personal money manager information system may reduce the time and effort to do the same thing. In addition, paying bills may actually be viewed as a more enjoyable task with the use of a personal money manager information system.

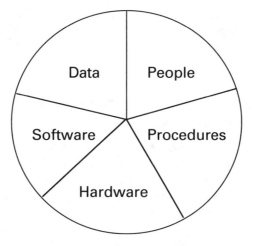

Figure 1.3   An Automated Information System

WHAT IS AN AUTOMATED INFORMATION SYSTEM?

An **automated information system** is an information system that incorporates the use of computer hardware and software as part of the system. Therefore, an automated information system adds two additional components, as in Figure 1.3, to the three associated with an information system. For example, a business may wish to send birthday cards to each of its employees just prior to their birthdays. This information system of course could be done manually with cards, envelopes, pen, and a handwritten roster of employee names, addresses, and dates of birth. But what if the business is AT&T and employs thousands of employees? Sending birthday cards could get expensive as well as take significant time for someone using a handwritten roster of employees. The more reasonable way to accomplish this birthday card task would be to have software read through a computerized database roster of employees and generate envelope address labels for those employees whose birthdays are approaching. A more elaborate automated information system might even create a customized birthday card for each employee or even send the birthday greeting to the employee through electronic mail.

Both large and small businesses have multiple information systems within them. For example, Blockbuster Video Stores probably have an automated payroll system, video rental/sale system, video purchasing system, as well as others. A small, independent video store a few blocks from your home may also have these same type of systems even though its systems may be manual. A manufacturing business with a few dozen employees has roughly the same information system needs as does a manufacturing business with thousands of employees. The difference is primarily the amount of data managed by these information systems.

From this point forward this book will be dealing with automated information systems. For simplicity, I will omit the word *automated* and refer to all automated information systems merely as information systems. This choice was made based on simplicity as well as my belief that the vast majority of systems analysis and design focuses on automated information systems, rendering the word *automated* unnecessary or implied.

WHAT ARE THE BASIC CHARACTERISTICS OF AN INFORMATION SYSTEM?

The basic characteristics that exist within an information system are data, functions, and behavior, as illustrated in Figure 1.4. Information systems do not necessarily have to have all three characteristics but most do. **Data** are either (1) input via some data entry device such as a keyboard, scanner, or mouse, (2) already in the system and stored on a storage device such as a magnetic disk or diskette, or (3) displayed or printed on an output medium, such as a display screen, paper, or microfilm. As part of the information system, data that are input and data that are stored are transformed or processed into meaningful output data called information. For example, in a payroll information system, the hours that employees work in a pay period and the hourly wage they are paid are important pieces of data for computing an employee's payroll check. The hours worked are data that need to be input each pay period, while the em-

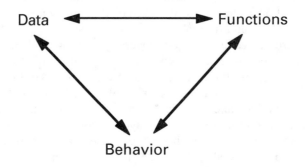

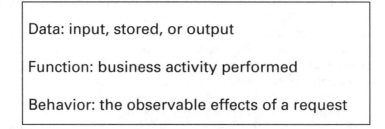

Data: input, stored, or output

Function: business activity performed

Behavior: the observable effects of a request

Figure 1.4   Basic Characteristics of an (Automated) Information System

ployee's hourly wage is stored data since it only changes when the employee gets a raise. The hours worked and the hourly wage are transformed by the payroll system into a paycheck and other meaningful payroll reports. Using another example, a simple multiplication can be a transformation of data into information, for example, three (data) times two (data) is transformed into six (information).

A **function** is a transformation or action taken by the information system. Information systems usually have many functions. Functions carry out and enforce business policies, rules, and procedures. Other synonyms for function are process, service, and method, the last two becoming more familiar with the popularization of object-oriented technologies. Examples of functions are printing a report, printing a paycheck, displaying the customer's account dollar balance, and dispensing cash from an ATM.

In information systems, **behavior** is the observable effects of a request. Behavior is present in all systems; however, it is more pronounced or accentuated in systems that have a human interface component (systems that respond to keyboard data entry, commands or mouse clicks, such as word processors, spreadsheets, e-mail, and other systems). In addition, a furnace motor turning on or off in response to a thermometer reaching a certain temperature, a bottle-filling device responding to messages like "fill bottle" and "full bottle," and an elevator responding to signals that its information system receives as it passes by actuators on the way up or down the elevator shaft are all examples of behavior in systems that have minimal human interaction.

Different types of information systems may emphasize one or more of these three characteristics over the others. In a typical business information system the primary emphasis is usually on the data component with secondary emphasis placed on the functional aspect, followed by the behavioral. Why is this? Because in most business information systems, such as accounts payable and receivable, purchasing, manufacturing bill of materials, material requirements planning, payroll, and order entry, the data are the primary component. In an elevator information system the primary emphasis may be on the behavioral component, while realizing that it must also give attention to the functional and data components.

Historically, business information systems were developed primarily from a functional perspective while being aware that these systems also had a strong data component. Since the mid-1970s and the introduction of database management systems and relational database technology, the functional perspective for developing business information systems has often been relegated to a lesser position in favor of a data perspective. However, the need to give primary consideration to the functional component may still be present in some types of information systems. Some businesses, no doubt, still prefer to practice using a functional perspective instead of a data or behavioral perspective.

With the strong acceptance of the windowing graphical user interface made popular by Apple's Macintosh Computer in the 1980s followed by Microsoft's Windows for most other PCs in the 1990s, behavior has become a dominant aspect of desktop business information systems. The iconic metaphor for user interaction has taken the human interface component of information systems to a new and improved level.

WHAT IS SYSTEMS ANALYSIS AND DESIGN?

Systems analysis and design is about developing software, but it is more about developing a complete automated information system, which includes hardware, software, people, procedures, and data as shown in Figure 1.3. These five components exist in virtually all automated information systems, although the amount of each will vary with respect to the specific system being developed. All of these components must be considered and addressed during systems analysis and design. If any one of them is slighted or overlooked, the system will more than likely not be successful. When you buy software at a retail store or the campus bookstore, you are buying more than the software. For example, you expect the software to work with certain hardware, have its own set of procedures for use, work with the data you want or need to give it, and interact with you; hence, the five components have been considered.

The culmination of the systems analysis and design process is to produce an acceptable automated information system for use in one or more of the following ways: (1) software to be used within the business that it was developed for (e.g., a payroll system developed by XYZ Corporation for its own internal use to produce paychecks for its employees), (2) software for sale via retail stores, mail-order catalogs, or direct from the company that created it, or (3) software to be used within products produced by a business (e.g., an automated teller machine sold by IBM). In all three of these scenarios the other four components of an automated information system—hardware, data, procedures, and people—have been analyzed and designed to work with the software.

WHAT MAKES SYSTEMS ANALYSIS AND DESIGN SUCH A DIFFICULT HUMAN ENDEAVOR?

Many times people have said that systems analysis and design activities are perhaps some of the most complex and involved activities ever conceived by and for humans. One significant study of information systems analysis by Vitalari and Dickson cited six reasons that contribute to the support of this belief, and each is discussed here.

1. Analysis problems, at their inception, have ill-defined boundaries and structure, and have a sufficient degree of uncertainty about the nature of the solution. What systems analysts constantly find at the start of systems development projects is that the users (remember, often the users can be many hundreds or thousands of people) are not certain of what they want (ill-defined boundaries and structure), and this contributes to the uncertainty about the potential solution to the problem. You may be asking, "how can people (users) not know what they want out of an information system?" Consider the analogy of many first-year college students being uncertain of their major field of study as they begin taking their general education courses.

2. The solutions systems analysts come up with to solve the problems are artificial, and since they are designed by humans with different backgrounds, experiences, biases, and so on, there exists an endless variety of potential solutions. What this could translate to mean is that there is *no single correct solution* to a problem. Any number of solutions could possibly address the problem in a satisfactory manner. The difficulty often is the fact that most systems solutions are compromises—a satisficing solution rather than the optimal solution. Why is this? Because of the diversity of needs that the user community may have for the impending information system. One user wants it to do something one way and another user wants it to do it a completely different way. So compromise needs to be reached in order to avoid win-lose situations with the users.

3. Analysis problems are dynamic. No business is standing still. A business is continually growing in one or more ways, shrinking in one or more ways, or growing in some ways and shrinking in others. In fact, the one thing that is constant about a business is **change**. Two common examples of this are the fast-food industry menu changes to stay abreast of consumer preferences and its competition, and the frequent shuffling of videos around the shelves in video stores to accommodate new arrivals, fast-moving, and slow-moving videos. As a result of this constant change, information system requirements and needs are like a moving target. The longer it takes to plan, analyze, design, program, test, and implement an information system, the greater the chances that it will not meet the business's needs simply due to the dynamics of change. A project scheduled to take five years to develop will either not meet the business's needs when implemented or will be in a constant state of change throughout its five-year development cycle in order to keep up with the changes that will be occurring within the business during that same time. An information system developed in six to nine months will probably have a greater chance of meeting the business's needs, which may have not changed too much in this shorter time period.

4. The solutions to analysis problems require interdisciplinary knowledge and skills, hence, the need for a team approach to information systems development. Today there is a strong emphasis on the partnership concept between the user community and the information systems developers.

5. The knowledge base of the systems analyst is continually evolving. As the apprentice systems analyst progresses through the junior, associate, and senior systems analyst ranks over time, he or she continues to learn more about business problem domains as well as improving his or her analytical skills and software development tool and technique skills.

6. The process of analysis is primarily a cognitive activity in that we are asked to (1) put structure to an abstract problem domain, (2) process diverse information from a variety of users, and (3) develop a logical and consistent specification that will lead to the creation of a successful information system.

With these thoughts in mind, it is no wonder that the men and women in the systems analysis and design profession consider systems analysis and design to be very difficult activities to perform successfully.

One additional factor that was not explicitly cited in the foregoing study, which many believe contributes to the complexity of the analysis and design activities, is that of working with people. Why is working with people so difficult? Information systems involve people and, whenever people are involved in an activity, difficulties can arise due to the myriad number of human behavior and communication issues that people bring with them into the project. Oh, they don't carry these issues in a bag slung over their shoulder. On the contrary, these issues are buried deep into the makeup and experiences of each human being.

Many information systems academics recommend to their students that they study human and organizational behavior as a compliment to systems analysis and design. The information learned in human and organizational behavior courses can go a long way to contributing to a more successful information systems development project and career.

STAKEHOLDERS OF AN INFORMATION SYSTEM

In a typical business, the systems analyst may need to interact with individuals who have different roles within that business, shown in Figure 1.5. The greater the number of interactions with individuals who have different roles calls for more experienced systems analysts. In situations where the systems analyst needs to be involved in many different human interactions, he or she may be functioning as a project manager.

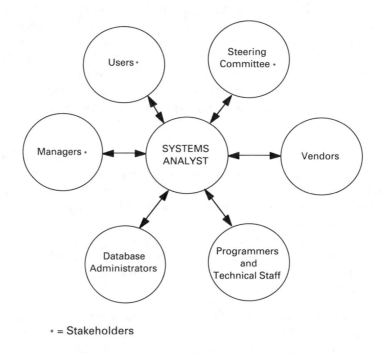

* = Stakeholders

Note: Systems analyst may be acting as a project
manager for some of these interactions

Figure 1.5   Systems Analyst Interactions with Individuals During Systems Development

What is a stakeholder? A **stakeholder** is a business unit, individual, or group of individuals that affects or is affected by an information system. Be careful not to extrapolate from this definition that everyone in the known universe is a stakeholder of every system. Most of the time business units, individuals, or groups of individuals are considered stakeholders if they are directly affected by a system. For example, employees are stakeholders of a payroll system because they receive payroll checks from the system. The stockholders of this company are not stakeholders of the payroll system even though stock values may be affected by payroll.

Often the majority of the stakeholders are the users of the information system, but they need not be. For example, a manager may be a stakeholder because he or she affects or is affected by the system but is not an actual user of the system.

Who are the users of an information system? To answer this question, each system must be independently considered based on its requirements. Users can be almost anyone. For example, you are the potential user of a student registration system, an automated teller machine, and a bank credit card system such as VISA or Mastercard. The astronauts in a space shuttle as well as the crew in the ground control rooms are users of a number of information systems during a space flight.

Referring to Figure 1.5, a **steering committee** is usually composed of cross-functional, senior managers within a business, such as vice presidents or directors. Included in this committee should be the senior information systems manager or a designate. The main role of this group is to conduct high-level reviews and evaluations of proposed information systems development projects and make recommendations for prioritization and resources for the projects.

Finally, **vendors** are those businesses which support the information systems development effort, such as consultants, hardware and software companies, training companies, telecommunications companies, documentation companies, and so on.

SYSTEMS ANALYSIS AND DESIGN AS A CAREER

Systems analyst is the title usually given to someone who chooses a career in systems analysis and design. Other synonymous titles can be found, such as software engineer, systems analyst/programmer, information systems engineer, and systems engineer. Systems analysis and design is a fascinating career choice for several reasons. First, you rarely, if ever, develop the same software or information system twice, which means that systems analysts are constantly being challenged to grow professionally, and rarely, if ever, are they bored with this discipline. Second, systems analysis and design is constantly changing and evolving, which points to a high degree of excitement for the men and women involved in this profession. Third, the systems analyst's learning and professional growth are never ending. Finally, organizational competitiveness and success often are dependent on the information systems that systems analysts help create, giving them a sense of significance and contribution to the business.

Even if you don't intend to make systems analysis and design a career choice, its study and understanding will be very useful to you as you pursue other careers, since most careers involve the use of computers and the software that makes them do what they do. More than likely you will be called upon to be a part of a systems analysis and design project sometime during your career, and having some appreciation of its process will allow you to contribute more effectively and hopefully be more understanding of the difficulties involved in developing quality software and information systems.

WHAT DOES A SYSTEMS ANALYST DO?

A systems analyst studies the problems and needs of a business in order to ascertain how hardware, software, people, procedures, and data can best accomplish improvements for the business. Improvements translate primarily into one or more of three areas: (1) increasing the revenue and/or profits of the business, (2) decreasing the costs of the business, and (3) increasing the quality of the service(s) of the business. A systems analyst should keep these objectives in mind when working on information systems projects because these are the material ways that the systems analyst can make a valuable contribution to the business. If the work of systems analysts stops contributing to one of these three objectives, then it is possible that the business may no longer need their services.

WHAT IS A SYSTEMS ANALYST RESPONSIBLE FOR?

A systems analyst is responsible for determining the most effective and efficient (1) capture of input data from the myriad of potential sources, (2) processing and storage of the data, and (3) delivery of timely and accurate information to the user or other information systems.

As technology evolves, it is clearly the responsibility of systems analysts to investigate and determine the most effective and efficient method for capturing input data from its source. For example, most major retail stores and supermarkets have moved from a cashier keying in the prices of each item purchased to a laser scanner device that they simply point at the product's barcode or move the product across or in front of the scanning device. Student course registration at some universities has migrated from students walking through a long line and pulling punch cards out of a box of cards for each class they want to students calling on the telephone at appointed times to sign up for a class using the keys on the telephone and getting immediate confirmation or denial of a seat in the class.

The processing and storage of data and delivery of timely and accurate information are continually investigated due to advances in technology. Businesses have migrated from information systems that were primarily run on the computer overnight (batch systems) to systems that the users directly interact with via a keyboard, mouse, scanner, and so on during their normal workday (on-line systems) and receive visual display of results or paper-based printouts on demand. These changes were handled by systems analysts and other technology professionals hoping to improve one of the three objectives discussed earlier.

SYSTEMS ANALYSIS AND DESIGN SKILLS AND ACTIVITIES

By now, you might be thinking, "What kind of skills are required to be a competent systems analyst?" Well, there are several systems analysis and design skills, depicted in Figure 1.6, required to consistently and successfully perform the role of a systems analyst. **Problem-solving ability and people skills** are at the core of the skills and competencies of an aspiring or seasoned systems analyst. You may be thinking, "I don't like to solve problems. In fact, I don't even like problems." That's okay. That's a normal human tendency. Perhaps it's just the word *problem* that is bothering you. A different word set, such as *situation-solving skill* or *condition-solving skill* may be more palatable to you. These word sets may actually be more representative of systems analysis and design work; nonetheless, most people refer to this as problem-solving skills. People skills refer to your ability to successfully work with diverse groups of people, such as those shown in Figure 1.5. Chances are, depending on your work experience, you have worked with diverse groups of people in some capacity. Did you mostly enjoy your interactions with them?

The next skill and competency layer, working outward, is a familiarity and understanding of systems analysis and design **concepts and principles**. These first two inner circles form the basis for the core competencies for a well-rounded understanding

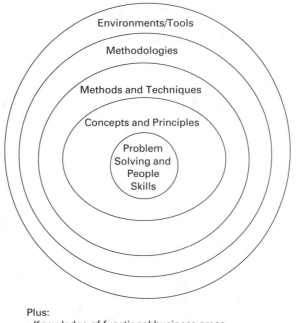

Plus:
• Knowledge of functional business areas
• Verbal and written communication skill
• Work experience in systems analysis and design

Figure 1.6   Skills and Competencies of a Systems Analyst

of the process of analyzing and designing automated information systems. Concepts and principles are general and abstract statements that describe desirable properties of software development processes and the resulting software. By themselves, concepts and principles are not sufficient to drive information systems development. An example of concepts and principles is the notion that the information system being developed is for the user and the user is most often the owner of the system. This may seem quite basic yet you would be surprised to find out that some systems professionals view the systems that they help develop as their own rather than belonging to the users. There are many other concepts and principles, some of which are scattered throughout this book.

Looking at the layers in Figure 1.6, the innermost layers generally change more slowly over time than do the outermost layers. In fact, some of the concepts and principles of systems analysis and design that were taught in undergraduate curricula in the late 1960s still apply today.

Methods and techniques operationalize concepts and principles. Systems analysts and programmers must learn the details of the prevailing methods and techniques in order to incorporate concept and principle properties into their information systems development process and the resulting products. Methods and techniques are sufficiently detailed, much like a cookbook full of recipes, to allow a certain amount of consistency for using the method or technique across a spectrum of systems analysts and programmers. There are methods and techniques for conducting user interviews, conducting surveys, doing feasibility studies, diagramming and documenting requirements, diagramming software, testing software, and so on. Other techniques include flowcharts, data flow diagrams, entity-relationship diagrams, and HIPO charts to name a few.

Since the mid-1980s, many of the techniques used in systems analysis and design have been automated. For example, drawing a data flow diagram can be done by hand, but is most often drawn with the assistance of data flow diagramming software, which is discussed later. Methods and techniques evolve more rapidly over time than do concepts and principles and are often synonymous with the automated tools that support them.

Packaging methods and techniques together form a **methodology**. The purpose of a methodology is to promote a certain problem-solving strategy by preselecting the methods and techniques to be used. Some methodologies can fill a bookshelf or two, similar to an encyclopedia. Some methodology examples are structured analysis, structured design, structured programming, object-oriented analysis, object-oriented design, object-oriented programming, and information engineering. Methodologies also evolve over time, a little slower perhaps than do methods and techniques. For example, during the 1960s traditional (informal) systems analysis and design was the dominant methodology; in the mid-1970s structured analysis and design methodologies became dominant; in the 1980s information engineering arose; today we are in another prolonged transition to object-oriented analysis, design, and programming methodologies. Keep in mind, however, that businesses will more than likely continue to use more than one of these methodologies for different projects due primarily to expertise in using the methodology as well as the strengths and weaknesses of the methodology when compared to another for a specific information systems project.

Environments and tools are available to support the application of methods, techniques, and methodologies. Tool examples include automated flowcharts, data flow diagrams, entity-relationship diagrams, HIPO charts, and others. Remember that these tools deliver automated support for the techniques they support. Environments are often labeled as being one of computer-aided software engineering (CASE), software development environments (SDE), or integrated project/program support environments (IPSE). As you might expect, environments and tools evolve or change more rapidly than do the methodologies, methods and techniques, and the concepts and principles.

Most of the initial use of techniques was done manually using pencil, paper, and an appropriate notation template for sketching out the tool's symbols. Since the mid-1980s a large number of automated technique support tools have become commercially available. In other words, personal computer graphical software is available that allows us to draw the technique symbol notations as well as apply some or all of the rules, guidelines, or heuristics of the technique. This class of software is commonly referred to as *C*omputer-*A*ided *S*oftware *E*ngineering, or CASE.

Empirical studies, professional trade journal articles, and discussions with information systems managers have suggested a few additional and complimentary skills necessary for a competent systems analyst. These include (1) general knowledge in the functional areas of business, such as accounting, marketing, finance, personnel, manufacturing, engineering, and so on, (2) verbal and written communication skills, and (3) work experience in systems analysis and design.

GENERAL MODEL OF SYSTEMS ANALYSIS AND DESIGN

Figure 1.7 is a general model of systems analysis and design (including implementation). Remember, systems analysis and design is a process consisting of many necessary activities in order to produce a successful information system. The general model identifies three major items: activities (analysis, design, and implementation), people involved in the activities (user, information technology staff), and inputs and outputs (all of the numbered areas on the arrows). The inputs and outputs are numbered in this figure in order to give you a visual queue for the standard sequence of events.

A "talk-through" this "partnership" general model would go something like this: A stakeholder(s) (user) has a problem or situation that he or she believes can be solved or addressed with the help of an information system. With the assistance of the *information technology staff* during the *analysis* activities, the user communicates the essence of the problem in written and verbal form. As the *analysis* activities progress, the *information technology staff* utilizes systems analysis *problem-definition skills* to document the user's *requirements* and produce a *requirements specification* document. The requirements specification document is often equated to architectural drawings used in the construction industry because it represents "what" the system should do when developed just as the architectural drawings represent "what" a building or

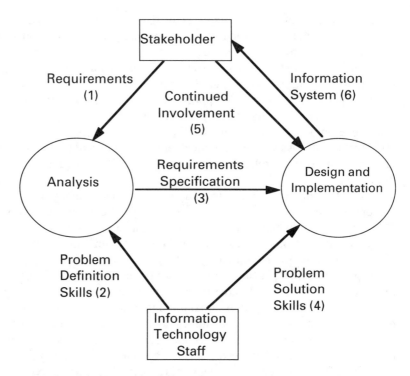

Figure 1.7   General Model of Systems Analysis, Design, and Implementation ("Partnership")

home should look like. The *requirements specification* becomes input to the *design* activities along with continued user *involvement* and information technology staff *problem-solution skills*. During *design*, the information technology staff develops detailed blueprints for the information system, similar to the detailed blueprints used in the construction industry and used by the construction site supervisor for detailed instructions regarding the construction of the building. Once *design* progresses through its activities, *implementation* activities begin. During implementation, software is constructed (although it may be purchased in some situations), tested, and finally implemented. The end result of the implementation is the completed *information system,* which is delivered to the stakeholder(s) (users) for his or her (their) use.

THE DETAILED ACTIVITIES OF ANALYSIS AND DESIGN

Zooming in on the two circles shown in Figure 1.7, Figure 1.8 "explodes" these two processes to reveal their detailed activities and deliverables. Activities, also referred to as tasks, are the work that is performed by systems analysts and others as part of analysis and design in order to complete the systems development process. Deliverables are the products that are produced during and at the conclusion of the systems development process. For example, drawings, documentation, training materials, requests for proposals, software, and of course the completed system are considered deliver-

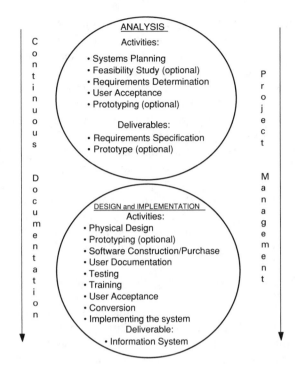

Figure 1.8 Systems Analysis, Design, and Implementation Activities and Deliverables

ables. Each of these activities and deliverables is briefly described here. Additional coverage of these activities and deliverables is found in subsequent chapters. Even though the discussion that follows appears to be highly linear and dogmatic, recognize that there is significant flexibility during projects to overlap, omit, and replace these activities and deliverables.

I believe one final point is necessary before discussing the analysis and design activities and deliverables. You need to understand that there is a variety of ways in which the details of analysis and design could be divided into detailed activities and deliverables. In fact, be aware that other seasoned professionals might choose to partition analysis and design somewhat differently than the way it is presented here—that's okay! Diversity of opinion usually based on years of experience may lead others to view analysis and design activities and deliverables somewhat differently than I do, and I want to recognize their views as being legitimate. Your instructor or another reference book may have a variation on what is presented here. I usually find no serious discrepancy or inconsistency in this. The software development community of professionals of which systems analysis and design is a part needs to focus on what we have in common more than where we disagree. Where there is disagreement, let us as professionals not allow these disagreements to keep us from moving forward with our efforts to find and communicate better ways to do systems analysis and design. Now on with the discussion of the activities and deliverables. Refer to Figure 1.8 when necessary.

Starting with the analysis circle, **systems planning** is an activity which seeks to identify and prioritize technologies and business applications that will yield high value to the business. A **feasibility study** is often done to determine the merits of developing or enhancing an information system. A feasibility study takes into consideration the technical, operational, and economic aspects of a proposed information systems project and compares them to goals, objectives, and limitations as set forth by the business. The more in line a proposed project is with the business's goals, objectives, and limitations, the more likely the project will be approved. Sometimes feasibility studies are bypassed during analysis simply because the information system is mandatory due to some internal or external regulation from labor unions, the government, or other situation.

Requirements determination is the most important and most difficult activity performed during systems analysis. It may also be the most important and most difficult activity performed during the entire systems analysis and design process. During this activity the systems analyst and the user work together to identify and document the true information system requirements. In larger projects there are usually many systems analysts and users involved in this activity. The systems analyst is constantly questioning: "What is this information system supposed to do?" Mistakes and oversights made during this activity become prohibitively expensive when they are eventually discovered later in the project. In fact, a mistake found late in the project can cost as much as 70 times more to fix than if it were found and corrected early in the project. The output or deliverable of this activity is a requirements specification document, which is analogous to architectural sketches and drawings of a home or building.

User acceptance is the informal and formal endorsement of the documented requirements by the users. User acceptance is more readily given if the users have been continuously involved in the work activities of the project that lead up to the formal acceptance point. **Prototyping**, an optional activity during analysis, may be useful to demonstrate the feasibility or some other aspects of the proposed information system in order to more fully understand the user's real requirements or to improve the chances of user acceptance of the proposed information system. The two primary deliverables from the analysis process are the **requirements specification**, mentioned earlier, and any **prototype** that may be developed.

During the design process, several activities are performed. The **physical design** activity produces a physical design document, sometimes called a detailed design, for the proposed information system, which is analogous to the construction industry's detailed blueprints used to guide the construction of homes and buildings. Another optional **prototyping** activity may be useful at this time to produce another model of the proposed information system.

Once physical design is completed or at least determined to be far enough along, implementation begins. This is when **software construction/purchase** commences. If a decision was made to purchase most of the information system from a commercial vendor, retail store, or mail-order house, then the detailed design document would be less complex than if the majority of the information system is to be developed by your own information systems staff. If the decision were made to mostly construct the software, then this is the activity for doing so.

User documentation, both in paper and electronic form, is usually created during design. **Testing** usually runs parallel with software construction and continues up to the point of implementation. User **training** is conducted as the information system approaches its usable state. *Just-in-time* training is advisable so that the user will get maximum benefit from it. **User acceptance** again is required before implementing the new or changed system. In many systems development projects, the user acceptance at this point in the project is the final checkpoint prior to accepting the system as being ready for implementation. Recall once again that user acceptance is more readily given if the users have been continuously involved in the work activities of the project that lead up to this final and formal acceptance point.

Conversion is the activity that transforms the stored data in the existing information system into the format required for use in the new or changed information system. Other activities may also be done during conversion. Conversions can be very simple or very complex. The final activity within design is **implementing the system**, which involves installing and getting the users to use the new or changed information system.

The single deliverable from design, as depicted in the figure, is the completed information system. This deliverable includes more than just software. All documentation, models, prototypes, training aids, conversion aids, and so on may be considered as part of the completed information system. More often than not, the systems professionals retain some or all of the system development models, system documentation, prototypes, and other analysis, design, or implementation-specific portions of the system. The reason for this is that the systems professionals will most likely be maintaining and enhancing the system over its lifetime and would prefer having these informational items for their support efforts.

Two ongoing activities, **documentation** and **project management**, apply continuously throughout both analysis and design. As much as possible, the systems analyst must document what is discussed, discarded, approved, and learned during the entire systems analysis and design process. Failure to do this could result in some organizational knowledge loss when the systems analyst is no longer available as part of the business's development team. Project management is necessary for almost all projects in order to keep them on track from both a financial and scheduling perspective.

SYSTEMS ANALYSIS AND DESIGN PROJECTS

Don't let anyone tell you differently, systems analysis and design is a highly labor-intensive activity involving groups of people working together as a project team. The smallest team possible is you creating software for yourself. You are the user of the software as well as the systems analyst and programmer who creates it. So you would be playing all of the roles on this project team—user, systems analyst, and programmer. The only one you would have discussions, disagreements, and compromises with on this project team is yourself!

The next smallest team possible is you creating an information system for one other person, say your roommate. He or she would be the user of the system and you would be the systems analyst and programmer. On this project team you only play the technical roles of systems analyst and programmer. Now you discuss, disagree, and compromise with the other team member who is the user of the system. Did you learn in a marketing class that the customer (user) is always right? Well, in systems analysis and design this notion is also true. The job of the systems analyst is to create an information system that will meet or exceed the user's requirements for the information system.

Users come in all sizes and shapes. They also have differing levels of understanding about the process of systems analysis and design and their role in that process. Compounding this situation is the complexity of the problem that they are trying to solve with the information system that you will help create for them. Likewise, systems analysts also come in all shapes and sizes. Systems analysts have varying amounts of experience with analysis and design activities as well as varying experiences with the specific problem that the user is trying to solve with the information system that will be created. This combination of user and systems analyst characteristics makes for very different project teams each and every time one is formed.

Most information systems are developed in projects that consist of more than two team members. This makes the communication between team members even more important and more difficult to coordinate than what was discussed previously. Project teams of ten, twenty, fifty, one hundred, and more are quite common in "industrial strength" systems analysis and design projects. Some of these members are managers and users, some are systems analysts, some programmers, and others are support staff.

More often than not, users have a difficult time articulating their real and full requirements of the problem to the systems analyst, not because they are incompetent or inarticulate, but because the problem being solved is not straightforward. For example, try listing all the necessary mental and physical tasks, in correct order, that you would go through to successfully walk yourself across a busy traffic light controlled intersection in your city. When this task has been done as a class exercise, the students identify as few as five steps up to a maximum of about fifty steps. Quite a wide variation in "user requirements"!

Another task that is easy to do but difficult to describe is tying your shoes. Try writing down the steps involved to do this. You are no doubt an expert at shoe tying, but trying to accurately articulate how you do it in writing will be very difficult. Have you ever tried to write down how you would apply cosmetics or tie a necktie? The same is true for users trying to tell you exactly how their job is performed or needs to be performed. They usually know their job very well and are competent at it but still have a difficult time telling someone how it is done.

WHERE DO INFORMATION SYSTEMS ANALYSIS AND DESIGN PROJECTS COME FROM?

New or changed information systems development projects come from problems, opportunities, and directives and are always subject to one or more constraints. Most maintenance of existing information systems projects come from users discovering real problems with existing information systems.

Problems, often called bugs, can arise at any time during the life of an information system. There is no such thing as a problem-free or bug-free information system. What exists are information systems that are waiting for the next problem or bug to arise. As soon as a problem/bug is suspected, a project to find and correct the problem should be initiated.

Opportunities are the most preferred way to kick off an information systems development project. This means that the business is hoping to create a system that will help it with increasing its revenue, profit, or services, or decreasing its costs. American Airlines capitalized in a very big way many years ago when it created its SABRE on-line airline reservation system, years ahead of its competition. In the past an on-line reservation system was an option or luxury. Today an airline reservation system is a basic business requirement just to be in the passenger airline business. Southwest Airlines, one of the most profitable U.S. airlines in the 1990s, has a much simpler on-line reservation system than does American Airlines. Their reservation system does not include the option of advance seat selection.

Directives are mandates that come from either an internal or an external source of the business. For example, the chief operating officer or president of a company (internal source) may issue a directive saying that the company will need to automate an information system during the next calendar year. A labor union (external source) may require certain reporting requirements for companies that have employees that are members of that labor union. The federal government (external source) may require that companies produce certain reports for equal employment opportunity statistics. In these situations the company has to comply or be penalized in some way.

Constraints are limitations and compromises that come with the soon to be developed information system. A video store rental system may only work on Intel Pentium-type computers; the word processing software you use may only work with Windows; the company's payroll system cannot produce an individual paycheck greater than $10,000, and so forth. Constraints and compromises are present in every information system and often exist in order to allow the system to be completed and implemented in a timely manner. For example, if a software development company, such as Microsoft, waited for its word processing software package to simultaneously run under Windows, OS/2, Unix, VM, CMS, and the NextStep operating systems before releasing it, it would take several additional years. Limiting it to one specific operating system, such as Windows, would allow that version of the word processing software to be released much sooner. Versions for the other operating systems could follow after additional time has lapsed.

INFORMATION SYSTEMS REQUIREMENTS SPECIFICATION

If you were going to go on a two-week vacation, would you plan ahead of time for several aspects about your vacation, such as where you will go, where you will stay, what you will do and see? Or would you just jump in your car on the first day of the vacation and start driving? This latter question seems more like running away from home than going on a vacation! Hopefully, you would do some planning, scheduling, and documenting of your vacation plans prior to the start of your vacation.

The same would apply if you decided to do a room addition to your home or build a swimming pool in your backyard. The city you live in would more than likely require drawings and blueprints of the anticipated additions prior to the actual work being done.

Likewise, information systems of any consequence are usually preceded by documenting in writing with words, drawings, and possibly pictures exactly what the information systems requirements are. Even changes made to enhance existing information systems should be preceded by documenting exactly what the change will be. An article in *Fortune* magazine suggested that an IBM checkout scanner has 90,000 lines of code in it, and a Citibank automated teller machine has 780,000 lines of code in it. Another report indicated that a space shuttle has 500,000 lines of code in it, and the ground support systems have 1.7 million lines of software code in them. It would be virtually impossible to create the software for a successful space shuttle flight without first documenting what the software is supposed to do. The document that contains the words, drawings, and pictures is often called the **user requirements specification** document. It becomes the blueprint for the information system waiting to be built or modified.

Historically, information systems development has been plagued with three significant user criticisms: (1) development takes too long, (2) development costs too much, and (3) the information systems that are developed do not truly meet the business's objectives for the system. These items are still a constant source of tension for systems developers, but there is a significant movement to integrate the user as a strong participant in the development project. Keep in mind that the use of the word *user* often refers to more than one person. Therefore, a partnership is established for the development project between the user and the developers. In some situations, the user is actually the project leader, which places a significant amount of project responsibility for overcoming the preceding three items directly on the user. With this type of project arrangement, should a system fail on any of the preceding three items (or others), the user is mostly responsible rather than the information systems development staff.

INFORMATION SYSTEMS LIFE CYCLE AND INFORMATION SYSTEMS DEVELOPMENT LIFE CYCLE (SDLC)

Similar to plants, animals, and humans, information systems have a life cycle of their very own. A typical human life cycle starts when a man and a woman plan to have a baby. The couple implements the plan, which leads to pregnancy, baby delivery, development from baby to child, teenager, young adult, adult, golden years, and death. Information systems have a similar life cycle in that they are planned for, then developed using an information systems engineering strategy, implemented for use, evolve as the needs of the business change, and replaced with some other information system at the end of their usefulness to the business.

The information systems development life cycle (SDLC) is the process by which an information system comes to life and maintains its usefulness to a business as it moves from inception to replacement. The SDLC is a much discussed topic among practitioners and researchers because there is so much variability in what it represents. What does this mean? Well, how many different "ways" can you think of for getting from Los Angeles, California to Boston, Massachusetts? You could fly, drive, take a bus or train. Within each of these transportation modes, there are several routing choices to get you from Los Angeles to Boston. The same holds true for the SDLC. We know where we are starting from and where we want to end up, but the process (routing) can be an infinite number of activities, choices, and decisions.

Systems analysis and design textbooks have historically listed as few as five SDLC activities and a maximum of about twelve SDLC activities necessary for a complete SDLC process. A close examination of these life cycles reveals that all necessary activities are covered in one way or another regardless of the actual number of activities identified in the text. So, for our purposes the SDLC consists of the following nine activities:

1. Systems planning—planning for an information system
2. Feasibility study (optional)
3. Requirements determination
4. Conceptual design
5. Physical design, prototyping (optional), construction, and testing, *or* purchase, testing, and integration
6. Conversion from current system to new or changed system
7. Training
8. Implementation
9. Evolution for enhancements and maintenance—this activity actually could involve doing activities 1–8 over and over again.

These same analysis and design textbooks have long presented the preceding or similar SDLC activities in pictorial form, but each has done so by presenting them in a variety of ways, some of which are shown in Figure 1.9. Recognizing that there are professional differences of opinion on the visual presentation and the interpretation of how to carry out each of the SDLC activities, we simply list the activities here for your consideration. In looking for some common ground, many practitioners agree that the SDLC process moves through its activities in mostly a sequential manner while allowing for significant overlap of activities. In addition, inclusion or exclusion of one or more of these activities is allowed as a particular project dictates. Finally, revisiting or looping back to a particular activity as the need arises is also supported and advocated. Perhaps the worst thing that could happen relative to the application of the SDLC is to have a project manager treat it as if its strict adherence must be followed no matter what the cost or consequences to the project.

1. Planning for an information system
2. Feasibility study (optional)
3. Requirements determination
4. Conceptual design
5. Physical design and/or purchase and/or prototyping
6. Conversion from current system to new or changed system
7. Training
8. Implementation
9. Evolution for enhancements and maintenance

a) Sequential or Traditional SDLC

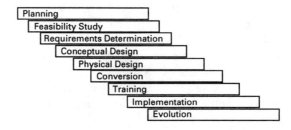

b) Waterfall or Staircase SDLC

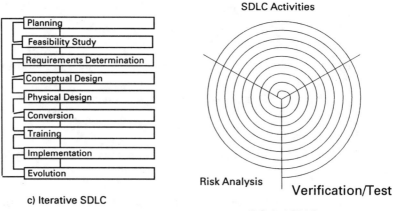

c) Iterative SDLC

d) Spiral SDLC

Figure 1.9   Systems Development Life Cycle (SDLC)

PRINCIPLES TO GUIDE INFORMATION SYSTEMS ANALYSIS AND DESIGN

Over the last three decades a number of systems analysis and design principles have been presented. Although not exhaustive, the list includes:

1. The system is for the user (it is not ours; we do not own the system just because we developed it).
2. A work breakdown structure such as an SDLC should be established for all information systems development projects.

3. Information systems development is *not* a sequential process; it allows for activity overlap, revisiting, inclusion/exclusion, and so on.
4. Information systems are capital investments for the business.
5. The project manager should not be afraid to cancel a project if its success is seriously challenged.
6. Documentation (manual and/or electronic) is a deliverable product during each activity of the SDLC.
7. Senior management support is necessary for the development project.

SUMMARY

This chapter has presented a number of introductory topics in order to set the framework for the remainder of this book. A system, an information system, and an automated information system have been defined and discussed along with the basic components of an automated information system. The three basic characteristics of an automated information system are data, functions, and behavior. The term *systems analysis and design* has many synonyms. An assertion was made that the systems analysis and design process is a difficult human endeavor.

Systems analyst skills, activities, and career opportunities were presented along with a general model of information systems development activities and deliverables. Information systems requirements specifications were discussed followed by discussion of information systems life cycles and the information systems development life cycle. Finally, a discussion and a list of some of the common principles that are used to guide systems analysis and design were presented.

QUESTIONS

1.1 With what other terminology is systems analysis and design synonymous?
1.2 What activities and deliverables are included in analysis?
1.3 What activities and deliverables are included in design and implementation?
1.4 Describe a system and the components of a systems model.
1.5 What two key components distinguish an information system from an automated information system?
1.6 How are data incorporated into an automated information system and what role do they play?
1.7 How do the components of an information system relate and what is the purpose of each component?
1.8 Name and describe each of the basic characteristics of an automated information system.
1.9 What are some of the problems associated with systems analysis and design?
1.10 What sociological and psychological factors play a role in systems analysis and design?

1.11 Who is affected most by an information system?

1.12 Given the previous question, explain how a systems analyst is more than just a programmer.

1.13 Briefly describe the components of the general model of systems analysis and design.

1.14 What are the activities involved in systems analysis and design?

1.15 What is the most critical element to keep in mind when developing an information system? Why?

1.16 How does your answer to the previous question affect a standardized application of a particular information system to all problems of a similar nature?

1.17 What are some of the causes that trigger new or updated information systems projects?

1.18 What is the purpose of the systems development life cycle (SDLC)?

1.19 Briefly describe the activities in an SDLC.

1.20 What are some of the guiding principles of information systems analysis and design?

REFERENCES

COAD, P., and E. YOURDON, *Object-Oriented Analysis* (2nd ed.). New York: Yourdon Press/Prentice Hall, 1991.

EL EMAM, KHALED (ed.), "IBM Federal Systems (Loral)—Space Shuttle Program," *Software Process Newsletter*, IEEE Computer Society—TCSE, no. 1 (September 1994), p. 6.

KOZAR, K.A., *Humanized Information Systems Analysis and Design*. New York: McGraw-Hill Inc., 1989.

MAGLITTA, JOSEPH, "Lean, mean flying machines," *Computerworld,* 28, no. 28 (July 11, 1994), 81–84.

MARCINIAK, J.J., (ed.), *Encyclopedia of Software Engineering*. New York: John Wiley & Sons, 1994.

SCHACH, S.R., *Software Engineering* (2nd ed.), p. 63. Homewood, IL: Irwin, Inc.

SCHLENDER, B.R., "How to Break the Software Logjam," *Fortune*, September 25, 1989, pps. 100–112.

VITALARI, N.P., and G.W. DICKSON, "Problem Solving for Effective Systems Analysis: An Experimental Exploration," *Communications of the ACM*, 26, no. 11 (November 1983), 948–956.

WHITTEN, J.L., L.D. BENTLEY, and V.M. BARLOW, *Systems Analysis & Design Methods* (3rd ed.). Boston: Irwin, 1994.

2

Feasibility Analysis and Requirements Determination

CHAPTER OBJECTIVES (YOU SHOULD BE ABLE TO)

1. Define information systems development feasibility.
2. Define feasibility analysis.
3. Discuss what feasibility analysis allows systems analysts to do.
4. Name and discuss three types of feasibility.
5. Identify the main challenges to requirements determination.
6. Describe the concept of problem domain.
7. Define the subactivities associated with the requirements determination activity.
8. Define and apply the PIECES framework for doing requirements determination.
9. Define and apply Kozar's Requirements Model framework for doing requirements determination.
10. List and discuss Coad's object-oriented framework for doing requirements determination.
11. Discuss techniques for gathering an information system's true requirements.
12. Identify the most common causes of requirements ambiguity.

As its name implies, systems analysis and design is comprised of two major components. This chapter concentrates on the first component—systems analysis. More specifically, it will investigate the feasibility analysis and the requirements determination activities central to systems analysis.

FEASIBILITY ANALYSIS

A major but optional activity within systems analysis is feasibility analysis. A wise person once said, "All things are possible, but not all things are profitable." Simply stated, this quote addresses feasibility. Systems analysts are often called upon to assist with feasibility analysis for proposed systems development projects. Therefore, let's take a brief look at this topic.

Consider your answer to the following questions. Can you ride a bicycle? Can you drive a car? Can you repair a car's transmission? Can you make lasagna? Can you snow ski? Can you earn an "A" in this course? Can you walk on the moon? As you considered your response to each of these questions, you quickly did some kind of feasibility analysis in your mind. Maybe your feasibility analysis and responses went something like this: Can you ride a bicycle? "Of course I can! I just went mountain bike riding last weekend with my best friend." Can you drive a car? "Naturally. I drove to school today and gasoline is sure expensive." Can you repair a car's transmission? "Are you kidding? I don't even know what a transmission is!" Can you make lasagna? "I never have, but with a recipe and directions I'm sure that I could. My mom makes the best lasagna, yum!" Can you snow ski? "I tried it once and hated it. It was so cold and it cost a lot of money." Can you earn an "A" in this course? "I think it would be easier to walk on the moon." Can you walk on the moon? "People have done it. With training, I think I could, and I would like to also."

Each of us do hundreds or thousands of feasibility analyses every day. Some of these are "no brainers" while others are more thorough. Every time we think words like "can I...?", we are assessing our feasibility to do something.

Information systems development projects are usually subjected to one or more feasibility analyses prior to and during their life. In an information systems development project context, **feasibility** is the measure of how beneficial the development or enhancement of an information system would be to the business. **Feasibility analysis** is the process by which feasibility is measured. It is an ongoing process done frequently during systems development projects in order to achieve a creeping commitment from the user and to continually assess the current status of the project. A creeping commitment is one that continues over time to reinforce the user's commitment and ownership of the information system being developed. Knowing a project's current status at strategic points in time gives us and the user the opportunity to (1) continue the project as planned, (2) make changes to the project, or (3) cancel the project.

Feasibility Types

Information systems development projects are subjected to at least three interrelated feasibility types—operational feasibility, technical feasibility, and economic feasibility. **Operational feasibility** is the measure of how well a particular information system will work in a given environment. Just because XYZ Corporation's payroll clerks all have PCs that can display and allow editing of payroll data doesn't necessarily mean that ABC Corporation's payroll clerks can do the same thing. Part of the feasibility analysis study would be to assess the current capability of ABC Corporation's payroll clerks in order to determine the next best transition for them. Depending on the current situation, it might take one or more interim upgrades prior to them actually getting the PCs for display and editing of payroll data. Historically, of the three types of feasibility, operational feasibility is the one that is most often overlooked, minimized, or assumed to be okay. For example, several years ago many supermarkets installed "talking" point-of-sale terminals only to discover that customers did not like having people all around them hearing the names of the products they were purchasing. Nor did the cashiers like to hear all of those talking point-of-sale terminals because they were very distracting. Now the point-of-sale terminals are once again mute.

Technical feasibility is the measure of the practicality of a specific technical information system solution and the availability of technical resources. Often new technologies are solutions looking for a problem to solve. As voice recognition systems become more sophisticated, many businesses will consider this technology as a possible solution for certain information systems applications. When CASE technology was first introduced in the mid-1980s, many businesses decided it was impractical for them to adopt it for a variety of reasons, among them being the limited availability of the technical expertise in the marketplace to use it. Adoption of Smalltalk, C++, and other object-oriented programming for business applications is slow for similar reasons.

Economic feasibility is the measure of the cost-effectiveness of an information system solution. Without a doubt, this measure is most often the most important one of the three. Information systems are often viewed as capital investments for the business, and, as such, should be subjected to the same type of investment analyses as other capital investments. Financial analyses such as return on investment (ROI), internal rate of return (IRR), cost/benefit, payback period, and the time value of money are utilized when considering information system development projects.

Cost/benefit analysis identifies the costs of developing the information system and operating it over a specified period of time. It also identifies the benefits in financial terms in order to compare them with the costs. Economically speaking, when the benefits exceed the costs, the system has economic value to the business. How much value is a function of management's perspective on investments.

Systems development and annual operating costs are the two primary components used to determine the cost estimates for a proposed information system. These two components are similar to the costs associated with constructing and operating a new building on the university campus. The building has a one-time construction cost—usually quite high. For example, a new library addition on campus recently costs $20 million to build. Once ready for occupancy and use, the library addition will incur operating costs, such as electricity, custodial care, maintenance, and library staff. The operating costs per year are probably a fraction of the construction costs. However, the operating costs continue for the life of the library addition and will more than likely exceed the construction costs at some time in the future.

Systems development costs are a one-time cost similar to the construction cost of the library addition. The annual operating costs are an ongoing cost once the information system is implemented. Figure 2.1 illustrates an example of these two types of costs. In this example, the annual operating costs are a very small fraction of the development costs. If the system is projected to have a useful life of ten years, the operational costs will still be significantly less than the development costs.

Two types of benefits are usually identified and quantified—tangible and intangible. **Tangible benefits** are those that can objectively be quantified in terms of dollars. Figure 2.2a lists several tangible benefits. **Intangible benefits** are those that cannot be objectively quantified in terms of dollars. These benefits must be subjectively quantified in terms of dollars. A list of several intangible benefits is shown in Figure 2.2b.

Comparing the benefit dollars to the cost dollars, one can tell if the proposed information system is going to break even, cost the business, or save the business money. Once a project is started, financial analyses should continue to be done at periodic intervals to determine if the information system still makes economic sense. Sometimes systems development projects are canceled before they become operational, many because they no longer make economic sense to the business. Opera-

1. <u>Systems Development Costs</u> (one-time; representative only)

Personnel:
- 2 Systems Analysts (450 hours/each @ $45/hour) $40,500
- 5 Software Developers (275 hours/each @ $36/hour)   49,500
- 1 Data Communications Specialist (60 hours @ $40/hour)   2,400
- 1 Database Administrator (30 hours @ $42/hour)   1,260
- 2 Technical Writers (120 hours/each @ $25/hour)   6,000
- 1 Secretary (160 hours @ $15/hour)   2,400
- 2 Data Entry Clerks during conversion (40 hours/each @ $12/hour)   960

Training:
- 3-day in-house course for developers   7,000
- User 3-day in-house course for 30 users   10,000

Supplies:
- Duplication   500
- Disks, tapes, paper, etc.   650

Purchased Hardware and Software:
- Windows for 20 workstations   1,000
- Memory upgrades in 20 workstations   8,000
- Mouse for 20 workstations   2,500
- Network Software   15,000
- Office productivity software for 20 workstations   20,000

Total Systems Development Costs: $161,670

2. <u>Annual Operating Costs</u> (ongoing each year)

Personnel:
- Maintenance Programmer/Analyst (250 hours/year @ $42/hour) $10,500
- Network Supervisor (300 hours/year @ $50/hour)   15,000

Purchased Hardware and Software Upgrades:
- Hardware   5,000
- Software   6,000

Supplies and Miscellaneous items   3,500

Total Annual Operating Costs: $40,000

Figure 2.1 Systems Development and Annual Operating Costs

a) Tangible Benefits

- Fewer Processing Errors
- Increased Throughput
- Decreased Response Time
- Elimination of Job Steps
- Reduced Expenses
- Increased Sales
- Faster Turnaround
- Better Credit
- Reduced Credit Losses
- Reduction of Receivables

b) Intangible Benefits

- Improved Customer Goodwill
- Improved Employee Morale
- Improved Employee Job Satisfaction
- Better Service to the Community
- Better Decision Making

Figure 2.2   Tangible and Intangible Benefits

tional and technical feasibility should also be continually assessed during the life of a systems development project in order to make adjustments when necessary.

REQUIREMENTS DETERMINATION

The requirements determination activity is the most difficult part of information systems analysis. **Requirements determination** addresses the gathering and documenting of the true and real requirements for the information system being developed. Many textbooks refer to this activity as the "what" portion of information systems development. In other words, the systems analyst is primarily thinking and trying to answer the question, "What must the system do?" during the requirements determination activity. Once information systems development progresses to the design activities, the systems analyst and the programmers focus their attention primarily on the question, "How does the system do what it is supposed to do?"

Why is requirements determination difficult? There are several reasons why this is true. Most are attributed to the fact that this is a highly cognitive and creative activity for all of the members of the development team, including the users. Requirements determination represents perhaps one of the last frontiers still awaiting significant automated and intelligent support. CASE technology, discussed later in the book, is making a contribution in this area, mainly as a documentation and communication aid.

The systems analyst's amount of functional understanding of the problem domain also contributes to the challenge. For example, an analyst who is gathering and documenting requirements for a student course registration problem domain would normally be more effective if he or she already had an understanding of how registration systems work. The reason for this is that the systems analyst with registration problem domain experience would be able to better relate to the user and the problem due to the systems analyst's familiarity and prior experience with registration systems. Therefore, the experienced systems analyst would ordinarily be able to ask more effective and complete questions. One very difficult and costly example of this happened to a colleague a few years ago. Even though he was an experienced systems analyst, he knew very little about financial information systems. His assignment was to gather requirements for a financial information system from the controller of the company. Due to the "question and answer" nature of requirements determination, the controller neglected to mention that he needed a general ledger report in the system. Since the systems analyst knew so little about financial systems, he could not even ask the controller if he needed this type of a report. Needless to say, the controller's "final" financial system was incomplete causing embarrassment, additional time to "fix" the system (i.e., add the general ledger report to the system), and additional cost to fix it. Now my colleague knows about financial systems.

Another challenge for the systems analyst doing requirements determination is the dynamic nature of the problem domain being investigated. The following situations illustrate this. Most businesses are either expanding or shrinking, which causes variability in their information systems needs. Government and labor union regulations change every few months, which might affect a system under development. Leadership within the business may change midstream in a systems development project, causing the business to want to rethink the system. Products that the information systems are to become part of are constantly changing due to technology improvements, manufacturing improvements, market demand for the products, and so on. A company that decides to acquire and merge with another company, such as the AT&T and NCR Corporation and EDS and General Motors mergers of a few years ago, can create some really interesting challenges for the systems analysts of both companies.

Life is dynamic. So is business. Try placing your life on hold for a few months or a year—no way! The same is true for almost all businesses. In spite of this, information systems development must coexist with the business dynamics. In effect, systems analysts are essentially shooting at a moving target during systems development. Today's requirements may be obsolete tomorrow or next week, month, or year. For example, today a company offering public seminars in U.S. cities only accepts U.S. dollars as payment; next month the company may decide to expand its seminars into the international market causing a change to the system requirement for only handling U.S. dollars for payment.

Communication among the information systems development project team members has traditionally been another challenge for requirements determination.

The larger the team, the larger the potential for communication inconsistencies and problems. The secret to success is developing a consistent understanding of the real requirements among all team members. This book is a good illustration of this challenge. The goal for the book is to effectively communicate the essence of systems analysis and design to you, the reader. How effective the book is at doing this is your judgment call. Sometimes information systems requirements are so involved that they could fill a book this size or larger. The user is expected to read through the requirements document (book) and discover inconsistencies, errors, oversights, and so on. This is a very difficult task, especially since the user more than likely already has a full day of work activities in addition to doing such a review.

Every problem domain has its own jargon. For example, computer technology has its own—RAM, ROM, RAID, BPS, Baud, VGA, EGA, SIMM, FDDI, and so on. Sometimes jargon gets interpreted differently by different people. This can cause communication problems, so caution is recommended when using it within any problem domain. For example, ATM is generally considered an automated teller machine, but recently ATM within the telecommunications discipline has emerged to mean asynchronous transmission mode.

Finally, there are still many other factors that can cause problems while doing requirements determination, for example, human factors, such as being tired, or not feeling well, distractions that occur inside a room or on the other side of a room's window during a meeting, stress of team members, and so forth. The probability of these types of challenges existing in every systems development project in varying degrees is quite high.

PROBLEM DOMAIN

A **problem domain** refers to any business area or function. In systems analysis and design, the problem domain refers to the business problem, area, or function being investigated and analyzed. The purpose of the investigation and analysis is to determine the need to purchase, make, or revise an information system to enhance the business activity of the problem domain. For example, an accounts payable problem domain refers to the activity that all companies perform to pay their bills to the businesses they buy materials and services from. Those of us individuals who pay household bills on a recurring basis, such as utilities, rent, cable television, car loans, and others, also are engaging in an accounts payable problem domain when we make our payments. The problem domain for developing the software and other systems components for an automated teller machine would be the automated teller machine problem domain. The problem domain for the software and other system components for a video arcade game would be the video game problem domain. The idea is to determine the scope or boundaries of the problem domain and then assign an appropriate name to it.

Rarely does everything within a problem domain become part of the information system. For example, there is a variety of information about you as a person, such as your name, your residence address, grade point average, courses taken, high school attended, social clubs you belong to, sports interests, hobbies, religion, your favorite actor, actress, sports figure, and so on. In your role as a student, your university's course registration information system problem domain probably has no need to know what high school you attended, your social clubs or sports interests, or who your favorite actor, actress, or sports figures are, so we can eliminate them from this system. On the other hand, in an information system of broader scope, say a student information system on campus which is integrated with the previously mentioned course registration system, there may very well be a need to have this information. Discussions with the information system's user will help determine when to include aspects of the problem domain and when to exclude them. Areas to be included within an information system's problem domain are often referred to as the **information system's responsibility or requirements.**

As mentioned earlier, requirements determination addresses the gathering and documenting of the true and real requirements for the information system being developed. To say this another way, it is the activity systems analysts and users engage in to determine the information system's responsibilities. This is illustrated in a general way in Figure 2.3. For example, you have hundreds of courses available to you at the university, yet you only take a small percentage of those courses (even though it seems like a lot of courses!). You go through some analysis to determine which courses to take. Some of your analysis is based on courses that are required in order for you to graduate, such as general education courses. Other courses may also be required in order for you to graduate with a specific major. Both of these examples are analogous to figuring out what the information system is required to do in order to be useful and successful to the user.

Other courses are selected based on personal preference (e.g., golf, racquetball, physics, and others) and are counted as electives. This could be analogous to including some desirable, but not mandatory, features in the proposed information system. Fi-

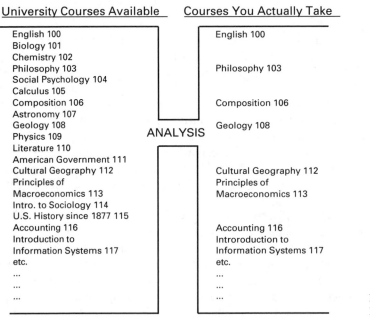

Figure 2.3   A Requirements Determination Process

nally, you may have to take one course from a group of required courses (e.g., take one of the following three courses . . .). This could be analogous to having the user pick one feature from a list of features that could be included in the information system. Determining the responsibilities of an information system involves the use of an analysis technique also, not exactly like the course example here, but certainly similar.

Determining the scope or boundaries of the problem domain is not always easy because it often involves trade-offs and compromises. Therefore, the definition of **requirements** for our purposes is the wants and/or needs of the user within a problem domain. Technically speaking, *requirements* refers to those items that are necessary or essential in the system. Within the systems context, however, requirements often includes those items that may not necessarily be essential but are, nonetheless, desired and, therefore, required.

Perhaps it is this confusion over what is essential, needed, desired, or required of the system that makes it so difficult to systematically articulate exactly what a systems analyst is to do during the requirements determination activity of the development project. In reality the requirements are defined at the beginning of the project along with the system's objectives. However, additional requirements are often identified during later activities of the systems development life cycle (SDLC). For example, new or changed requirements may surface during the testing activity, an activity where it is ultimately determined how the system will be best implemented in the environment. Therefore, while it is nice to think about requirements determination being completed very early during systems development, it is somewhat artificial to close off the definition and gathering of requirements as we move from analysis activities to design and implementation activities.

Using an object-oriented approach to systems analysis and design to analyze the informational, functional, and behavioral requirements of the system helps eliminate the need to artificially close one activity of the SDLC as we move to another activity. Regardless of the approach used for analysis and design, systems analysts need to continuously decide throughout their careers what to glean from current thinking about requirements determination techniques. In most cases articles and books on the subject fall into two broad categories: (1) frameworks or ways to classify requirements into subject areas so that categories of requirements are not overlooked by the systems analyst, and (2) guidelines or heuristics (rules of thumb) that guide the systems analyst toward specific kinds of questions to ask users during the requirements determination activity of the project. Each of these is addressed separately next.

FRAMEWORKS FOR UNDERSTANDING AND DOING REQUIREMENTS DETERMINATION

While there are many frameworks that can be discussed in this section, four have been selected that represent different perspectives of the problem and are discussed here:

1. requirements determination subactivities
2. PIECES framework

3. Kozar's Requirements Model
4. object-oriented requirements modeling activities

Requirements Determination Subactivities

Requirements determination is the general data-gathering activity done during analysis. It has four subactivities—requirements anticipation, requirements elicitation, requirements assurance, and requirements specification. One of the earliest research articles to deal with understanding the requirements determination activity was presented by Naumann, Davis, and McKeen and later expanded by Vitalari's work. Together they have identified four subactivities within the requirements determination activity as listed previously and described in more detail here:

1. Requirements Anticipation. The systems analyst hypothesizes that particular requirements are relevant based on his or her previous experience with other similar systems and knowledge about the problem domain. As you progress through college, you continue to anticipate instructors' requirements for passing their courses. If you have the same instructor twice, you are even more able to anticipate his or her course requirements.

2. Requirements Elicitation. The systems analyst uses this activity to gather the essential requirements from the user employing a variety of techniques, such as interviews, questionnaires, group brainstorming meetings, and voice and e-mail.

3. Requirements Assurance. The systems analyst uses the activity of requirements assurance to validate and verify the requirements with the user as being the real and essential requirements. User walk-throughs in which the systems analyst and the user review documented requirements in detail is one assurance technique.

4. Requirements Specification. This is the activity that the systems analyst uses to explicitly catalog and document the requirements either during or after the elicitation and assurance activities. This is the activity most often associated with automated computer-aided software engineering (CASE) technology, which is discussed later in the book.

The preceding four subactivities are tightly coupled with each other and highly iterative in nature. Systems analysts have often commented that it is difficult to isolate one subactivity from the others because they are so interrelated. Nevertheless, these same systems analysts believe that having a more complete understanding of the detailed subactivities within the requirements determination activity makes them more effective as they gather requirements for a proposed information system.

The PIECES Framework

The second framework to discuss, called the **PIECES** model and first presented by Wetherbe, focuses on the actual work of doing requirements determination. This model is used to classify identified requirements into one of six subject areas—**Performance, Information, Economy, Control, Efficiency,** and **Services.** The goal of the model is to assure the systems analyst and the user that questions will be included during analysis about each of these six essential subjects as it relates to the problem domain. The responses to the questions for each of these subject areas significantly contribute to the definition of the system's requirements. What follows is a brief summary of each of the six subject areas.

Performance questions address how the system needs to perform for the user. Issues of throughput (the amount of work performed over some period of time) and response time (the average delay between a transaction or user request and the response to that transaction or user request) are considered. For example, the systems analyst may ask questions about the needed response time or throughput required on the network, the quality of print needed, or the need to have a graphical user interface or a menu or text type of interface. In other words, the question the systems analyst asks is, "How does the system need to perform in this environment?" Its answer can be multifaceted depending on the needs of the user.

The **information** category provides the basis for the information or data model that the system needs to maintain. Issues dealing with input data, output data, and stored data are considered. The systems analyst may ask the following questions: "What information is required by the users of the system?" or "What outputs are required?" and "What do these outputs need to look like?" These questions need to be addressed and answered while the systems analyst is interacting with the user to define output or report definitions.

Similarly, questions related to input data required in order to produce the outputs are also included in this category, for example, "What input screens are needed?" or "What is the source for the input (where does it come from)?" and "Can the input enter the system with source data acquisition equipment such as barcode scanners, laser guns, mouse, and so on?" Ultimately, the data need to be defined with a high degree of detail, which is discussed further in a later chapter of this book.

The third category in this framework is **economy.** This subject area addresses project development and operational cost information along with any objectives that may relate to economy or savings associated with the system. For example, the systems analyst may ask, "What is the budget for this project?" or "What is a workable solution to the problem worth to the user of this system?" Other questions can include: "What are some anticipated cost savings associated with this system?" and "Are there current manual activities that an automated solution to the problem may affect?" If so, "How will the automated system transform the role of these workers?"

The **control** category is closely associated with system security issues as well as the editing required on the incoming data. For example, questions may be asked related to needed accounting controls for some processes, or at what levels (workstation, user, screen, file, data element, and so on) security is needed. Any issue related to controlling the use of the system, its outputs and inputs, or required controls over the data can be included in this category.

Somewhat related to economy, the other "E" in the PIECES framework refers to **efficiency.** Efficiency is a measure of method correctness. In other words, "Are things being done right?" Efficiency's impact is usually measured at least at one of three levels—corporatewide, department, or individual. Questions related to efficiency are primarily directed toward the impact that any solution must have on the environment. For example, "How can the operations in the office be improved by this system?" and "What values can be added to the environment by using an automated solution to the problem?" are two questions that the analyst can ask in this subject area.

The final category in Wetherbe's PIECES framework is essentially the functional requirements of the system which he associates with **services.** "What does the system need to do in order to solve the problem?" and "What processes need to be performed?" or "How are the objects expected to perform?" and "What do the objects need to be able to do?" are typical questions the analyst asks for this subject area. In addition to functional requirements, services may also include implementation concerns, such as ease of use and needed support for ongoing use of the system, maintenance of the system, and training and documentation requirements.

Kozar's Requirements Model

Kozar's Requirements Model is the third framework and is shown in Figure 2.4. It too focuses on a technique useful to a systems analyst doing requirements determination.

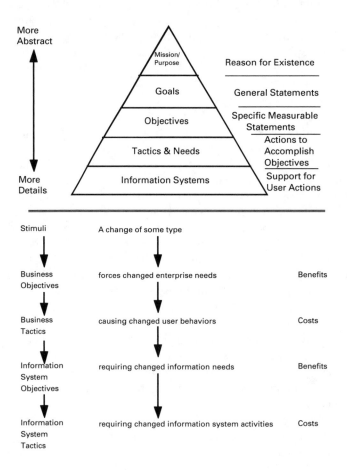

Figure 2.4   The Requirements Model [Adapted from *Humanized Information Systems Analysis and Design,* by K. A. Kozar, 1989, p. 61, McGraw-Hill, NY.]

Instead of classifying requirements into one of six categories as in the PIECES model, the requirements model associates the established business objectives and tactics to information system objectives and tactics. The key to this model is to establish relationships between what the business wants to accomplish and what the information system can do to help.

The model presents five tiers starting with some internal or external stimuli (e.g., problems, opportunities, or directives as discussed in Chapter 1) representing the need or desire for some type of change. Sometimes the change affects the mission of the business, but most often it affects the business objectives that have already been documented. Business objectives are often action-oriented, measurable statements that could lead to one or more ways of increasing a business's revenue or profit, or reducing the business's expenses. For example, "Increase sales of fishing equipment by 10 percent this year" could be a business objective for a sporting goods store. Once new or changed business objectives are established, business tactics to support these objectives can be identified. Business tactics are usually what people need to do to support the business objectives. The same sporting goods store could have "develop five or more new television advertisements" as a business tactic to support the foregoing business objective example. Information system objectives can be identified to support the business tactics followed by identifying the information system tactics necessary to carry out the information system objectives.

Most often, successful use of the requirements model expects that the business has documented specific goal and mission statements, often in a document that is called an **enterprise or business model.** Briefly, an enterprise or business model attempts to answer the question, "Why do we exist?" This type of question can be asked and answered at every organizational level—corporate, division, region, department, section, and so on. The important point for the requirements model is that a general business direction must be provided before the information system requirements can be identified. Although desirable, it is not absolutely essential to have an entire organization's business model defined before applying the requirements model. It is necessary, however, for the business model to be established for the business unit (e.g., division, region, department, section, and so on) that is doing requirements determination using the requirements model approach.

Cost/benefit analysis, described in more detail later in the book, is usually associated with the justification for doing an information systems development project. Using the requirements model, it is often possible to equate both business and information system objectives with benefits and business and information system tactics with costs as shown in the bottom half of Figure 2.4.

In the requirements model, **business objectives** are specific statements of how the organizational goals can be achieved. For example, "increase profit 10 percent each year" and "reduce customer complaints 15 percent each year" could be examples of business objectives. The objectives are business directed, always measurable, usually stated in terms of time and/or money, and in the spirit of total quality management (TQM) often do not have an ending point so that continuous improvement and excellence becomes an ongoing objective.

Business tactics are specific actions that can be taken to realize the business objectives. The business tactics may or may not specifically involve the information systems of the business. Often other noninformation systems activities are associated with business tactics, such as "hire two new order entry clerks" and "install a new telephone system." The business tactics that involve the information system are used to form the final two tiers of the model, the information system objectives and the information system tactics.

Information system objectives are the information system accomplishments, such as using scanners to enter sales data, displaying calculations such as total price and sales tax, printing reports of sales summary information, and so on. These objectives are in direct support of and correlate directly to one or more business tactics. Quite often the information system objectives represent "what the user of the information system sees" when interacting with the information system.

Finally, the **information system tactics** are the information system actions done "behind the scenes" by the information system in order to accomplish the information system objectives or "what the user of the information system sees." For example, doing a gross pay calculation in a payroll information system may be an information system tactic to support the information system objective of producing a payroll check. Information system tactics usually represent the work done by systems analysts and other technical professionals to accomplish the information system objectives.

A good guideline for developing business tactics, information system objectives, and information system tactics is that each business objective leads to one or more business tactics; each business tactic leads to zero or more information system objectives (remember not all business tactics equate directly to the information system); each information system objective will create one or more information system tactics. A partial example of mission, business objectives, business tactics, information system objectives, and information system tactics is shown in Figure 2.5. This list was developed in the classroom with students working on a video sale/rental store information system project.

The requirements model for eliciting and developing requirements has the advantage of building on good business strategic modeling. If the last two tiers of the model, information system objectives and information system tactics, were expanded to consider the PIECES framework, then both views are consolidated into one framework.

The biggest drawback to the requirements model is that in practice businesses often do not have well articulated and documented business models, mission statements, or goal statements. Even when the business realizes that it needs to have these in order to get the most out of information systems, it typically continues with information systems development projects that may not have anything to do with the desired, but undocumented, business goals and objectives.

Ideally, the requirements model provides a good framework for new system development as well as identifying areas that need reengineering or constant reevaluation. In practice, however, the PIECES model may be more practical when the

<u>Mission Statement</u>

To be the video store of choice by successfully providing a generous selection of home video products for sale or rental at competitive prices.

<u>Goals</u>

1. Increase market share and maintain profitability.
2. Offer superior customer assistance and browsing environment.

<u>Business Objectives</u>

(what we want to accomplish for the business)

1. Decrease checkout time for customers by at least 50%.
2. Improve membership management by 50%.
3. Increase memberships by 75% each year for the next two years.
4. Improve inventory management by 60%.
5. Purchase at least one new store each calendar year for the next three years and then begin acquiring several stores each year thereafter.

<u>Business Tactics</u>

(how we plan to accomplish the business objectives)

1.1 Revise the checkout method for rentals and sales to be more efficient and effective.
2.1 Revise the membership management method to be more effective and efficient.
3.1 Implement a marketing strategy to increase membership.
4.1 Revise inventory management to be more effective and efficient.
5.1 Replace/implement accounting and financial systems.

<u>INFORMATION SYSTEMS OBJECTIVES</u>

General Objectives:

A. Provide just-in-time (JIT) training.
B. The systems we implement must be friendly and easy to learn and use.
C. The systems we implement must give considerations to security issues.

Specific Objectives:

1.1.1 Provide an automated system to assist with customer sales/rental check-outs.

2.1.1 Provide and maintain an automated membership database.
    a. provide current (up to date) membership information on demand.
    b. capability to add, change, and delete (remove) membership information.

2.1.2 Provide membership information reports such as (but not limited to):
    a. least used memberships.
    b. most used memberships.
    c. delinquent memberships (both money owing and outstanding rentals).

4.1.1 Provide and maintain an inventory database for both sales and rental items.
    a. provide current (up to date) inventory information on demand.
    b. capability to add, change, and delete (remove) inventory information (sales and rental).

4.1.2 Provide inventory information reports such as (but not limited to):
    a. least popular rentals.
    b. most popular rentals.
    c. delinquent tape rentals outstanding.
    d. products "on order" (purchasing report) for sale and for rental items.

5.1.1 Provide sales reports such as (but not limited to):
    a. sales for a time period (day, days, week, weeks, month, etc.) by product code.
    b. rentals for a time period (same as above).

Figure 2.5   Video Store Requirements Model

business model is not well defined and/or the business objectives lack specific tactics that can direct information system activities.

Object-Oriented Requirements Determination Modeling Activities

Gathering requirements using an object-oriented perspective emphasizes objects, patterns, responsibilities, and scenarios. Be aware that there are several object-oriented methodologies competing in the commercial marketplace, and each of these methodologies has its own term, synonym, or variation for these four generic terms. Each of these is described in significant detail in a later chapter along with the specific object notation used in this book. However, a simple definition for each of these terms may be useful at this time.

An **object** is a person, place, or thing, such as student, faculty, sales clerk, city hall, famous park, ATM machine, and video tape. A **pattern** is a template of objects with stereotypical responsibilities and interactions; the template may be applied again and again, by analogy. Pattern instances are building blocks used to assemble effective object models. For example, a transaction object and transaction line item objects are a familiar pattern or template in business information systems. An actual instance of the transaction to transaction line item pattern is a sales order (transaction) with its associated sales order line items (transaction line item).

Responsibility is associated with objects and has three aspects to it:

1. **What the object knows about itself.** The things that an object knows about itself are called attributes. An **attribute** is a characteristic associated with a person, place, or thing object. Each characteristic has a value or state. For example, the following are attributes: person's name, person's telephone number, person's grade point average, place name, place location, vehicle name, and vehicle type. The following are values or states for the preceding attributes: Ronald Norman, 619.594.3734, 3.75 (pretty good GPA huh?), Central Park, New York City, Mercedes Benz, automobile.

2. **Who the object knows.** A problem domain has many objects within it. Who the object knows identifies relationships between objects. A standard **relationship** is a connection between different types of objects, such as students and courses (relationship: students take courses; courses are taken by students), sales order and line item (relationship: sales orders have line items on them; line items are found on sales orders), and video tape and customer (relationship: video tapes are rented by customers; customers rent video tapes).

3. **What the object does.** This translates into a list of services for each object. A **service** is some functionality that an object is responsible for performing, such as registering for a course, dropping a course, checking out a video tape, purchasing products at the supermarket, and so on.

The last term to be discussed here is a scenario. A **scenario** is a time-ordered sequence of object interactions to fulfill a specific responsibility. A scenario would be developed for each of the preceding services. For example, just as making a cake and changing the oil in your car have detailed steps to accomplish the task, registering for a course also involves several detailed steps to actually accomplish the service. These steps would make up the scenario for this service.

With these simplified definitions in mind, Coad's object-oriented requirements determination modeling approach would involve the following four major activities:

1. Identify purpose and features of the information system.
2. Identify objects and patterns.
3. Establish object responsibilities: "what I know," "who I know," and "what I do."
4. Work out the system's dynamics using scenarios.

Each of these four major activities would be considered within four major model components. Each is discussed in greater detail in a later chapter so they are just listed here for preliminary exposure:

1. Problem domain—activities related primarily to the problem domain under consideration.
2. Human interaction—activities related primarily to the human-computer interface, such as displays (windows) and reports.
3. Data management—activities related primarily to the persistent storage of data, such as databases.
4. System interaction—activities related primarily to the interaction of this system with other systems.

The details of object-oriented requirements determination and the resulting object-oriented model of the problem domain is left to later chapters in this book.

METHODS USED TO GATHER AN INFORMATION SYSTEM'S REQUIREMENTS

Assuming a systems analyst understands what information system's requirements mean in a general sense, the first step in deciding how to gather, document, and validate the requirements is deciding which method(s) to use to gather and document them. There are several methods to pick from. Generally speaking, the methods for gathering requirements can be viewed from global, individual, or collective (group) perspectives.

Starting with the global view of the system, the requirements can be gathered by (1) reviewing current or past reports, forms, files, and so on, (2) conducting research into what other companies have done for the same problem domain, and (3) conducting site visits to similar system installations. The drawback to the global view, and thus all three of its requirements-gathering methods, is that it focuses on what has already been done and may overlook innovations needed for the future. The benefits of the global methods are that they all help to familiarize the systems analyst with the environment that the new system is being proposed for, and they can help acquaint the systems analyst with at least minimum, already established, requirements.

To customize the system requirements to the problems at hand, however, individual requirements are always necessary. Within the individual category, common methods include: (1) **interviews,** (2) **observation,** (3) **questionnaires** or surveys, and (4) **creating a prototype** of the information system in order to obtain feedback from the potential users of the system. Interviews involve at least one systems analyst and at least one user conversing about the information system's requirements. Interviewing for requirements is similar to your interviewing for a job position or a television talk show host interviewing a guest. The purpose is to gather information, hopefully in an interesting way. Observation is the act of the systems analyst going to a specific location to observe the activities of the people and machinery involved in the problem domain of interest. Hopefully, firsthand knowledge of the problem domain's process can be seen by the systems analyst. Questionnaires are feedback forms designed to gather information from large groups of people. No doubt you have responded to at least one questionnaire or survey in your lifetime.

Creating a prototype of the information system can be done on an individual level or in a group setting. The idea is to explore system alternatives by developing small working models of the proposed system so that user reactions can be gathered. It goes along with the notion that "users don't know what they want until the users see what they don't want." Therefore, quite often the value of prototyping at the requirements level is to eliminate the unwanted features of the system, as well as define the desired features.

The main objective at this level of gathering requirements is to find out what the individual user needs or wants from the system. This includes identifying current problems, current needs, future wants, needs and expectations, and getting to know the user well enough to determine what organizational requirements may be necessary in order to make the system functional in the user's environment. One-on-one interviewing coupled with observation is perhaps the most popular method for gathering these requirements but has the drawback of often taking the longest time to accomplish.

As mentioned earlier, prototyping can also be done in the group or collective user setting and has proven to be quite effective. Sometimes prototyping is done in conjunction with **joint application development** (JAD) or rapid analysis techniques of any type. Essentially, JAD and rapid analysis techniques are facilitated groups of users that collaborate in concentrated work sessions to define needed system functions, screens, reports, expectations, and data elements. Often the results of each session can be translated into a prototype, which can be reviewed by the user in subsequent sessions and used to communicate the systems analyst's understanding of what the system needs to be or do.

In addition to prototyping, rapid analysis techniques, and JAD, other group techniques include group brainstorming, electronic JAD (called EJAD), and the use of group systems software often referred to as groupware. Regardless of whether a computer is used to gather the results obtained from meetings, as is the case with EJAD and groupware, the meetings consist of facilitated sessions in which multiple users interact with each other in order to produce an agreed-upon list of requirements.

The facilitator of the group meetings needs to have a clear and precise idea of what the objectives for each group session need to be. For example, if the objectives of a session are to identify potential problems with implementing a new information system and rank order them in terms of severity, the group might brainstorm barriers to implementation in the business followed by developing a rank-ordered list of barriers to implementation. Follow-up sessions can be held to brainstorm tactics that can be used to overcome or address each of the top barriers mentioned.

Gathering requirements as group interactions has several advantages over individual interviewing and observation. First, the group develops its own synergy. Conflicts between or among the users can be readily identified, and a global view of the system is possible. Next the user can see that others are affected by what he or she does and if the group is well facilitated with a way to effectively resolve conflict, communication among the users improves (at least with respect to the requirements of the system at hand). Finally, because the individual ideas of the users are gathered in one place at one time, group methods are more efficient for the project team and provide a natural way to synthesize and consolidate results.

Gathering requirements as group interactions has a few disadvantages as well. Group interactions can be a disaster for the project team when the meetings are poorly facilitated, have no way to resolve conflict, and/or consist of people who are not directly and/or potentially involved with the system. Attention needs to be paid to each of these issues so that they can be minimized or eliminated from group interactions.

Facilitated groups can be used to effectively brainstorm and rank order system preferences, solution attributes or the characteristics the solution needs to consider or include, constraints or limitations to implementation, expectations, and evaluation criteria. System preferences just mentioned include defining system information, data, and functional requirements. The rank ordering can be as simple as classifying each of these items as mandatory, highly desirable, desirable, nice to have, or not necessary.

The steps to use in facilitating groups are to brainstorm with the expectation that as many ideas will be generated as possible. When ideas begin to be repeated or no new ideas come forth, or if the preassigned time is up, idea generation is closed off. All ideas, regardless of "value" or seriousness, are recorded so that everyone in the room can see the list. The list is then reviewed and consolidated where needed. Depending on the length of the list and its intended use, the next step can either consist of rank ordering the items into a priority list, classifying each item into one of several categories, or rated in importance to the system. Categories that are typically used to classify requirements include: (1) essential or required now, (2) nice to have or deferred until later, or (3) nonessential.

The systems analysts can then review the user-generated lists for feasibility and classify the essential and nice-to-have items into categories based on user visibility and technical feasibility. Visibility refers to whether the item needs to be visible to or hidden from the user. The technical feasibility refers to an inclusion possibility with relatively little additional cost, or an impossibility with present cost constraints. Feedback to the user for any modified items needs to be communicated promptly because the output from the group sessions forms the foundation of user expectations.

Recently and coinciding with the development of computer-based group decision support systems (GDSS), the manual group facilitation techniques such as JAD, RAD (rapid application development), and other rapid analysis techniques take on a new dimension as the groups use software to assist in the gathering and documenting of the ideas generated from brainstorming, consolidating the ideas into categories, voting on the ideas using several different voting techniques, and preparing some of the final documentation. The computer-assisted use of these techniques is often referred to as EJAD (electronic JAD). Sometimes CASE technology is used to assist with the documentation of requirements during facilitated, group-based meetings.

The jury is still out on the overall effectiveness of automated support used to assist with gathering and documenting subactivities of requirements determination. Some businesses have had significant success in their use while others have found that the technology seriously inhibits the objectives of the facilitated, group-based meetings. Few practitioners would argue with the fact that computer-based support for these meetings is beneficial when the technology is unobtrusive to the group.

Regardless of whether requirements are gathered electronically or manually at the global, individual, or group level, there are three common threads: (1) feedback to the user for verification is necessary, (2) context-free content is desired, and (3) good communication skills are required. Context-free content means that the solution is not part of the question. For example, the systems analyst might ask: "What problems are likely to be encountered in the environment where the system will be used?" instead of "What problems are we likely to encounter using a database to solve this problem?"

Feedback to the User

In some ways the feedback to the user is what drives the development of many of the new automated software tools used to document systems development. For example, flowcharts and pseudocode were replaced by context and data flow diagrams as primary tools to diagram system processes because these two diagrams are often more easily understood by the user. Similarly, object technology is currently exploring notation and symbols that can be used both by the systems analyst to document the essence of the system and by the user to verify that the systems analyst has indeed gathered the essence of the information system accurately.

Other documentation techniques such as the pages of system narrative and signed-off input and output screen designs are also feedback to the user from the systems analyst. All of these techniques fall short of providing a fail-safe way to assure that the systems analyst has gathered all the requirements. Certainly records and minutes of meetings held, decisions made in prototyping sessions, facilitated group session results, and user interview responses are all available to provide multiple ways to verify that the systems analyst has not misrepresented what the user has indicated is important. No doubt much more research is needed to find improved techniques to document and verify requirements.

Requirements Ambiguity

Gathering requirements is filled with potential problems due to the uncertainty and ambiguity of this highly cognitive activity. Figure 2.6a shows that the goal for gathering requirements is to determine exactly what the user wants and exactly what the user does not want. Systems analysts often assume that what the user does not want is everything not mentioned in what the user does want. When systems analysts start the requirements-gathering activity, they often vacillate back and forth with the user as depicted in Figure 2.6b. Users may ask for something to be in the system when in fact they really do not want it (but genuinely don't know at this time that they don't want it). This is kind of like wishing you could switch places with some sports hero or movie star and then, after doing so, realize how complex and public such a life would be. Switching back to your real self again, you are thankful that you are not Mr. hero or Ms. movie star.

During the requirements-gathering activity, which could span several hours, days, or weeks, the systems analyst and the user explore and iterate around the real requirements as in Figure 2.6c, hoping to come as close to the real requirements goal as possible. The larger and more complex the information system that the systems analyst is working on, the less likely he or she will be to exactly match the goal as shown in Figure 2.6a.

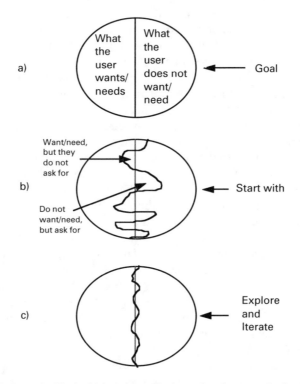

Figure 2.6   Requirements Ambiguity [Adapted from *Exploring Requirements: Quality Before Design,* by D. C. Gause and G. M. Weinberg, 1989, p. 20, Dorset House Publishing, NY.]

The three most common sources of requirements ambiguity are (1) missing requirements, (2) ambiguous words, and (3) introduced elements. Missing requirements are simply those things that are needed and necessary for the success of the information system but are not included for a variety of reasons, such as the user forgetting to mention them, politics within the business, cost, additional time required to include them, systems analyst did not think to ask about them, and so on. In most situations missing requirements happen for legitimate and understandable reasons; however, on a less frequent basis, requirements are missing for prejudicial or illegitimate reasons. In such a situation the user either overtly or covertly attempts to sabotage the system because of personal reasons. This may seem hard to believe, but it happens. Haven't you ever tried to sabotage something that you did not want to do or participate in?

The second most common problem that leads to requirements ambiguity is ambiguous words. Words such as *large, small, inventory, service, user, overnight, weekend, net pay, going out, and inexpensive* have a significant amount of flexibility as to their interpretation and exact meaning. There are thousands more just like these, and all of them leave the use of the word in a particular situation open to a great deal of subjectivity. For example, a requirement like "build me a large bedroom addition to my home" has an incredible amount of subjectivity to it. The builder's interpretation of a large bedroom may be significantly different than that of the person who wants the large bedroom to be built. With that statement as the requirement, even the cost and time estimates to complete the bedroom submitted from different home building contractors bidding on the job could be all over the map. Information systems are no different. The requirements must be exactly spelled out or significant interpretation and subjectivity creep into the project. This often leads to user dissatisfaction with the outcome because it differs from his or her expectations of what the information system would be like and would accomplish.

The third most common problem that leads to requirements ambiguity is introduced elements. Simply put, introduced elements are liberties taken by the systems development team with the hope that "the user will like it" (I call this the "Mikey likes it" syndrome). With best intentions, the systems analyst and even the programmer may make some decisions that affect the system without first getting the approval of the user. It could be something as simple as printing the date at the beginning of a report or as complex as interpreting an ambiguous word such as *overnight* to mean that a report is to be delivered to the user by 10:30 a.m., since that's the time most overnight delivery services advertise. The user may really need the report by 8:00 a.m. and so will be unhappy with the liberty taken by the systems analyst.

As with most sports teams, a systems development team develops synergy as it works together. The team members begin to anticipate responses from other team members based on interaction over time with each other. This is often good, but it can also lead to problems if our anticipated response is different than what would be the actual response had we asked the individual.

Requirements ambiguity is further exacerbated by a systems analyst making observational and recall errors or interpretation errors, and can also be attributed to the complex nature of human interaction. Observational errors occur when the systems analyst observes some operation within the business, say the canning of tuna fish, but misses some aspect of the process for one reason or another. Recall errors occur primarily when a systems analyst tries to either remember something that was said during a meeting or looks at his or her notes of a meeting and has difficulty recalling the correct interpretation of the notes. Have you ever had one of these problems when looking at your own class notes while studying for a test?

Interpretation errors arise when you thought an idea was expressed in one way when in actuality it was expressed or intended to be expressed differently.

The bottom line for requirements gathering and ambiguity is that of time, money, and information systems that do not meet the needs of the user. Studies over many years have suggested that the cost to fix errors, oversights, and other problems that exist in requirements documents escalates the further you are into the project when the problem is detected. For example, the cost of fixing a requirements problem that is discovered a few days prior to actual system usage by the user can be anywhere between 30 and 70 times the cost of discovering and fixing the same problem during the requirements determination activity of the project. A cost of $1,000 to fix a problem during the requirements activity could wind up costing as much as $70,000 to fix the same problem when it is finally discovered very late in the project. Based on this alone, the skills and abilities of the systems analyst to perform requirements gathering successfully is critical to any information systems project.

SUMMARY

Information systems feasibility analysis should take into account operational, technical, and financial feasibility prior to starting and during a systems development project. Information systems are viewed as capital investments and are therefore subjected to the same kinds of investment analyses as are other capital resources of the business. All three feasibility types should be monitored during the life of a systems development project. Information systems have two major cost components, the systems development costs and the ongoing operational costs. Information systems also have benefits, some of them are tangible and others are intangible. The intangible ones may be the most difficult to assess in economic terms, but often these benefits are the ones that can make an information system which looks to be a bad economic investment become a good one for the business.

To summarize the requirements determination activity, one needs to keep in mind that of all the activities in information systems development, this one is perhaps the most cognitive and the least understood. Consequently, very little automated support is available for the cognitive portion of the work done. Most systems analysts and users agree on what the outcome of requirements determination needs to be: a definitive list of things (data, information, functions or services, expectations, and so on) required to build the system. What is usually vague or missing is a well-articulated methodology for arriving at the definitive list.

The PIECES framework and the requirements model provide a way to categorize questions that need to be asked so that dimensions to the problem are not overlooked.

The requirements model actually subsumes the PIECES framework by addressing the need for the system or project from a business perspective. This is desirable so that the information system being developed can enhance some aspect of the business. Coad's method for gathering requirements was given in order to present an object-oriented approach to gathering requirements.

Regardless of the framework, model, or methodology used to gather requirements, there are three generally accepted ways to answer the questions needed to build the requirements list: (1) global research, such as reviewing reports, forms, and files, and reviewing the performance of other companies by contacting or visiting them; (2) individual interviews, surveys, observation, research, site visits, and so on; and (3) group sessions in the form of JAD, EJAD, and/or rapid analysis techniques. Each of these approaches requires that the systems analyst ask appropriate questions, provide feedback to the user, and have good communication skills.

Requirements ambiguity needs to be avoided when gathering and documenting requirements. The three most common sources of requirements ambiguity are missing requirements, ambiguous words, and introduced elements. The systems analyst is responsible for eliminating ambiguity in the final requirements specification document.

QUESTIONS

2.1 What is the meaning of feasibility in an information systems development project context?

2.2 What is feasibility analysis and what is its purpose in information systems development?

2.3 Discuss operational feasibility in an information systems development project.

2.4 What does technical feasibility measure?

2.5 What are some of the implications of economic feasibility (or nonfeasibility) of an information systems development project?

2.6 What are some of the different elements that make up systems development costs and annual operating costs?

2.7 In information systems development, what does the requirements determination process accomplish?

2.8 List several of the problems and difficulties associated with requirements determination.

2.9 Describe how a business's environment can have an effect on the requirements determination process.

2.10 Discuss a particular problem that may result when a systems analyst is working with an unfamiliar topic or functional business area.

2.11 Contrast the problem domain with an information system as a whole. What differentiates the two? And what or who determines what each will consist of?

2.12 What do requirements mean in an information systems context? How does this differ from a common definition of requirements?

2.13 Why is requirements determination regarded as a perpetual activity?

2.14 What are the four subactivities within requirements determination and what is the role of each?

2.15 How do these subactivities relate to each other?

2.16 Briefly describe the main idea behind the PIECES method of requirements determination.

2.17 What are the components of the PIECES model? Briefly describe each.

2.18 What is the first part of Kozar's Requirements Model? Discuss its importance with regard to the model as a whole.

2.19 In what case would the PIECES model rather than the requirements model be a more practical method of requirements determination?

2.20 Name and briefly describe the four major activities in Coad's object-oriented method for gathering and modeling problem domain requirements.

2.21 Name and briefly describe the four major model components in Coad's object-oriented method for gathering and modeling problem domain requirements.

2.22 What are the global, individual, and group methods available for gathering requirements? What are some of the problems with these particular methods?

2.23 What is most important to keep in mind during the requirements determination process?

2.24 What is prototyping and what advantages does it present during requirements determination?

2.25 What are some of the other group-level techniques and how can they be used to enhance the requirements determination process?

2.26 Give an example of how group-level interaction in requirements determination can fail.

2.27 Describe the steps that facilitated groups take during requirements determination.

2.28 What are the three common elements essential to requirements determination, regardless of the method used to gather?

2.29 What is the main goal behind requirements determination?

2.30 Discuss some of the problems that lead to requirements ambiguity. How can these problems be alleviated?

2.31 Why are correcting errors and realizing deficiencies so important during the requirements determination process?

REFERENCES

CHONOLES, M.J., "Success with Object Analysis: Lessons Learned on How to Chart Your Course," *Object Magazine* (February 1994), 50–58.

COAD, P., D., D. NORTH and M. MAYFIELD, *Object Models: Strategies, Patterns, and Applications*. Englewood Cliffs, NJ: Prentice Hall, 1995.

DAVIS, A.M., "Requirements Engineering," in *Encyclopedia of Software Engineering,* ed. J.J. Marciniak. New York: John Wiley & Sons, Inc., 1994, pps. 1043–1054, Vol. II.

DAVIS, A.M., *Software Requirements Analysis and Specification.* Englewood Cliffs, NJ: Prentice Hall, 1990.

DAVIS, GORDON B., "Strategies for Information Requirements Determination," *IBM Systems Journal,* 21, no. 1 (1982), 4–29.

GAUSE, D.C., and G.M. WEINBERG, *Exploring Requirements: Quality Before Design.* New York: Dorset House Publishing, 1989.

HSIA, P., A. DAVIS, and D. KUNG, "Status Report: Requirements Engineering," *IEEE Software,* 10, no. 6 (November 1993), 75–80.

IEEE Software, vol. 11, no. 2 (March 1994), theme issue on requirements engineering.

KOZAR, K.A., *Humanized Information Systems Analysis and Design.* New York: McGraw-Hill Inc., 1989.

NAUMANN, D.J., G.B. DAVIS, and J.D. MCKEEN, "Determining Information Requirements: A Contingency Method for Selection of a Requirements Assurance Strategy," *The Journal of Systems and Software,* 1 (1980), 273–281.

NORMAN, R.J. and G.F. CORBITT, "The Operational Feasibility Perspective," *Journal of Systems Management,* vol. 42, no. 10 (October 1991).

VESSEY, I., and S.A. CONGER, "Requirements Specification: Learning Object, Process, and Data Methodologies," *Communications of the ACM,* 37, no. 5 (May 1994), 102–113.

VITALARI, N.P., "An Investigation of the Problem Solving Behavior of Systems Analysts," unpublished Ph.D. dissertation, University of Minnesota, 1981.

VITALARI, N.P., "A Critical Assessment of Structured Analysis Methods: A Psychological Perspective," in *Beyond Productivity: Information Systems Development for Organizational Effectiveness,* ed. Th.M.A. Bemelmans: Elsevier Science Publishers B.V. (North Holland), 1984, pp. 421–433.

WOOD, J. and D. SILVER, *Joint Application Design.* New York: John Wiley & Sons, 1989.

WETHERBE, JAMES C., *Systems Analysis and Design* (2nd ed.). St. Paul, MN: West Publishing Company, 1984.

WHITTEN, J.L., L.D. BENTLEY, and V.M. BARLOW, *Systems Analysis & Design Methods* (3rd ed.). Burr Ridge, IL: Irwin, 1994.

3

An Object-Oriented Methodology and Model

CHAPTER OBJECTIVES (YOU SHOULD BE ABLE TO)

1. Define and describe a methodology.
2. Define and describe traditional, structured, information modeling, and object-oriented methodology classifications.
3. Discuss classification theory and its relationship with object-oriented methodologies.
4. Define and discuss key characteristics of an object-oriented methodology.
5. Define and describe Coad's object-oriented methodology.
6. Define and describe Coad's object model activities.
7. Define and describe Coad's object model components.
8. Describe Coad's object-oriented notation for creating an object model of the problem domain.
9. Demonstrate general familiarity with Coad's object-oriented notation used to depict class, class-with-object, and responsibilities.
10. Demonstrate general familiarity with Coad's object-oriented notation for attributes and services.

METHODOLOGIES

How do you change the oil in your car? How do you bake a cake? How do you study for a test? No doubt, you have an answer for all of these questions. Assuming you were going to do one of these three tasks (baking the cake sounds best—yum!), you probably have "a way" of doing it. Is your way the correct way? Is your way the only way? Is your way the best way? If you answered yes to all three of these questions, you are in trouble! Why? Because you may be suffering from a severe case of a big ego! Someone once said, "Systems analysts with huge egos to satisfy should be extinct." I tend to agree with this statement.

Don't get me wrong. There's nothing wrong with having confidence in "your way" of doing things, but maturity and professionalism as a systems analyst demand that you recognize and acknowledge that there are other ways of accomplishing the same task. The way something is done in systems analysis and design is usually called a **methodology.** When talking about methodologies, some people refer to them as the strategy, steps, directions, or actions they choose to do or think about something.

Methodologies can be (1) obtained or purchased from another business (you purchase a methodology for baking a cake when you buy the cake mix—it's on the box), (2) created by your own business or team (you may have created your own way to study for a test), or (3) a combination of the two. For example, you may use the cake mix methodology generally but take shortcuts (like not using measuring cups and spoons), or do things a little differently along the way (like putting the eggs in the bowl before the cake mix even though the methodology says to put the cake mix in the bowl first). Making changes to an existing methodology creates a new one—your own—and is often referred to as a hybrid methodology because of the changes that were made.

There are literally thousands of methodologies for developing information systems because most businesses prefer to create a hybrid methodology in order to maximize the methodology's utility to its own organizational culture. Over the last 40 years of information systems development, four general methodology classifications have evolved—traditional, structured, information modeling, and object oriented. Figure 3.1 presents each of these and highlights some of the more commonly associated techniques and tools used as part of each methodology.

Some colleagues believe that prototyping should be thought of as yet a fifth general methodology classification. On the other hand, prototyping and its myriad of variations can often be introduced as part of any of the four general methodology classifications presented here. Therefore, prototyping will be considered a technique used as part of a methodology and will be presented as such in this book. Many systems analysis and design projects incorporate some aspect of prototyping, since there are many benefits associated with prototyping. These benefits are discussed in the prototyping section of the book.

The Traditional Methodology

Today classifying one's methodology as being traditional is often a kind way of saying that you have no methodology at all or, at best, an unstructured, unrepeatable, unmeasurable, ad hoc way of doing something. Can you develop a system using a traditional methodology? Of course! It's done by someone every day! A team consisting of exactly one member (you developing a system for yourself) probably could use and get away very nicely with using this methodology. The traditional methodology becomes

	---------------Techniques and Tools Representing---------------			
	System Flows	Data	Communication with Users	Process Logic
Traditional	System Flowchart	Forms, Layouts, Grid Charts	Interviews	English Narrative, Playscript, Program Flowcharts, HIPO
Structured	Data Flow Diagram	Data Dictionary Data Structure Diagrams, E-R Diagrams	Interviews, User Reviews, JAD Sessions	Decision Tree/Table, Structured English, Structure Charts, Warnier-Orr Diagram
Data Modeling (Info Rmation Engineering)	Business Area Analysis, Process Model	Business Area Analysis, E-R Diagrams	Interviews, User Reviews, JAD Sessions, Brainstorming	Business Systems Design
Object-Oriented	Object Model	Object Model Attributes	Interviews, User Reviews, JAD Sessions, Brainstorming	Object Model Services, Scenarios, Decision Tree/Tables, Structured Englilsh

Note: Prototyping as a technique with associated tools can be introduced into any of the four methodologies where deemed appropriate by the project manager

Figure 3.1   Systems Development Methodology Techniques and Tools

much less effective as the size of the team grows or as the complexity of the information system grows. For reference purposes, some tools used in conjunction with a traditional methodology are system flowcharts and hierarchical input-process-output charts (HIPO). The implication being made here with respect to a traditional methodology is that this methodology has serious problems in most information systems development situations; however, the tools used to support it have applicability in certain development situations.

Structured Analysis and Design Methodology

The structured analysis and design methodology classification, often referred to as the data flow modeling methodology, emerged in the mid-1970s to complement structured programming, and today is one of the dominant methodologies being utilized for systems development. In this methodology, the real world is described by the flow of data through an information system and the transformations of the data into information as the data move. A structured analysis and design methodology, as its name implies, adds dimensions that allow it to be rigorous (structured), repeatable, and measurable. In addition, a new dimension has been added starting in the early 1980s—it can be enhanced by automated support called computer-aided software engineering (CASE). The predominant focus of this methodology is to do the analysis and design from a functional perspective. In other words, the analyst uses a "what does the system have to do?" problem-solving approach. With the introduction of CASE tools around 1984, computerized software support has emerged to provide assistance for drawing the methodology notations, validating and verifying the methodology models that the notations represent, and allowing management to monitor the progress of a project via computerized project management support.

For reference purposes, names such as Constantine, DeMarco, Gane, Hatley, Myers, Orr, Paige-Jones, Palmer, Sarson, Stevens, Ward, Warnier, Yourdon, and others have made significant and documented contributions to the structured analysis and design methodology classification. Tools such as data flow diagrams (DFDs), structure charts, Warnier-Orr diagrams, Petri nets, and data dictionaries are often used in conjunction with a structured methodology. Remember, conceptually there are only two dominant generic structured analysis methodologies (Yourdon, and Gane and Sarson), but businesses have in most cases created their own hybrid versions of the generic ones.

Information Modeling Methodology

The information modeling methodology classification, often referred to as the data modeling methodology or information engineering methodology, emerged in the early 1980s as businesses began to embrace database management systems. Today it too is a dominant methodology used to assist in the development of information systems. An information modeling methodology subscribes to the notion that good engineering is simple engineering. It approaches the development of information systems predominantly from an information perspective rather than a functional perspective as does the structured methodology. The real world is described by its data and the data's attributes and their relationships to each other. As with the structured methodology, information modeling adds dimensions that allow it, too, to be rigorous (structured), repeatable, and measurable, and automatable. The predominant focus of this methodology is to do the analysis and design from an information perspective. In other words, the analyst uses a "what information does the system have to be able to provide the user?" problem-solving approach. There is only a handful of purchasable CASE technology products that fully support an information modeling methodology, most notably from Texas Instruments and Knowledgeware Corporation.

For reference purposes, names such as Peter Chen, James Martin, and Ian Palmer among others have made significant and documented contributions to the information modeling methodology classification. Tools and techniques such as an entity-relationship diagram, business area analysis (BAA), and information models are often used in conjunction with the information modeling methodology.

Please note that the problem-solving strategy used in this methodology is different from that used in the structured methodology, for therein lies the fundamental difference between them. Each methodology must address both function and information in order to successfully address the information systems needs. It's really a matter of the problem-solving strategy used to address the functions and information. Do you begin with functions or information? The structured methodology advocates would say "functions." The information modeling methodology advocates would say "information or data." Both sides would agree that they need both functions and data, but they disagree with the fundamental problem-solving strategy used to discover the functions and data. In reality, both the structured and information engineering methodologies have strengths and weaknesses.

Object-Oriented Methodology

The object-oriented analysis and design methodology classification emerged in the mid- to late 1980s as businesses began to seriously consider object-oriented programming languages for developing systems. Even though Simula, circa 1967, is credited as being the first object-oriented language, popular object-oriented languages such as Smalltalk, C++, Objective C, and Eiffel came into their own in the 1980s. All of these object-oriented languages approach programming from a significantly different paradigm than previous programming languages. Rather than follow the structured, deterministic, and sequential programming paradigm associated with languages such as COBOL, Fortran, C, Basic, PL/1, Ada, and others, these languages follow the approach pioneered by Simula based on objects, attributes, responsibilities, and messages.

Often history repeats itself. Such is the case with object-oriented analysis and design. Just as structured analysis and design emerged as a complement to structured programming in the 1970s, object-oriented analysis and design emerged as a complement to object-oriented programming in the 1980s. The problem-solving strategy inherent in object-oriented programming is to approach the problem from an object (e.g., person, place, or thing) perspective rather than the functional perspective of traditional and structured methodologies or the information perspective of an information engineering methodology. Because of this, object-oriented programming is becoming the dominant programming strategy for behavioral graphical user interface (GUI) software, and software that runs on distributed and heterogeneous, client-server computer hardware platforms.

Since object-oriented programming's problem-solving strategy is unique, the need for an object-oriented systems analysis and design approach leading up to the programming task makes sense. The older, more mature methodologies have limitations in representing a model of the information system that can readily be implemented using an object-oriented programming language primarily due to their focus on functions or data, not objects. This is not a statement that the older methodologies are inferior. Rather, they are just different and well suited for the job they were intended for.

Object-oriented analysis and design has good and bad news associated with it. The bad news is that there is yet another new methodology on the market. The good news is that much can be borrowed from the other pervading methodologies and applied to the object-oriented methodology, which some believe makes it more evolutionary than revolutionary. In other words, experienced systems analysts need not throw out all their knowledge and experience with the other methodologies, which would tend to demoralize the veterans in the field. Rather than the object-oriented analysis and design methodology being revolutionary, it is evolutionary in the "borrowing from what we already know" sense. The methodology notations and subsequent models it represents are new, yet similar to other notations.

The most difficult aspect of learning an object-oriented methodology and notation for experienced systems analysts is the transition from a functional or information (data) centered problem-solving strategy to that of an object problem-solving strategy. Moving from a "function think" or "data think" problem-solving strategy to an "object think" strategy is not an easy one. For example, just think of the difficulty you would have retraining yourself to type using a Divorak keyboard layout, which is significantly different from the standard Qwerty keyboard layout that we are all familiar with! Study, practice, experience, and time will all be important for experienced systems analysts to make a successful transition in their thinking about the problem domain and then determining requirements and documenting them using an object-oriented notation and methodology. The use of the object-oriented methodology (or any other for that matter) for the novice systems analyst or trainee should be less difficult due to the limited experience he or she has with other methodologies. In other words, novices have less experience to retrain. Someone who does not know how to type using a keyboard has little or no keyboarding memory to retrain.

KEY CHARACTERISTICS OF AN OBJECT-ORIENTED METHODOLOGY

As discussed in Chapter 1, systems analysis and design is a complex process. Accommodating this complex process requires systems analysts to exploit commonly understood characteristics or principles for managing complexity. As cited earlier, today's information systems are complex and involve tens of thousands, hundreds of thousands, or millions of lines of programming code. In fact the systems being developed today are far more sophisticated than those of just five years ago. A personal computer word processor was "feature poor" back then compared to the ones currently available for purchase. Shrink-wrap software has moved into the competitive arena of hundreds of thousands or even millions of units in commercial sales. As a result, each new version or release of the software must exceed the features and benefits of the older version. Not only that, new releases often must exceed the capability of their strongest software competitors.

Some common characteristics for managing complexity during systems analysis and design are available for systems analysts to use to one degree or another regardless of the SDLC used to develop the information system. At least eight characteristics or principles for managing complexity are also considered foundational and generally accepted characteristics of object-oriented analysis, design, and programming. Each of these common characteristics are presented here for your consideration. You may already be familiar with many of them because they often have a much broader applicability to your own personal life, not just to software development. They are:

1. common methods of organization
2. abstraction
3. encapsulation or information hiding
4. inheritance
5. polymorphism
6. message communication
7. associations
8. reuse

Not all object-oriented books discuss all eight of these characteristics. However, because I believe they are important, I have included them here. These eight concepts are not necessarily new to object-oriented systems development as other methodologies incorporate one or more of them in some way. However, together all eight form the fundamental building blocks for object-oriented technologies. Each is discussed next.

1. Common methods of organization assist with organizing an information systems model as well as the software that is ultimately written. The common methods of organization that are applicable here are:

a. Objects and attributes or characteristics. For example, you could be the object and your name, address, height, weight, color of eyes, date of birth, and so on are the characteristics.

b. Wholes and parts. A desktop computer system—the whole—usually consists of a box with all kinds of hardware in it, a printer, a monitor, a keyboard, and a mouse—the parts, or your television set is the whole and the individual parts that make up the television set are the parts.

c. Classes and members. Conceptually this is similar to wholes and parts, but its examples are often different and do not seem to fit neatly into a "wholes and parts" classification. For example, a computer club would be a class and the individual members of the computer club would be the members, or your analysis and design course would be the class and the students in the class would be the members.

2. Abstraction is the principle of ignoring those aspects of a problem domain that are not relevant to the current purpose in order to concentrate more fully on those that are. Abstraction is a concept that you use every day. It has to do with the amount of detail you care to get involved in. In systems analysis and design, this is called levels of abstraction. Let's say that your friend decides to bake a cake. He offers you the choices of either helping him bake the cake and then helping him eat it or just helping him eat it. If you choose the first option, you are demonstrating a detailed level of abstraction because you have chosen to get involved with the details of making a cake. If you choose the second option, you are demonstrating a generalized level of abstraction because you chose not to deal with the details of baking the cake, only eating it. Telling your mother that you love her is abstract, or generalized, while telling her one or more of the reasons why you love her is specific and thus a more detailed level of abstraction.

In systems analysis and design, abstraction is used to identify essential information system requirements while simultaneously eliminating nonessential aspects. An abstraction intentionally ignores some qualities, attributes, or functions of an information system in order to focus attention on others. An abstraction is also a summary, covering the highlights and leaving out the details. It also omits the pieces of the system that are not necessary for understanding the system at a given level of detail. Maps typically represent a good example of abstraction. You can get maps of the United States, individual states, counties, cities, zip codes, and so on. Each map cited here in succession is less abstract and hence more detailed than the prior one. In addition, a city map intentionally leaves off the details of the surrounding county, state, and country which it is a part of in order to allow you to focus on the details of the city. Another example of abstraction would be to view an aerial picture of your entire university campus. Such a picture would not have detailed information regarding a specific office on a specific floor within a specific building on the campus. Abstraction has been a common characteristic in all of the prevailing information systems methodologies. Therefore, experienced systems analysts can continue to utilize this system-building characteristic as they transition to an object-oriented methodology.

3. Encapsulation or **information hiding** is the notion that a software component (module, subroutine, method, and so on) should isolate or hide a single design decision. This notion is based on the work of David Parnas dating back to the early 1970s. An example of programming encapsulation may illustrate this notion best. The use of a single software component to look up student account numbers in a student database isolates this lookup feature. Each time the information system needs to look up a student account number, this routine is used, since the logic to look up student account numbers has been encapsulated within this one routine. If a new lookup routine is created, tested, and used each time there is a new need to look up a student account in the database, two potentially negative things happen. First, prior work—software construction and testing as a minimum—is replicated. Second, additional maintenance activity will be required for the information system should it ever be necessary to alter the student lookup routine(s).

In systems analysis and design, systems analysts decompose the problem domain into small encapsulated units. These analysis and design decisions eventually become software modules. Encapsulation helps to localize their volatility when changes and maintenance are required after they have become software modules.

A further elaboration of encapsulation or information hiding is any mechanism that allows certain portions of the information system to be "hidden." This can be useful in at least two different situations:

a. When it is important that people only use or have access to a certain subset of the complete information system. For example, while an information system is being developed, development team members may be assigned to work on a certain part of the system and not allowed access to other parts of the system being worked on by other team members. Once the information system is operational, certain users may only be allowed to use a few of the features of an information system while other users may have access to many more features.

b. To purposely prevent other components of the information system from being aware of or unable to take advantage of certain other components in the system. This situation addresses another aspect of encapsulation—assignment of responsibilities. Just as in real life you have certain responsibilities, a specific component of an information system has its responsibility, say dispensing cash in an ATM, while other components of the same system have their responsibilities, which exclude the dispensing of cash since that function is done by another component.

Another way to think about encapsulation is from a programming perspective, which has long since advocated the creation and use of small blocks of function-specific code (e.g., procedure, subroutine, paragraph, and so on) that can be assembled into a working computer program in a modular fashion, much like the plug-compatible stereo components in homes today. Each of these blocks has encapsulated within it one or a few design functions. For example, each of the following functions could be represented by a small block of computer programming code and then bundled together to perform a higher-level function, such as registering for courses for this coming school year—"get student identification (Id) number," "validate student Id number," "check student Id number for outstanding fines," "get course Id to add to student's schedule," "determine if a seat is available for this course Id," "determine student Id number's eligibility to take this course Id," and so on.

Finally, encapsulation has been a common characteristic in all of the prevailing information systems methodologies. Therefore, experienced systems analysts can continue to utilize this system-building characteristic as they transition to an object-oriented methodology. There is one subtle difference. With the older, more mature methodologies, encapsulation was usually limited to encapsulation of functions separately from encapsulation of data. With the object-oriented methodology, encapsulation incorporates both functions and data together into objects. This notion will be discussed in more detail later in the book.

4. Inheritance is a mechanism for expressing similarity. Just as you inherited certain physical features from your mother and certain ones from your father, so also can information system components inherit certain things from related components. For example, Figure 3.2 shows that a hierarchical or parent–child relationship exists between a group called Person and groups of Faculty, Administrator, and Student. We purposely are using the singular for each of these words in keeping with the notation

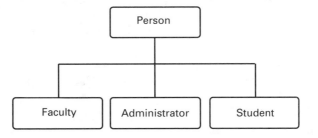

Figure 3.2   Inheritance Hierarchy

that is used throughout the book, but keep in mind that Administrator means one or more administrators, as does student and faculty.

A sibling relationship exists between Faculty, Administrator, and Student. There are certain characteristics that can be associated with Person and thus inherited by all the children in the hierarchy—Faculty, Administrator, and Student—such as name, address, phone number, and date of birth to name a few. Other characteristics may be unique to a single sibling. For example, office hours may only be applicable to Faculty, job title may only be applicable to Administrator, and grade point average may only be applicable to Student. These characteristics would be associated with the specific child node in the hierarchy. Similarly, there are certain things that a Person can do that would be common to Faculty, Administrator, and Student, such as eat at the cafeteria and walk on campus. Likewise, there are certain things that are unique to each of the children nodes. A Faculty may assign grades and add students to the class, whereas an Administrator may manage people and key in administrative data to the computer, while a Student could register for classes and take an exam. These actions would be associated with the specific child node that they belong to. As is usually the case in real life, inheritance in systems development is a one-way street—down the hierarchy.

5. Polymorphism in the general sense is the ability to take on different forms. For example H_2O can take on three forms—water, steam, or ice. In a sense, you could say that you are polymorphic with respect to approaching a traffic signal in your car. You respond differently depending on the status of the signal light. In software, for example, a print routine may be available that can print textual characters and numbers and graphic symbols and images on a printer. Because the print routine knows how to print both text and graphics, it is considered polymorphic. Polymorphism in computer programming isn't new, but its use during object-oriented analysis and design by systems analysts probably is.

6. Message communication is the way objects in an object-oriented methodology communicate with each other. It is similar to a subroutine call with parameters or a paragraph or procedure call in some other conventional programming languages such as Fortran, COBOL, C, or FoxPro. This concept is widely understood by practicing systems analysts so it, too, can be directly transferred into an object-oriented methodology by them.

7. Associations are useful to assist with relating things in an information system to each other. Things that happen at the same time could be associated as well as things that happen under similar conditions. For example, VISA, Mastercard, Discover, and other credit card companies send out periodic statements showing us our balance owing. These companies often take this opportunity to insert a number of advertisements along with the statement, since the postage is often the same with or without the advertisements.

8. Reuse. Everywhere you look in your world you see reuse in one form or another. The same power cord can often be used for either your computer, your monitor, or your printer. Apartment windows often come in standard sizes, and dozens of a certain size window are used in a large apartment complex. The can of orange juice you bought last time probably looks just like the one you bought today. How many shirts or blouses do you have that reuse the same type of button? How often do you reuse a textbook that was used by some other student before you? Have you ever reused a video cassette to record over some other event that you recorded at a previous time?

Reuse is a much talked about concept in systems analysis and design, but one which is still struggling to make a significant contribution to the development process due primarily to two factors. The first factor is the retraining of the mindset of the analysts and programmers who are accustomed to creating their own systems models and code. This translates to the "not invented here" mentality held by some systems analysts and programmers. In other words, they want to create and debug their own "whatever" rather than use someone else's "whatever" that does the same thing (and is already debugged I might add!). Management can address and overcome this problem given proper amounts of time, retraining, and other motivators for the systems analysts and programmers who are resistant to reuse.

The second factor relates to the administrative factors that must be established for setting up a library of reusable models and code. This is no easy task for a variety of reasons including strategy to accomplish, personal, organizational, cultural, and legal aspects.

As used in our world, reuse can take one of three forms: (1) sharing, (2) copying or cloning, or (3) adjusting. Each is discussed here.

An example of **sharing** is when you use a textbook and then sell it back to the bookstore, which in turn resells it to another student. Or everyone living in or visiting your apartment or home shares (reuses) the telephone you have. In software development, a subroutine, module, paragraph, procedure, service, or file definition are all potentially reusable. A small procedure could be created that only looks up student identification numbers in a student database and then responds with valid or invalid. Such a procedure could be used over and over again, intact, within an information system.

An example of **copying** or **cloning** is what is done every day in the manufacturing industry. An automobile model is prototyped and then moved into production to produce thousands of that model. In software development, we could have a paragraph of a COBOL program that we copy and put into another COBOL program, giving two paragraphs that are identical.

An example of **adjusting** is a variation on the copying or cloning. Using the automobile example, as a certain model moves down the manufacturing assembly line, standard features for this automobile are assembled (copying) as well as different combinations of options are assembled such as air conditioner, CD player, spoiler, tinted glass, and so on (adjusting). Granted that these options are standard reusable components just assembled in different combinations in the particular automobile. However, the final product, the automobile, is a variation on the standard make and model.

A variation of the adjusting type of reuse is the situation in which the standard item is actually changed and then used in some other application. For example, a computer modem cable can have two of its wires physically swapped within the cable housing, and then this cable becomes a different kind of modem cable called a null modem cable—an adjustment to the standard cable.

An example of adjusting in systems analysis and design is when a programmer locates a subroutine, module, paragraph, and so on from a library of reusable components. The subroutine becomes the programmer's beginning point for creating a new subroutine, somewhat similar to the original library version. Then the programmer begins making adjustments to the subroutine, such as removing some of the code in the subroutine, changing other parts of the code, and adding new code to this new subroutine to meet a specific purpose.

TWO CLASSIC PROBLEMS RESOLVED WITH OBJECT-ORIENTED ANALYSIS AND DESIGN

Two of the most significant problems that have continuously been associated with the pervading analysis and design methodologies appear to be solved or at least significantly reduced with an object-oriented analysis and design notation and methodology. The first of the two problems is that most other methodologies construct separate models for the functional and information views of the proposed information system as shown in Figure 3.3a. In addition, a third component—behavior—is becoming more important in today's information systems and needs to be modeled as well. The popularity of interactive, graphical user interfaces and object-oriented operating systems, such as Windows, Macintosh, OS/2, Motif, NextStep, Taligent, and others has shifted programming from a deterministic approach to an event-driven approach. This introduces the need for a third view of the system—a behavioral view, and this view is rapidly becoming quite important.

The function and data multiple models have value in that each depicts some portion of the same information system. The difficulty lies in the fact that neither systems analysts nor CASE software have been able to completely validate and verify the consistency and accuracy of the integration of the two models simply because they each represent a different perspective of the information system. Coad's object-oriented analysis and design notation and methodology propose the use of one model that is capable of representing all three views—function, information, and behavior. This is a very significant move forward because more rigorous validation and verification of the single model is now possible.

The second problem that has continuously plagued information systems development is the proverbial transition from analysis to design. Once again, in most methodologies, a completely new model or set of models is drawn that communicates more specifically to the programmers for the impending programming effort. Analysis models have historically been a communication tool between the systems analyst and the user. The design models have historically been a communication tool between the systems analyst and the programmer. With Coad's object-oriented analysis and design notation and methodology there is no transition to design. What happens instead is a progressive expansion of the analysis model as it progresses through de-

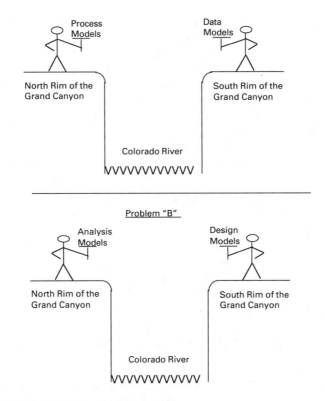

Figure 3.3   Two Classic Systems Development Problems

sign to include additional refinement details that address the programming task that is approaching. So systems analysts, users, and programmers deal with one model of the information system from analysis all the way through design, coding, testing, and implementation. The resulting information system can be validated back to the original analysis model because it is still the most valid model of the system even after the system is implemented.

For reference purposes, names such as Booch, Coad, Cox, Henderson-Sellers, Jacobson, Mellor, Meyer, Rumbaugh, Shlaer, Wirfs-Brock, and Yourdon among others have made significant and documented contributions to the object-oriented methodology classification. Tools and techniques such as dynamic and static object-oriented models, state-transition diagrams, and use case scenarios are often used in conjunction with object-oriented methodologies put forth by methodologists other than Coad.

CLASSIFICATION THEORY

Object-oriented methodologies and programming are based on classification theory, a fundamental concept most of us learned prior to or during our first few years of elementary school. Classification theory simply states that people constantly and consistently use three methods of organization to help clarify their thinking about something. The three methods are:

1. The differentiation of experience into particular objects and their characteristics—**objects and their characteristics.** An example is being able to distinguish between a tree and its size or spatial relationship to other things in our world, such as a rock, flower, bush, bicycle, and so on.
2. The distinction between whole objects and their component parts—**wholes and parts.** For example, a tree is made up of leaves, branches, and roots, while a bicycle is made up of wheels, rims, spokes, pedals, seat, and handlebars.
3. The formation of and the distinction between different classes of objects—**classes or groups of objects.** Examples include grouping trees (maple, elm, palm, eucalyptus) into a tree class; grouping buildings (homes, apartments, offices, churches, schools) into a building class.

People's common understanding and use of classification theory is one strong reason why an object-oriented perspective is a viable approach to analyzing, designing, and building information systems. People are already familiar with it. Learning something totally new is not required.

COAD'S OBJECT-ORIENTED METHODOLOGY

Several authors and researchers have put forth specific object-oriented notations and methodologies (refer to an earlier section of this chapter for some names; also the chapter references). For the purpose of this book, only Coad's notation and methodology will be discussed, presented, and used. The decision to do so does not mean that it is necessarily better than the other object-oriented methodologies. The motivation to present one succinct methodology in significant detail rather than several in somewhat more abstract and potentially confusing terms is a pedagogical one. I believe an understanding of one object-oriented methodology and notation will prepare you to understand and readily use any of the other object-oriented methodologies and accompanying notations.

A second, albeit weaker reason for presenting Coad's notation and methodology has to do with the refinement of notation and methodologies that continues to occur within the object-oriented community. Two examples come to mind. First, Coad's notation and methodology is an extension and refinement of what was initially documented as the Coad/Yourdon notation and methodology (refer to chapter references Coad and Yourdon, and Yourdon). Second, Booch and Rumbaugh have consolidated their individual notations and methodologies.

COAD'S OBJECT-ORIENTED ANALYSIS AND DESIGN METHODOLOGY AND NOTATION

Figure 3.4 presents Coad's object-oriented methodology in summary form. Referring to the figure, notice that there are four major activities, each of which may be per-

Activities

Model Component	1 Identify Purpose and Features	2 Identify Objects and Patterns (behavior, data)	3 Establish Object Responsibilities (behavior, data, functions)	4 Define Service Scenarios (behavior, data, functions)
Problem Domain (PD)	✔	✔	✔	✔
Human Interaction (HI)	✔	✔	✔	✔
Data Management (DM)	✔	✔	✔	✔
System Interaction (SI)	✔	✔	✔	✔

✔ Indicates that the activity has been performed
for the model component

Figure 3.4   Coad's Object-Oriented Methodology Summary (Adapted from Coad, North, and Mayfield)

formed with each of four major model components. The grid boxes in the figure all have check marks in them. A check mark indicates that the activity has been performed for the indicated model component. During a systems development project, a project manager could refer to the grid to get a visual perspective of what activities have and have not been performed as of a particular point in time.

Figure 3.5 presents Coad's object-oriented methodology in detail as a series of sequential activities that the systems analyst performs, hopefully with significant user participation where appropriate. Notice that activities 2 through 4 are performed for each of the four model components.

Figure 3.6 presents the summary activities using conventional SDLC approaches to show the iterative and spiral nature of the methodology. What is not shown in any of these three figures are the notions of parallelism, substitution, and omission. Parallelism gives the project manager license to perform activities in parallel when a systems development situation permits. Substitution gives the project manager license to substitute activities out of the standard order presented in Figure 3.5 when a systems development situation permits. Likewise, omission gives the project manager license to omit one or more detail activities presented in Figure 3.5 when the situation permits.

A simple explanation of each of the four methodology activities is presented here, since each one is discussed in much more detail in a later section of this chapter as well as in the following chapters.

1. Identify the information system's purpose and features.

Use four object model components (problem domain [PD], human interaction [HI], data management [DM], and system interaction [SI]) to guide and organize the work.

For each of PD, HI, DM, and SI repeat the following:

2. Select the model component's objects and organize them by applying patterns.

3. Establish responsibilities for model component's objects:
   - what the object knows
   - who the object knows
   - what the object does

4. Work out model component's dynamics using scenarios.

Variation notes for activities 2, 3, and 4:
1. They may be done in any sequence that is appropriate.
2. One or more of them may be omitted.
3. One or more of them may be done in parallel.
4. Model components may be done in any order that is appropriate .

Figure 3.5   Coad's Object-Oriented Methodology Standard Sequence [Adapted from *Object Models: Strategies, Patterns, and Applications,* by P. Coad, D. North, and M. Mayfield, 1995, Prentice Hall, Englewood Cliffs, NJ.]

Activity 1: Identify information system purpose and features. This activity needs to be done prior to doing any object modeling of the problem domain. In fact, this activity should be done before any type of modeling is done. The user and the systems analyst need to identify, understand, and document the system's intended purpose and its features. This is analogous to identifying the system's goals and objectives as discussed in Chapter 2.

Activity 2: Identify model component (PD, HI, DM, SI) objects and patterns. The systems analyst and the user working together discover through a variety of techniques the candidate objects and associated classes that exist in the problem domain. Once identified, they refine and challenge the discovered classes and objects in order to obtain the true class and object requirements. There could be dozens of class and objects in an average-sized information system. These classes and objects become the basis for the object-oriented information systems model that is created by the systems analyst. As objects are identified, patterns of objects are considered and utilized when appropriate. As defined in Chapter 2, a pattern is a template of objects with stereotypical responsibilities and interactions; the template may be applied again and again by analogy. Pattern instances are building blocks used to assemble effective object models. Object patterns allow objects to be related to each

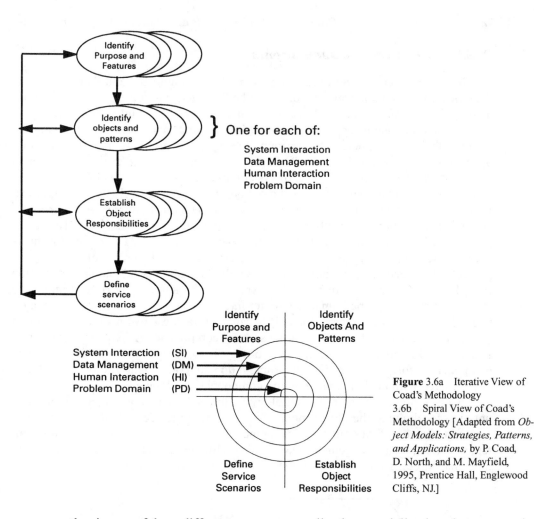

Figure 3.6a Iterative View of Coad's Methodology
3.6b   Spiral View of Coad's Methodology [Adapted from *Object Models: Strategies, Patterns, and Applications,* by P. Coad, D. North, and M. Mayfield, 1995, Prentice Hall, Englewood Cliffs, NJ.]

other in one of three different ways—generalization-specialization class connection, whole-part object connection, and object connection. Each of these is defined and described later.

Activity 3: Establish object responsibilities. All objects have up to three types of responsibility. The first is what the object knows about itself. The second is what the object knows about other objects, and the third responsibility is what the object does.

Activity 4: Define service scenarios. A scenario is a time-ordered sequence of object interactions needed to fulfill a specific service responsibility.

As discussed in Chapter 1, systems have three basic characteristics—data, function, and behavior. Notice in Figure 3.4 that the activities identify the information system characteristic(s) that they emphasize or support. Activity 1 does not have a direct emphasis on any of the three. Activity 2 supports both behavior and data. Activities 3 and 4 each support all three—function, data, and behavior.

Coad's Object Model Components

An **object model component** is a grouping of classes. Figure 3.4 and Figure 3.7 show the object model components in Coad's methodology. The groupings are problem domain (PD), human interaction (HI), data management (DM), and system interaction (SI). These components enable the **separation of concerns** concept by allowing the systems analyst the opportunity to group objects according to problem domain (PD) or technology area (HI, DM, SI). Doing so could enable a higher degree of object reuse and a higher (improved) level of system maintainability over its life.

Coad's methodology advances the notion of one progressively expanding model from analysis through design and implementation. The object model components facilitate such a notion. The problem domain component (PD) is usually the first component to which the four methodology activities may be applied. The reason for this is because the PD is what the user is most interested in—modeling his or her business problem. The second component that is usually developed is the human interaction (HI). The reason for this is that the HI is what the user is usually next most interested in, since it represents the human-computer interaction for the system. Data management (DM) and system interaction (SI) usually follow these two, as the user is often the least interested in these "behind the scenes" components. A frequent variation on this standard component development strategy is to develop the human interaction component before or along with the problem domain component. Prototyping is usually involved with the human interaction component development.

Coad's object-oriented methodology includes activities, an object model, and notation for drawing the object model. The four activities described earlier are performed for each of four object model components—problem domain (PD), human interaction (HI), data management (DM), and system interaction (SI). It is interesting to note that three of the four components—HI, PD, and SI—are very similar to the

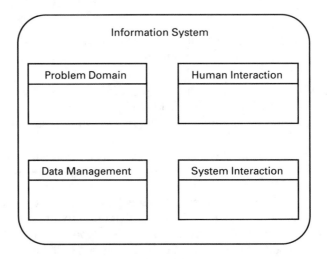

Figure 3.7   Coad's Object Model Components

Smalltalk programming language concept called model-view controller (MVC). Each component is briefly described next. More detail will follow in subsequent chapters.

The **problem domain component** contains objects that directly correspond to the problem being modeled. Objects in this component are technology-neutral and have little or no knowledge about objects in the human interaction, data management, and system interaction components.

The **human interaction component** contains objects that provide an interface between the problem domain objects and people. In an object model, such objects most often correspond to specific windows and reports.

The **data management component** contains objects that provide an interface between problem domain objects and a database or file management system. In an object model, such objects most often correspond to specific problem domain objects that need support for persistence and searching.

The **system interaction component** contains objects that provide an interface between problem domain objects and other systems or devices. A systems interaction object encapsulates communication protocol, keeping its companion problem domain object free of such low-level, implementation-specific detail.

An Object-Oriented Model

If you were asked to draw an office building or a home, could you do it? Probably so. Why could you draw an office building or a home? You could do so for at least two reasons: (1) You have a mental picture of what an office building or home looks like (even though your mental picture may be quite different from someone else's mental picture of an office building or home), and (2) you already know the basic notation needed to draw an office building or home, such as lines, circles, rectangles, triangles, squares, windows, doors, patios, porches, roofs, and so on.

Now if you were asked to draw a picture of a payroll information system or even a student course registration information system using an object-oriented notation, could you do it? Probably not, unless you (1) have some understanding of how a payroll information system or a student course registration system functions, and (2) know what an object-oriented notation looks like and have some minimal understanding of how to use the notation to draw a picture of an information system.

What's really required here? At least two things come to mind: (1) familiarity and understanding with a problem domain (e.g., office building, home, payroll system, student course registration system), and (2) familiarity and understanding of a specific notation. Familiarity with and understanding a problem domain means that you have knowledge of a specific problem domain ranging from minimal knowledge all the way to being a **subject matter expert** (SME), one who knows a great deal about a problem domain. **Notation** is a set of symbols used to communicate or represent something. An alphabet is a universal example of a notation. Familiarity with a notation implies that you know which notation symbols are available to use, and understanding of a notation implies that you know the methodology for integrating and combining the notation symbols. As discussed before, **methodology** refers to the packaging of methods and techniques together. In this situation methodology refers to the packaging of the notation symbols and their underlying meanings.

The details of Coad's object-oriented systems analysis and design methodology and its associated notation are presented in many of the following chapters in this book. However, before walking you through all of the details of the methodology and its notation in a step-by-step fashion, this section of the chapter presents a simplistic, partially completed user requirements problem domain component (PD) model using Coad's object-oriented notation. The intent is to show you in advance what an object-oriented problem domain component model looks like so you can see "the big picture" before diving into all the details in the following chapters. A colleague once made the following comment about textbooks. He said, "Textbooks are very good at explaining all of the ingredients of a tossed salad (lettuce, tomatoes, onions, carrots, celery, and so on), but few textbooks actually show you a completed tossed salad." The intent of this section is to show you the completed "tossed salad" allowing for some license with respect to the object-oriented model really being complete. In addition, this model only contains the problem domain component.

A video sale/rental store information system is used for presenting the object-oriented user requirements model example. The example is incomplete from two perspectives. First, the details of the video store information system problem domain are omitted for simplification purposes. Second, the object-oriented user requirements model presented is only showing the basics of the notation used through the systems analysis portion of systems analysis and design. This means that the HI, DM, and SI object model components usually added as part of systems design are not presented here, again for simplification purposes. As you might expect, the difficulty of presenting an example at this point in the book is that you need to focus your attention on the object-oriented model's notation rather than on the details of the specific video store problem domain being depicted by the model. Hopefully, you have purchased or rented a video from a video store and, by doing so, have at least minimal knowledge of the video store problem domain. Some of you may work at a video store or have done so in the past, making you somewhat of an SME. If you fit into this classification, realize that the model presented here probably does not exactly represent the video store information system you have personal expertise with. This is in keeping with the opening comments in this chapter that suggested that there are different solutions to a problem domain.

Since this is a model of a fabricated automated information system, please recognize that there are many other variations to the way this model is being presented. This video store model (and much more detail) was created by a large team of three dozen or so "aspiring systems analysts" (undergraduate students) working in the classroom for about two months during a semester with the instructor acting as the senior systems analyst and as the user. These requirements may or may not be the same as the ones you would come up with working with a different user in a similar problem domain. Remember, the intent here is not to make you subject matter experts for the video store problem domain, but to expose you to a partially completed object-oriented user requirements model so that you will know what one looks like early in the game.

Figure 3.8 presents the seven basic object-oriented methodology notation symbols as defined by Coad. The Coad notation is simplistic in that there are only a few symbols, yet it is sophisticated enough to depict large and complex information systems, ones exhibiting functions, data, and behavior. Referring to Figure 3.8, the symbols in the notation are the model component, class-with-objects, class, generalization-specialization connection, whole-part object connection, object connection, and message. Each of these is briefly defined here because they will be defined and described in greater detail in later chapters.

To assist with the discussion of Coad's object-oriented methodology notation, Figure 3.9 will be used. Figure 3.9 is the problem domain component (PD) object model of a video store information system for purchasing, selling, and renting inventory items, such as video movies and games, VCRs, and concession items. In addition, the information system keeps track of these same inventory items when they need to be ordered so that the store will have enough of them to sell or rent to its customers. The PD component of an object model contains objects that reflect the user requirements for the information system. The video store's problem domain requirements

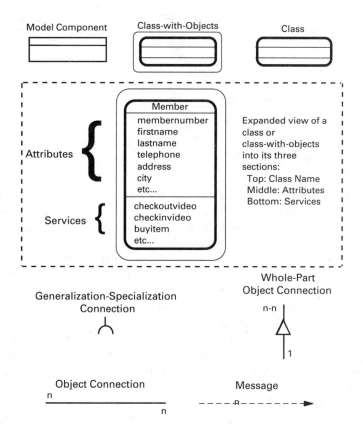

Figure 3.8   Coad's Object Model Notation

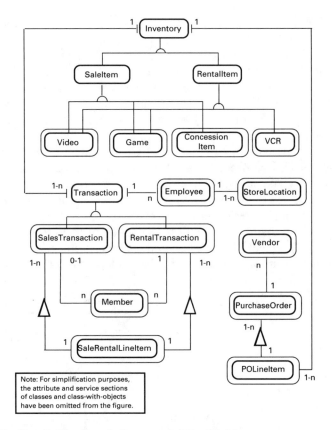

Figure 3.9   Video Store Problem Domain Component's Object Model

will be defined in the next chapter. The intent here is to show you the resulting object model based on the user requirements. Recall that a requirements model of the video store's mission, goals, objectives, and tactics was presented earlier in Chapter 2. The object model uses Coad's notation.

Simply stated, class-with-objects are the things that an information system needs to know within a specific problem domain. **Class-with-objects** are people, places, and things, such as employee, vendor, store location, rental transaction, video, VCR, and so on. **Classes** are abstract class-with-objects used to show generalization-specialization connections. Because they are abstract, classes have no objects. **Objects** are the actual instances of people, places, and things and are always associated with a class-with-objects, as its name implies. For example, individual video objects associated with the Video class-with-objects would be each specific movie video that is available for either rent or sale.

The core of an object model of user requirements is the class and the class-with-objects. In Figure 3.9's object model, Inventory, SaleItem, RentalItem, and Transaction are depicted as classes because they represent generalization-specialization connections. None of them have (or will ever have) object instances. Video, Game, ConcessionItem, VCR, Employee, StoreLocation, SalesTransaction, RentalTransaction, Vendor, Member, PurchaseOrder, SaleRentalLineItem, and POLineItem are all examples of class-with-objects because each one will more than likely have one or more object instances. The determination of classes versus class-with-objects surfaces during requirements discussions with the users. Classes and class-with-objects are treated similarly in most situations, since classes are simply class-with-objects that are and always will be void of objects.

Classes and class-with-objects usually have **attributes** or characteristics which describe them in more detail. For example, in Figure 3.8, the Member class-with-objects has attributes MemberNumber, FirstName, LastName, Telephone, Address, City, etc. Attributes are discussed in much more detail in a later chapter.

Classes and class-with-objects usually have **services** that the class or class-with-objects is responsible for accomplishing. In Figure 3.8, for example, the Member class-with-objects has CheckOutVideo, CheckInVideo, and BuyItem as services. Services are discussed in much more detail in a later chapter.

The **model component** notation symbol, which is not included in Figure 3.9, is used to partition the object model into four distinct partitioning groupings—problem domain, human interaction, data management, and system interaction. Refer to an earlier section of this chapter for a brief discussion of model components. All of the classes and class-with-objects in Figure 3.9 would be listed in the problem domain (PD) model component. Since each model component simply lists the classes and class-with-objects contained within it, it can be viewed as a more abstract view of the object model.

The next three notation symbols are the generalization-specialization connection, the whole-part object connection, and the object connection. All three of these notation symbols are used to associate classes and class-with-objects to other classes and class-with-objects. All classes and class-with-objects have responsibilities. One of their responsibilities is "who they know." Connections represent this responsibility in the object model. Each of the three connection types is briefly discussed next.

The **generalization-specialization connection** symbol is used to associate one generalization class or class-with-objects with one or more specialization classes or class-with-objects symbol together in a hierarchical parent-child pattern type of relationship. For example, a SaleItem class, the generalization portion of the parent-child pattern, may have several specialization class-with-objects, such as Video, Game, Concession Item, and VCR. Refer to Figure 3.2 for another example of the generalization-specialization "parent-child" pattern. Often specializations can be thought of as being a "type" or "kind" of the generalization. For example, a game is a type of sale item as well as a type of rental item.

The **whole-part object connection** notation symbol is used to associate two class-with-objects symbols together in a whole-part pattern type of relationship. For example, an Automobile class-with-objects could be the "whole" and have several "part" type class-with-objects such as Door, Window, Wheel, Seat, Gauge, and so on.

In Figure 3.9, the PurchaseOrder class-with-objects represents a "whole" portion of the whole-part pattern and the POLineItem class-with-objects represents the "part" portion of the pattern. Also the SaleRentalLineItem class-with-objects is the "part" portion of both the SalesTransaction and the RentalTransaction class-with-objects. The "n-n," "1-n," and "1" notation associated with the whole-part symbol and pattern in Figures 3.8 and 3.9 will be discussed in detail in a later chapter. Basically, this notation represents constraints placed on the connection relationship. For example, the "n-n" notation at the top of the whole-part symbol in Figure 3.8 is to indicate how many "part" objects are associated with one "whole" object. The "1" notation at the bottom of the whole-part symbol in Figure 3.8 is to indicate how many "whole" objects are associated with one "part" object.

The **object connection** notation symbol, like the whole-part object connection, is also used to associate two class-with-objects symbols together. However, this connection is not considered as strong a connection as whole-part. For example, both SalesTransaction and RentalTransaction class-with-objects in Figure 3.9 are connected to the Member class-with-objects because these two types of transactions are performed by members of the video store. From a video store application standpoint, members are not part of transactions so the connection is not shown as a whole-part object connection. The Vendor class-with-objects has an object connection with the PurchaseOrder class-with-objects. Vendor represents businesses or individuals that the video store purchases inventory from. PurchaseOrder represents a business transaction and identifies inventory items that are purchased from a vendor. The reason for the Vendor to PurchaseOrder relationship being an object connection rather than a generalization-specialization or whole-part connection is because PurchaseOrder is not a "type" of Vendor and because PurchaseOrder is not a "part" of Vendor. Constraints are also indicated on an object connection symbol. For example, referring to the Vendor to PurchaseOrder object connection in Figure 3.9, a specific Vendor may have zero or more (represented by "n") PurchaseOrders and a specific PurchaseOrder is related to only one specific Vendor.

A **message** is a request for a service to be performed, sent from a sender to a receiver. It is used to help a class or class-with-objects carry out its intended responsibilities. Classes and class-with-objects usually have one or more services belonging to them. Often a service needs help from another service in the same or another class or class-with-objects in order to perform its responsibility. For example, a createPaycheck service may send a message to getEmployeeSalary, computeGrossPay, and computeFedTax services (among others) in order to fully prepare a paycheck for printing. In other words, the createPaycheck service does appear to do all of the work necessary to prepare a paycheck; however, it does the work by delegating portions of it out to other services located somewhere within the object model. Because a typical object model has so many, messages can significantly complicate the "look" of the object model. Therefore, they are only shown in the object model's **scenario** view. Simply stated, scenarios are the actions taken by services to perform work. Figure 3.9 does not include a scenario view, since presentation of service details is not necessary at this point in time. Scenarios are discussed in more detail in a later chapter.

Continuing the discussion of the video store object model shown in Figure 3.9, at some subjective point additional details become important and necessary. The object model has two more levels of detail: (1) attributes and services, and (2) scenarios. Figure 3.10 expands Figure 3.9 by including attributes and services. Attributes and services help communicate more about what data (attributes) are associated with the information system, and what functions (services) are performed by the information system. As you look at Figure 3.10, keep in mind that this is the exact model as shown in Figure 3.9 with additional details. Due to the extent of the details, the model was divided into three sections in order to fit on book pages. Using a computer-based CASE tool to draw this model, you would have it all connected even though you might only see sections of the model at any moment in time on a computer screen.

As described earlier, attributes are characteristics which describe a class or class-with-objects in more detail. Sometimes the name of a class, such as StoreLocation in Figure 3.9, is not sufficient to communicate its intended purpose in the system. Looking at this class's attributes might help communicate its intended purpose. For example, StoreLocation's attributes are storeNumber, storeAddress, storeCity, storeState, storeZipcode, and storePhone. Data associated with these attributes would indicate that the system needs to keep track of the video store's identification information for various reasons. One reason could be to print it on certain reports. Another

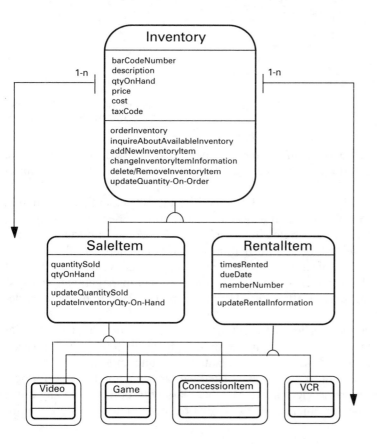

Figure 3.10a   Video Store Object Model with Attributes and Services

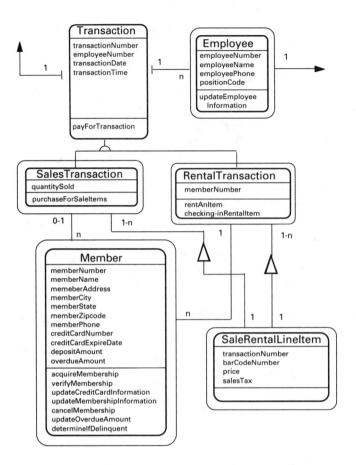

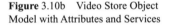

Figure 3.10b   Video Store Object
Model with Attributes and Services

reason could be the need to consolidate two or more video store's data together in the information system and still be able to electronically separate it into the data belonging to each store.

Are the attributes shown in Figure 3.10 all of the attributes that this information system will have? Probably not. As the systems analyst progresses through analysis and into design, additional attributes may surface and become important to keep track of and would be added to the information system via a change to the problem domain component of the object model with the user's approval. Remember also that the other three model components—human interaction, data management, and system interaction—may also have attributes. Attributes are discussed in much more detail in a later chapter.

Services, as described earlier, represent the actions or functions that a particular class or class-with-objects must accomplish in order to help fulfill the overall mission of the information system. A service is associated with one and only one class or class-with-objects, the one most closely associated with the function to be performed. That is why there are more services in the Member class-with-objects than any of the other classes and class-with-objects. Much of what happens in a video store is initiated by a member.

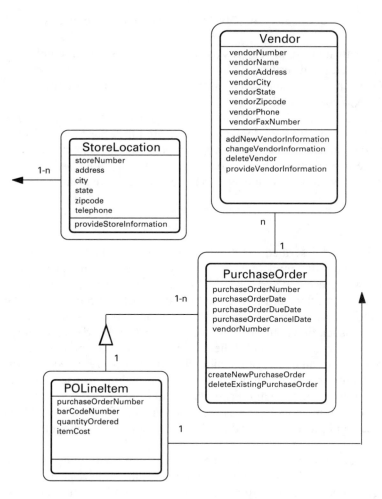

Figure 3.10c   Video Store Object Model with Attributes and Services

Are these all of the services that this information system will have? Probably not. As the systems analyst progresses through analysis and into design, additional services may surface and become important and required for the information system to accomplish its mission. These new services would be added to the information system via a change to the problem domain component of the object model, similar to the way attributes are added, with the user's approval. Remember also that the other three model components—human interaction, data management, and system interaction—may also have services. Services are discussed in much more detail in a later chapter.

As you look at the details in Figure 3.10's object model, you will notice that several class-with-objects have no attributes and services shown in them. For example, Video, Game, ConcessionItem, and VCR have no attributes and no services listed in their class-with-objects symbol. This is okay since each of these class-with-objects are the specialization's part of a generalization-specialization. In a generalization-specialization, all specialization classes and class-with-objects inherit all the attributes and all the services from the generalization class or class-with-objects. Therefore, ConcessionItem has the same attributes and services as Inventory and SaleItem combined. VCR has the same attributes and services as Inventory and RentalItem combined. Finally, Video and Game have the same exact attributes and services as Inventory, SaleItem, and RentalItem combined.

This concludes the review of the video store object model. Remember, Figures 3.9 and 3.10 only represent the problem domain component of the object model. There are still three others that make up the complete object model—human interaction, data management, and system interaction. The next few chapters will explore Coad's object-oriented methodology, object model components, and notation in more detail.

SUMMARY

This chapter has defined and described a methodology as the way one thinks about or performs an activity in order to complete it successfully. Each of four major methodology classifications was presented and discussed—traditional, structured, information engineering, and object oriented. Eight basic building block concepts—common methods of organization, abstraction, encapsulation, inheritance, polymorphism, message communication, associations, and reuse—for an object-oriented methodology were also presented and discussed. Not all of them are new, which is a good thing. Another fundamental concept of an object-oriented methodology was presented—classification theory. This theory is common to most people, thus making its use within an object-oriented methodology a more natural activity. Finally, the Coad object-oriented systems analysis and design methodology was presented and briefly discussed. Much more will be discussed and presented regarding this methodology throughout the remainder of the book.

Understanding and knowledge of a problem domain along with understanding and experience using a methodology and notation are important for successfully documenting user requirements for an information system. Coad's object-oriented methodology notation consists of seven symbols—model component, class, class-with-objects, generalization-specialization connection, whole-part object connection, object connection, and message. A partially completed problem domain component object model was presented to show how the notation symbols are combined to depict the user's requirements for an information system. Finally, the object model was expanded to reveal attributes and services along with a discussion of these model features.

QUESTIONS

3.1 What is a methodology?

3.2 Briefly discuss the different ways of acquiring a methodology.

3.3 How would you best describe the traditional methodology classification?

3.4 Given your answer to the previous question, what is the major disadvantage of this type of methodology?

3.5 Why is the structured analysis and design methodology sometimes called the data flow modeling methodology?

3.6 How has computer-aided software engineering (CASE) changed the structured methodology?

3.7 What problem-solving approach is at the heart of an information modeling methodology?

3.8 How does an information modeling methodology differ from the approach used in a structured methodology?

3.9 How does the approach taken in an object-oriented methodology differ from strategies employed by the information modeling methodology?

3.10 List and describe the eight characteristics of an object-oriented methodology.

3.11 What role does abstraction play in an object-oriented methodology?

3.12 What is another term for encapsulation, and what is its purpose within an information system?

3.13 What is one of the main advantages of reuse within an object-oriented methodology?

3.14 Discuss a few of the roadblocks that are standing in the way of industry acceptance of reuse.

3.15 What significant aspect of object-oriented methodology resolves problems associated with the older methodologies?

3.16 What are the three concepts within classification theory and how do they relate to object-oriented methodology?

3.17 What is meant by reuse? What role does it play in systems analysis and design?

3.18 Name and discuss the different types of reuse that are possible.

3.19 List and briefly define the four model components in Coad's object-oriented methodology.

3.20 List and briefly describe Coad's object-oriented analysis notation symbols.

3.21 If you wanted to draw a picture of an information system, what would you need, at minimum, to undertake this task?

3.22 Define notation and briefly explain its importance in an object-oriented information system.

3.23 Define and give a few examples of the class and class-with-objects notations.

3.24 What is the importance of the class and class-with-objects symbol notations within the problem domain component of the object model?

3.25 What are the two additional elements usually found within the class and class-with-objects notations and what is their purpose?

3.26 What is the purpose of the generalization-specialization connection within Coad's object-oriented methodology notation?

3.27 Distinguish the whole-part object connection symbol from the generalization-specialization connection symbol.

3.28 What is the main purpose of the "n-n" and "1" notation in the whole-part object connection?

3.29 What is the nature of relationships as symbolized by the object connection?

3.30 What is the purpose of a message? Describe how one works.

3.31 Why is it sometimes helpful to identify attributes within a particular class or class-with-objects?

3.32 What distinguishes a service from an attribute within a particular class or class-with-objects?

REFERENCES

BOOCH, G., *Object-Oriented Analysis and Design with Applications* (2nd ed). Menlo Park, CA: Benjamin/Cummings, 1994.

"Classification Theory," *Encyclopedia Britannica,* 1986.

COAD, P., D. NORTH, and M. MAYFIELD, *Object Models: Strategies, Patterns, and Applications.* Prentice Hall, Englewood Cliffs, NJ, 1995.

COAD, P., and E. YOURDON, *Object-Oriented Analysis* (2nd ed.). Englewood Cliffs, NJ: Yourdon Press/Prentice Hall, 1991.

COAD, P., and E. YOURDON, *Object-Oriented Design.* Englewood Cliffs, NJ: Yourdon Press/Prentice Hall, 1991.

COX, B.J., *Object-Oriented Programming: An Evolutionary Approach.* Reading, MA: Addison-Wesley, 1986.

FIRESMITH, D.G., "Basic Object-Oriented Concepts and Terminology: A Comparison of Methods and Languages," White Paper from Advanced Software Technology Specialists (ASTS), Fort Wayne, IN, January 3, 1994, 37 pages.

HENDERSON-SELLERS, B., *A Book of Object-Oriented Knowledge.* New York: Prentice-Hall, 1992.

JACOBSON, I., M. CHRISTERSON, P. JONSSON, and G. OVERGAARD, *Object-Oriented Software Engineering.* New York: Addison-Wesley, 1992.

MEYER, B., *Object-Oriented Software Construction.* Hemel Hempstead, United Kingdom: Prentice Hall, 1988.

OLLE, T.W., et al., *Information Systems Methodologies: A Framework for Understanding.* Wokingham, England: Addison-Wesley, 1988.

OLLE, T.W., H.G. SOL, and C.J. TULLY, (eds.), *Information Systems Design Methodologies: A Feature Analysis.* Amsterdam: North-Holland Publishing Co., 1983.

OLLE, T.W., H.G. SOL, and A.A. VERRIJN-STUART, (eds.), *Information Systems Design Methodologies: A Comparative Review.* Amsterdam: North-Holland Publishing Co., 1982.

RUMBAUGH, J., M. BLAHA, W. PREMERLANI, F. EDDY, and W. LORENSEN, *Object-Oriented Modeling and Design.* New York: Prentice Hall, 1991.

SHLAER, S., and S.J. MELLOR, *Object-Oriented Systems Analysis: Modeling the World in Data.* Englewood Cliffs, NJ: Yourdon Press/Prentice Hall, 1988.

WIRFS-BROCK, R.I., B. WILKERSON, and L. WIENER, *Designing Object-Oriented Software.* New York: Prentice Hall, 1990.

YOURDON, E., *Object-Oriented Systems Design: An Integrated Approach.* New York: Prentice Hall, 1994.

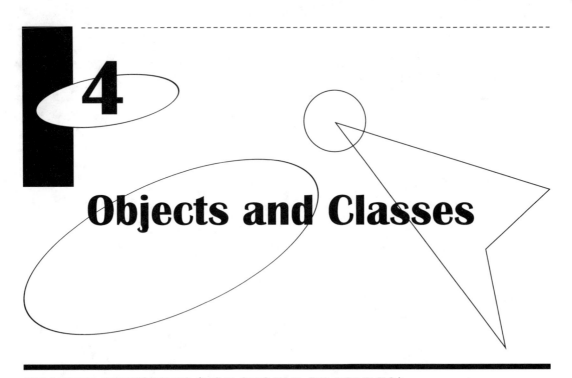

Objects and Classes

CHAPTER OBJECTIVES (YOU SHOULD BE ABLE TO)

1. Define and give examples of class and objects.
2. Describe and use several strategies used to find objects.
3. Describe a technique for keeping track of future enhancements for the information system.

Getting started! For many people, these two words represent an enormous hurdle or mountain to climb over. That term paper. That big test to study for. The whole apartment needs cleaning. Writing all those "thank you" letters. Many projects seem overwhelming as we think about just getting started.

Getting started with an information systems development project can also be difficult because there are so many ways in which one could actually begin a project. You may recall that "ways to get started on a project" was discussed in the chapter dealing with requirements determination. In that chapter three dominant problem-solving strategies were identified—function, data, and object. To review briefly, the functional strategy approaches a problem by first identifying the actions the system must accomplish, such as register for a course, drop a course, check out a movie video, and so on. The data strategy approaches a problem by first identifying the data that the system is responsible for, such as student information, course information, schedule information, movie video information, and so on. Finally, the object-strategy approaches a problem by simultaneously focusing on objects, such as people, places, and things, and the object's data and services. The object problem-solving strategy best fits with an object-oriented systems analysis and design methodology. Because of this, it will be used here to identify the objects, classes, and class-with-objects.

The requirements model, also discussed in that same chapter, is one strategy used for defining the information system requirements in terms of business goals, objectives, and tactics as well as information system objectives and tactics. Once this is done or some other approach is used to accomplish this, Coad's object-oriented systems analysis and design methodology suggests finding objects and relating them according to some well-established patterns. The methodology further suggests that we start looking for objects within the problem domain component (PD) of the object model.

OBJECTS AND CLASSES

What is an object? What is a class? Simply stated, an **object** is defined as an abstraction of a person, place, or thing within the problem domain of which the information system must be aware. A **class** is a set or collection of abstracted objects that share common characteristics. Coad's object model notation differentiates a class that will never have objects from a class that has or has the potential to have objects. The way Coad does this is by having two distinct notation symbols, one for class (with no objects) and another for classes with objects. Further clarification of these two will be presented when necessary. For now, the discussion regarding classes will mean classes with objects. In keeping with the reference materials on Coad's methodology, the first letter of any identified class in an object model is always capitalized. For example, a class of bicycles would be called Bicycle.

Examples of objects, or what is often referred to as instances, could be your bicycle, your backpack, your automobile, your telephone, each shoe you own, each key on your key chain, each slice of bread in a loaf of bread, the paycheck you received last pay day, and so on. These physical objects and others in the real world are the easiest type of objects to identify. Later in the chapter we will discuss the more difficult types of objects to identify—the ones that are not so obvious.

Examples of classes could be Bicycle, Backpack, Automobile, Telephone, Shoe, Key, Bread, and Paycheck. Classes are always referred to in the singular even though they typically represent a collection of one or more objects with similar characteristics. So, the Bicycle class could contain instances of your bicycle, my bicycle, your roommate's bicycle, and so on. A class is also partitioned into three parts—its name, its attributes, and its services—as shown in Figure 4.1. Attributes and services are discussed later in the book.

Be sure to use terminology and names the user of the information system will be comfortable with, otherwise user acceptance of the information system may be difficult to obtain. For example, in a recent requirements-gathering session with a client, several synonyms were discovered to mean basically the same thing. Of the seven people in the meeting, five users and two systems analysts, four different terms were favored—*origin code, priority code, source code,* and *key code*—by different individuals. Further discussion revealed that all four were in fact synonyms. The group finally reached a consensus to use one of the four terms—origin code—universally from that point on in order to eliminate confusion and misunderstanding. Sometimes

Class (with no objects)

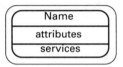

Class-with-Objects

Examples:

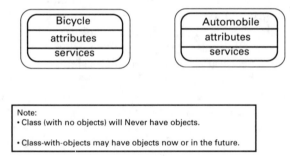

Note:
• Class (with no objects) will Never have objects.

• Class-with-objects may have objects now or in the future.

Figure 4.1   Class and Object Notation

use of the synonyms is acceptable and agreed to by the users, but, over time, organizational dynamics can negatively affect the use of synonyms as in the preceding example. Sometimes it is necessary to use an adjective along with a noun for the class name such as Rental Agreement instead of Agreement in order to further clarify the class and to make it more meaningful to the user. In Coad's notation, class names with more than one word are compressed together with no spaces. For example, Rental Agreement would become RentalAgreement.

Object and Class Rules and Guidelines

There are a few simple rules and guidelines for creating classes. These rules and guidelines about the class notation symbol are listed in Figure 4.2 and summarized here. First, an object must always belong to a class. There is no provision in the notation for a stand-alone object. Next all class names should begin with a capital letter, such as Bicycle instead of bicycle. Class names should also be singular, such as Bicycle instead of Bicycles. Class names should also be meaningful to the user. A class is partitioned into three parts—its name, its attributes, and its services.

1. Objects must always belong to a class.

2. The first letter of class names and each word in the class name should be capitalized. Examples: Students, Bicycle, DataEntryClerk.

3. All names of classes, attributes and services should be singular. Examples: Student instead of Students; Bicycle instead of Bicycles; DataEntryClerk instead of DataEntryClerks.

4. All names of classes, attributes, and services should be meaningful. Examples: SalesDept, AccountsPayable, Cashier, SaleLineItem.

5. The class notation symbol is partitioned into three parts: name, attributes, and services.

6. Attribute and service names should start with a lower case letter; each additional word used in the name should begin with a capital letter. Examples: studentIdNumber, inventoryControlNumber, gradePointAverage, requestTranscript, calculateSalesTax.

Figure 4.2   Rules and Guidelines for Coad's Object Model Notation

Some people refer to both class symbols as *squircles* (pronounced like squirkles). Referring to Figure 4.1, notice the only difference between the class and class-with-objects squircle is the outer, thin squircle that surrounds the inner, thick class squircle. When you see the class-with-objects squircle, you can infer that this class has the potential to have zero, one, two, or more object instances associated with it. Even if the class-with-objects squircle has thousands of objects, it is drawn with only one thin squircle surrounding the class's thick squircle. You could think of this like a deck of playing cards having 52 objects, but when the deck is ready for dealing, all cards are organized and brought together and hidden behind the initial card as in Figure 4.3. Nearly all classes created within an object model will contain objects. A few will not. Knowing when to create a class with no objects versus a class-with-objects squircle is important and is discussed in a later section of this chapter.

Class Attributes and Services Defined

Figure 4.4a illustrates a template of a class. The top section is reserved for the name of the class such as Bicycle, Paycheck, Student, Video, and so on. The center section is reserved for any associated characteristics of the class and is referred to as attributes. Examples of attributes that are all related to a student could be a student's last name, first name, address, city, state, and grade point average. Attributes related to a course at a university could be course number, course name, course units, and course prerequisites. Attributes are more fully discussed in the next chapter.

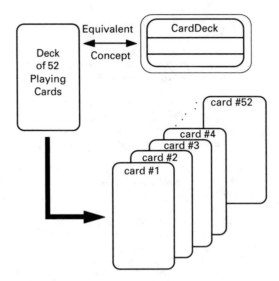

Figure 4.3   Playing Card Illustration of Class Notation Symbol

a) Class Template

b) Class Example

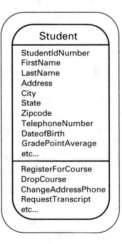

Figure 4.4   Class Template and Example

The bottom section of the class symbol is reserved for services, often referred to as methods in the object-oriented programming world. Services are actions that the class is responsible for exhibiting or performing. A few examples of potential services would be withdraw cash from ATM, scan product code, select course number, print paycheck, and rent a movie video. Services, too, are more fully explored in a later chapter.

An example of a class squircle with attributes and services for a student registration system is illustrated in Figure 4.4b. Not all possible attributes and services are listed due to space considerations, but hopefully enough are shown to give you the idea of this concept. The class notation is flexible enough to fit the needs of the problem domain. You may have a situation in which a class only has one or more services and no attributes. For our purposes, however, most classes that you will be seeing and reading about in this book will have both attributes and services.

In business information systems, many of the classes will represent some data such as students, courses, faculty, products, rooms, video movies, and so on. Within these data-oriented classes, each object must have one or more attributes whose values make the object unique from all other objects within the class. For example, a class called Automobile that has unique instances of automobiles, say yours, your roommate's, your brother's, your parent's, and so on, must have some way of identifying one specific automobile object from all the other automobile objects within the class. This is no different than an ATM machine being able to read your bank card and tell your bank account number from any other of the tens of thousands it keeps track of, or the fact that your fingerprints are like no other on the face of this earth (so we are told).

The way business information systems identify one object as being unique from all other objects in the class is to identify some attribute or combination of attributes that would distinguish it from all other objects. For example, how is one automobile object instance identified uniquely from all other automobiles in the Automobile class? Well, in the real world, each automobile has a unique vehicle identification number (VIN) associated with it and assigned by the vehicle's manufacturer. There should not be two VIN numbers alike on the face of the earth (so we are told). Therefore, this attribute could be used to uniquely identify one automobile from all the others.

Comment on Object-Oriented Problem-Solving Strategy

An important characteristic of the object-oriented problem-solving strategy is that classes and their definitions including attributes and services are relatively stable over time. Even though some of the attribute values of a class may change (e.g., you move so your address and telephone number must be changed in your object), or the way the information system performs one of the services that belongs to a class may change down the road (e.g., an ATM machine's display screen changes from monochrome to color), the class will remain a constant and an integral part of the overall problem domain. This is good news! Stability over time is very important when it comes to the amount of time, resources, and cost to maintain the information system throughout its life. Like a building whose foundation is instrumental in the longevity of the building, an information system's foundation is important. In general, the more stable the information system's foundation, the less time, resources, and cost to maintain it over time.

FINDING OBJECTS

Now that you have an understanding of objects and classes, we will discuss three strategies for helping you identify these in the problem domain of interest to you and your user. This section is going to present strategies for finding objects. As you know, no object may stand alone; therefore, when you find an object, you automatically find the class that will group all objects of that type. For example, in a student course registration system, a student object is deemed a candidate object. Since there are lots of student objects (one for each student registering for courses), you automatically find a class called Student, which will be the grouping of all student objects.

The strategy or strategies you select for finding objects are often dependent on several factors. First, the type of requirements documentation you have to work with will play a significant role in determining this. If you have something like Kozar's requirements model or Wetherbe's PIECES framework as the requirements document for the problem domain, as discussed in an earlier chapter, or some other well-defined document, you may choose freely from the strategies discussed later. If you are getting together with your user in a brainstorming or JAD session in lieu of having a requirements document, you may need to use the third strategy discussed later.

Second, your user may have a preferred way of initially looking at the problem domain. Discussed in an earlier chapter were the data, functional, and behavioral characteristics of an information system. Your user may have a preference for one or more of these as the basis for discussing a model of the problem domain. For example, a recent client of mine had prepared a simplified user requirements document prior to meeting with me. His document focused primarily on the data aspects of an information system. He had drawn samples of half a dozen or more reports that he wanted from the information system. An object-oriented model could still be constructed starting from a data perspective such as this.

Third, you may have a bias toward one or more of these three aspects of information systems and feel more comfortable using a certain one. With over 25 years of systems analysis and design consulting experience, I have a predisposition to use the functional aspect but have tempered this to use both the data and behavioral when necessary. Lastly, the information system itself may cry out for a specific aspect emphasis—data, functional, or behavioral.

Wirfs-Brock Noun Phrase Strategy

The first strategy, discussed in more detail in the Wirfs-Brock reference book, approaches the problem of finding objects by focusing on the noun phrases that exist in the requirements specification document, such as a requirements model document, PIECES framework document, or some other requirements document. These are the suggested steps and are quite often done with discussion and assistance from the user:

1. Read and understand the requirements document because the goal of the "finding objects" activity is to create a model that very closely parallels the real-world problem domain in question.
2. Reread the document looking specifically for noun phrases. Make a preliminary list of these phrases and change all plurals in the list to singular, such as bicycles to bicycle.
3. Divide the noun phrase list into three categories: obvious objects, obvious nonsense objects, and "not sure of" objects.
4. Discard the nonsense noun phrase list.
5. Discuss the "not sure" noun phrase list in more detail until each phrase can be moved to either the obvious or nonsense list.

Figure 4.5 is a copy of part of the video store's requirements model from an earlier chapter. It is presented here for illustration of step 2 in the Wirfs-Brock approach to finding objects. The pertinent noun phrases in the list have been underlined as object candidates. I need to point out that normally the systems analyst would be working with the information system tactics portion of the requirements model, not the

INFORMATION SYSTEMS OBJECTIVES

General Objectives:

A. Provide just-in-time (JIT) training.
B. The systems we implement must be friendly and easy to learn and use.
C. The systems we implement must give considerations to security issues.

Specific Objectives:

1.1.1 Provide an automated system to assist with customer sales/rental checkouts.

2.1.1 Provide and maintain an automated membership database.
    a. provide current (up to date) membership information on demand.
    b. capability to add, change, and delete (remove) membership information.

2.1.2 Provide membership information reports such as (but not limited to):
    a. least used memberships.
    b. most used memberships.
    c. delinquent memberships (both money owing and outstanding rentals).

4.1.1 Provide and maintain an inventory database for both sales and rental items:
    a. provide current (up to date) inventory information on demand.
    b. capability to add, change, and delete (remove) inventory information (sales and rental).

4.1.2 Provide inventory information reports such as (but not limited to):
    a. least popular rentals.
    b. most popular rentals.
    c. delinquent tape rentals outstanding.
    d. products "on order" (purchasing report) for sale and for rental items.

5.1.1 Provide sales reports such as (but not limited to):
    a. sales for a time period (day, days, week, weeks, month, etc.) by product code.
    b. rentals for a time period (same as above).

Figure 4.5 Video Store Information Systems Objectives (partial) with noun phrases underlined [Wirfs-Brock]

objectives as done here. Due to the large number of tactics that would derive from the information system objectives in the video store problem domain, we are using objectives for this example.

Figure 4.6, which represents the results of step 2, lists the noun phrases from the information system objectives in Figure 4.5. These are the candidate objects. Note that plurals have been converted to singular at this point, and that all words have been capitalized. Steps 3 through 5 would be performed by the systems analyst working with the user. Finally, the systems analyst would end up with a refined list of candidate objects that may undergo additional changes as the analysis and design proceed.

Wirfs-Brock CRC Strategy

The second strategy looks at the same beginning requirements document as did the Wirfs-Brock strategy discussed earlier. However, instead of looking for nouns, this strategy looks for verbs. The noun strategy tends to find objects based on "what the object is" (e.g., a person, place, or thing) while the verb strategy is based on "what the object does" (e.g., it prints, it computes, it displays, it dispenses cash, and so on). This strategy is generally referred to as **class-responsibility-collaboration** (CRC) and is discussed in detail in the Wirfs-Brock reference book also.

Conglomeration Strategy

The third strategy is a combination of suggestions documented in several of the chapter references. The ideas are similar and compatible with each other so each is being presented here for your consideration. Later in the chapter these suggestions will be

Security Issue

Automated System

Customer Sales/Rental Checkout

Membership Database

Membership Information

Membership Information Report

Inventory Database

Sales and Rental Items

Inventory Information

Inventory Information Report

Sales Report

Figure 4.6   Candidate List of Objects Using Wirfs-Brock's Noun Phrase Strategy

applied to finding objects in the video store information system example. As you begin to find objects, you should look for the following:

1. Tangible items such as vehicles, furniture, credit/debit cards, insurance policies, buildings, animals, scanners, keyboards, displays, conveyors, bins, invoices, shipping documents, sales receipts, ATM transaction receipts, paychecks, patient charts, maps, and so on. Since we are starting to look for objects in the problem domain component of the object model first, some of these tangible objects may be deferred until later when we are ready to address either the human interaction component, data management component, or system interaction component of the object model.

2. Roles played by people or organizations such as student, teacher, administrator, patient, clerk, employee, nurse, homeowner, department, region, sales office, division, and so on.

3. Incidents/interactions. Incidents happen at a specific point in time such as paying bills, an airplane flight, payday for employees, register to vote, register for classes, football or other game, maintenance call, telephone call, and so on. Interactions are of a transactionlike nature and similar to incidents, such as making a purchase at a store, checking out a book at the library, withdrawing money from an ATM, making a deposit through the ATM, and so on. It is not necessarily important to distinguish between an incident object and an interaction object, nor is it always even possible to do. The important point here is to give you two concepts to use to help you identify objects.

4. Specifications which have a tabular (row/column) quality such as a list of sales offices, list of U.S. state abbreviations, list of Standard Industry Codes (SIC), tax rate table, and so on.

As you create the list of candidate objects or after you have a list, each object on the list should be challenged based on:

1. Needed remembrance—does the information system need to keep track of this object? Why?

2. Needed services–does the information system need to have certain services performed by this object? Why?

3. Does the object have more than one identifiable attribute? If none or only one can be identified, then seriously consider the need to keep this object. If it is still justifiable, then by all means, keep it.

4. Will there be more than one of this type of object that could be grouped into a class? If none or only one object can be identified for the class, then seriously consider the need for this object. If it is still justifiable, then keep it. This challenge can often be similar to the foregoing needed remembrance challenge. An example of this challenge could be a class-with-objects called University, which contains instance objects such as Harvard, Yale, Stanford, UCLA, Michigan, Notre Dame, Arizona, Minnesota, and so on. This candidate class would pass this challenge. In another information system situation having University as a class-with-objects, you may find that the information system need only have one instance object, such as the object for your specific university. This should be challenged. During the challenge, your user tells you that the information system needs to keep track of specific information about your university because it is needed to allow the information system to perform its duties. Based on the user's explanation, the need for the University class would override this challenge and be kept as a candidate. If the user could not justify the need for keeping the University class, then it should be dropped from the candidate list of objects.

5. Avoid derived results as objects. Often these appear in the form of reports or displays. For example, an employee report listing employee names, addresses, and telephone numbers is initially identified as a candidate object and is called EmployeeReport. This candidate object is a derived report based on data from one or more other objects such as one called Employee. Therefore, eliminate EmployeeReport as an object. Many of these derived result type objects are identified initially, but they end up being eliminated as objects. The systems analyst must retain the fact that the report is still an output of the information system. With this in mind, derived result type objects become the output of one or more services in some other object, such as the Employee object having a service that creates an employee report.

No single strategy has emerged as the definitive way to find objects; therefore, it is common for different project teams to create somewhat different candidate objects for the same problem domain. This situation continues to make the modeling activity more of an art rather than a science. As object-oriented technology matures, strategies for finding objects will be refined, merged, invented, and so forth until one day when there may only be a very few, tried-and-true, algorithmic strategies for finding objects.

The Video Store Example—Finding Objects

The video store's problem domain component object model presented in Chapter 3 will continue to be used to develop an object-oriented information system example from the ground up. Here the emphasis is on finding objects. The video store information system model, as mentioned earlier, was developed in the classroom working with about three dozen students over several weeks of a semester. The intent was not to build the ultimate video store information system but to practice using an object-oriented methodology and notation to create a problem domain component object model of a simple business information system.

The students helped develop the video store mission statement, goals, business objectives, business tactics, information system objectives, and information system tactics as shown in Chapter 2 and partially shown here as Figure 4.5. Several brainstorming sessions were held to create the object-oriented video store model, shown in Chapter 3, working from these lists intermixed with classroom discussion.

The instructor chose to use strategy 3 discussed earlier, the conglomeration strategy, as the strategy for finding objects. The primary reason for doing so was that almost all of the students had some personal interaction with video store movie rentals and, therefore, could be considered knowledgeable in the problem domain. Strategies

1 and 2, as discussed earlier, would have perhaps reduced the amount of free-flowing discussion of the problem domain, a situation that the instructor did not want to inhibit for pedagogical and learning reasons. What the students discovered through the brainstorming sessions was that there were many different opinions of what happens in a video store. Many of the differences were subtle but some were significant. The significant ones were mostly due to the variety of types of video stores that the students had visited to check out movie videos.

During an initial class period of brainstorming in order to find candidate objects, Figure 4.7 was developed. This list was referred to as the Video Store Information System Candidate List of Objects—Pass One. No attempt was made during the session to discuss details of these candidate objects. In fact, in the spirit of brainstorming, any comments about the positive or negative merits of each were expressly discouraged by the instructor. No attempt was made to identify whether a candidate item was an object or not. In fact, we referred to all of the items discussed as just objects. The discussion was lively with many of the students offering their object suggestions.

After a few brainstorming and discussion sessions, the student/instructor team refined the first-pass list of candidate objects to come up with the second-pass list of

- Members
- Customers
- Movies
- Reports
- Concessions
- Sales
- Games
- Inventory
- VCR Rentals
- Movie Rental
- Employees
- Automated System
- Hardware
- Software
- Backup System
- Bonus Plan
- P.O.S. Terminal
- Clothing
- Transaction
- Marketing
- Popcorn
- Prices
- Physical Environment
- Scheduling
- Location
- Accounting
- Advertising
- Overhead
- Database
- Payroll

Figure 4.7   Video Store Information System Candidate List of Objects—Pass 1

candidate objects grouped into classes as shown in Figure 4.8. Note that the class names are the same as the original object names. This is not required but often ends up being the case. Some of the candidate objects were removed from the problem domain component because they were better suited for one of the other three components—human interaction, data management, system interaction—such as P.O.S. Terminal. The students indicated that they felt pretty good about their first few attempts at finding objects during the brainstorming and discussion sessions. They did admit, however, that having experience with video stores allowed them to find more of the important objects early in the process.

Having completed pass 2 in the "finding objects" activity for the video store information system, the students believed that the core objects for the problem domain component of the object model had been identified. More objects may be identified as the methodology's activities progress. Figure 4.9 shows the grouping of the identified objects into classes, each having the same name as the object it is a collection of. This is a common phenomena in object modeling—class names being the same as the object names—since a class is a collection of objects.

Attributes and services are still needed for each of the identified classes. This activity, called responsibilities, could be done at this time if the systems analyst or user would prefer it. However, for purposes of this book, identifying responsibilities is deferred until the next chapter so that a discussion of a future enhancements strategy and patterns can be presented here first.

A FUTURE ENHANCEMENTS STRATEGY

In a real problem domain, additional object identification brainstorming and discussion sessions may take place. As they do, it is quite likely that the user will suggest additional ideas and features for the information system. Many of these ideas and features can be deferred until a later time or a later version of the proposed information system. Comments like, "It would be nice if the system could do . . ." tend to re-

- Video
- Game
- ConcessionItem
- VCR
- SalesTransaction
- RentalTransaction
- Member
- Employee
- StoreLocation
- Vendor
- PurchaseOrder

Figure 4.8   Video Store Information System Candidate List of Classes—Pass 2

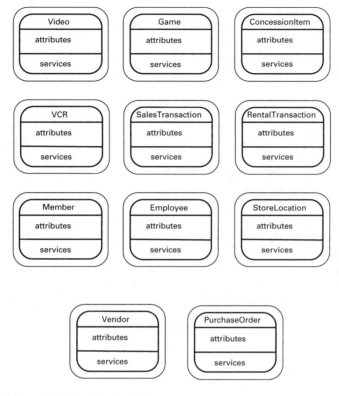

Figure 4.9   Video Store Candidate Class Symbols

sult in ideas or features that can be put on a future enhancements list to be dealt with after the information system has been implemented. Also time to develop and cost factors may shift current features into a future enhancements list in order to reduce the overall development time and/or cost of implementing the first iteration of the information system.

A future enhancements list was created and expanded throughout the brainstorming and discussion sessions surrounding the video store example. The final list is presented here as Figure 4.10 even though the actual future enhancements list at this point in the project would have included less items.

SUMMARY

This chapter has defined object, class, and class-with-objects and given examples of each. Several strategies were presented as ways to help the systems analyst find objects and classes in a problem domain. These strategies and others are helpful given the facts that the problem domains are quite diverse and the object-oriented methodology is still emerging and needs to mature. One of the strategies was discussed in a step-by-step fashion in order to illustrate its approach. The third strategy was used to

1. Open-ended memberships (indefinite; i.e., no cancellations) with credit card or $50 deposit (P).

2. Establish bonus point system for rentals (i.e., rent nine get one rental free, etc.) (E).

3. Handle special orders for videos (E).

4. Ability to reserve a movie ahead of time (E).

5. Automatic billing to credit card at end of month for month's charges (E).

6. Establish a membership debit card (E).

7. Allow a monthly tab to be paid at the end of the month (see #5) (E).

8. No physical security devices on videos (P).

9. System will not include payroll system, accounts payable, or financial systems such as general ledger and income statements (P).

10. The system will not keep information on nonmember customer sales (P).

11. Game inventory can be for sale or rent (but only one of these at a time) (P).

12. The system will not include any work shift scheduling of employees (P).

13. The video store will rent VCRs.

14. Payment for rental transactions will be at time of transaction (not when item is returned).

(P) = Policy decision   (E) = Potential Future Enhancement

Figure 4.10   Video Store Future Enhancements List—Final Pass

begin the development of the video store information system object-oriented user requirements model. As requirements are being gathered for an information system, it is helpful to keep a list of proposed future enhancements for possible addition to the information system at a later time.

QUESTIONS

4.1 Briefly define and give an example of a class and an object.

4.2 What is the difference between a class and class-with-objects?

4.3 What is the major distinction between classes and objects?

4.4 How many different notation symbols are used for class and object? Draw each of them.

4.5 In object-oriented notation, how is a class distinguished from a class-with-objects?

4.6 What are the three sections within the class template? Give examples of each.

4.7 What are the steps in a noun phrase strategy of identifying objects?

4.8 In general, what is a noun phrase strategy trying to find?

4.9 How does your answer to the previous question change if you are using a verb strategy?

4.10 When starting a search for objects, what are some of the items you should be especially aware of?

4.11 What is meant by challenging a list of possible objects?

4.12 What is generally done with derived results such as an employee report? Why?

4.13 What is done with requirements that can be deferred?

REFERENCES

BOOCH, G., *Object-Oriented Analysis and Design with Applications* (2nd ed.). Menlo Park, CA: Benjamin/Cummings, 1994.

COAD, P., D. NORTH, and M. MAYFIELD, *Object Models: Strategies, Patterns, and Applications.* Englewood Cliffs, NJ: Prentice Hall, 1995.

COAD, P., and E. YOURDON, *Object-Oriented Analysis.* (2nd ed.). New York: Yourdon Press/ Prentice Hall, 1991.

FIRESMITH, D.G., *Object-Oriented Requirements Analysis and Logical Design.* New York: John Wiley & Sons, 1993.

HENDERSON-SELLERS, B., *A Book of Object-Oriented Knowledge.* New York: Prentice-Hall, 1992.

RUMBAUGH, J., M. BLAHA, W. PREMERLANI, F. EDDY, and W. LORENSEN, *Object-Oriented Modeling and Design.* New York: Prentice Hall, 1991.

SHLAER, S., and S.J. MELLOR, *Object-Oriented Systems Analysis: Modeling the World in Data.* New York: Yourdon Press/Prentice Hall, 1988.

WINBLAD, A.L., S.D. EDWARDS, and D.R. KING, *Object-Oriented Software.* Reading, MA: Addison-Wesley, 1990.

WIRFS-BROCK, R.I., B. WILKERSON, and L. WIENER, *Designing Object-Oriented Software.* New York: Prentice Hall, 1990.

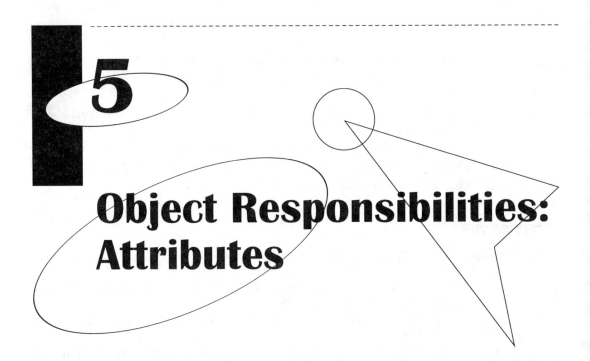

5
Object Responsibilities: Attributes

CHAPTER OBJECTIVES (YOU SHOULD BE ABLE TO)

1. Define attributes and their purpose.
2. Identify several questions to ask to help discover attributes.
3. Define three types of attributes.
4. Describe one attribute data dictionary technique.

What's your name? What size shoe do you wear? What is your address? What color are your eyes? How much do you weigh? What is your date of birth? These questions are somewhat intrusive, aren't they? Characteristics such as name, shoe size, address, eye color, weight, and date of birth are called attributes in the object-oriented methodology. The responses you give regarding the preceding questions represent the data values of the attributes for you. Your best friend's answers to these questions represent his or her data values for these attributes, and your instructor's answers represent his or her data values for these attributes.

Objects (and classes) have responsibilities that they are expected to exhibit in an object-oriented information system. There are three basic types of object responsibilities: (1) what "I" know, (2) who "I" know, and (3) what "I" do. Each object knows things about itself, called attributes; each object knows about other objects, called object connections; and each object knows what functions it has to perform, called services. This chapter addresses the "what 'I' know" responsibility of an object. The other two basic responsibilities are covered in the next two chapters.

ATTRIBUTES

Attribute names (e.g., name, shoe size, eye color, and so on) become the template or pattern that can be applied to all object instances within a class that has attributes associated with it. Remember, infrequently in business information systems, some classes will not have any attributes. So if the object model of an information system has a Student class with the foregoing attributes as in Figure 5.1, then each Student object instance (e.g., you and each of your classmates) has its own personal data values for each of the attributes that makes up the attribute template.

Sometimes the values in one object instance may be the same as the values in another object instance, but that happens most often by chance. For example, the eye color attribute probably has only a few possible values such as blue, brown, green, and so on, whereas a date of birth attribute could have tens of thousands of possible values.

Another name for an attribute value is an attribute state, and we will consider these two terms as interchangeable. **State** represents the condition of the attribute at any moment in time for a specific object instance. Once established, some attributes' state rarely or never changes. For example, your value for the Date of Birth attribute should never change; your value for the Name attribute might change a few times in your lifetime. For

a) Template

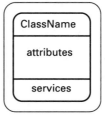

b) Example

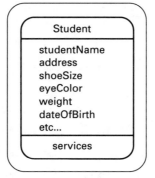

Figure 5.1   Class Showing Attributes

example, Hillary Rodham's name changed to Hillary Rodham Clinton. Some attributes' state changes frequently. For example, your value for the shoe Size attribute would have changed several times as you were growing up; your value for the weight attribute could change every day or more frequently; your value for your body temperature could also change frequently. Some information systems do not need to be as precise as knowing that your weight changes daily. For example, weight values on driver's licenses are only changed, if at all, at renewal time, which may be every couple or more years.

Candidate classes often have one or more attributes. In both the problem domain component (PD) and the data management component (DM) of the object model, more often the case is that many of the classes have dozens of attributes. Attributes have already been mentioned in earlier chapters because they are an integral and necessary component of object-oriented models and often contribute to the identification and grouping of classes. Attributes add more detail to classes, which in turn reveals more of how the model should be constructed.

Over the years professional men and women and students have commented that when they were first learning about objects, attributes, and states, they found it helpful to make a mental picture of some examples. This may help you also, so Figure 5.2 represents a mental picture that has been helpful to others.

Identifying and defining attributes is an ongoing and iterative activity that involves the systems analyst interacting with the user. A specific problem domain may potentially have thousands of attributes, but chances are that only a portion of them

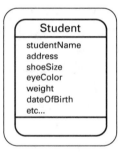

studentName	address	shoeSize	eyeColor	weight	dateOfBirth
Susan McIntyre	123 Franklin St. San Diego CA	11	Blue	175	4-12-74
Greg Fisher	765 Park Ave. San Diego CA	10	Brown	170	12-2-73
Minder Chen	222 Dallas St. La Mesa CA	9.5	Brown	140	10-5-76
Sally Athey	862 Grand Ave. Pacific Beach CA	6	Brown	125	6-28-75
Laura Applegate	914 Garnett La Jolla CA	5	Blue	110	3-15-74
Margie Heltne	479 55th St. El Cajon CA	5.5	Brown	105	5-22-75
Bill Martz	876 Balboa Mission Beach CA	10.5	Blue	190	1-26-70
Anna Easton	309 Del Mar Hts. Del Mar CA	6	Brown	120	8-14-74
Gene Easton	309 Del Mar Hts. Del Mar CA	10	Brown	165	9-19-72

Figure 5.2　Class Showing Attribute Values

are a necessary requirement of the system. In an earlier chapter, the necessary subset was referred to as the information system's responsibility. Once again, the analysis activity is performed to prune the list of potential attributes down to the list of those that are necessary for the information system to fulfill its responsibilities, as in Figure 5.3.

Attributes are manipulated by services, the topic of a later chapter. In general, attributes in a specific class may only be manipulated by their own services. This feature of object orientation supports the encapsulation or information-hiding principle for managing complexity within a problem domain. In other words, the services in a specific class have the responsibility of managing and changing the state or data values of their attributes. Some object-oriented programming languages have supported the notion of *friend* objects. Friend objects are allowed to manipulate other friend-designated objects.

A change made to an attribute state can be anything that is allowed for the specific attribute. For example, a new object instance may not have a specific value for any of its attributes when created; then one or more services within the class are called upon to populate the attributes with data values for each attribute. The service that changes the colorOfEyes attribute value to "blue" had better know the policy or rule pertaining to allowable data values for this attribute. If the request was made for this service to change the colorOfEyes attribute data value to "purple," it should know

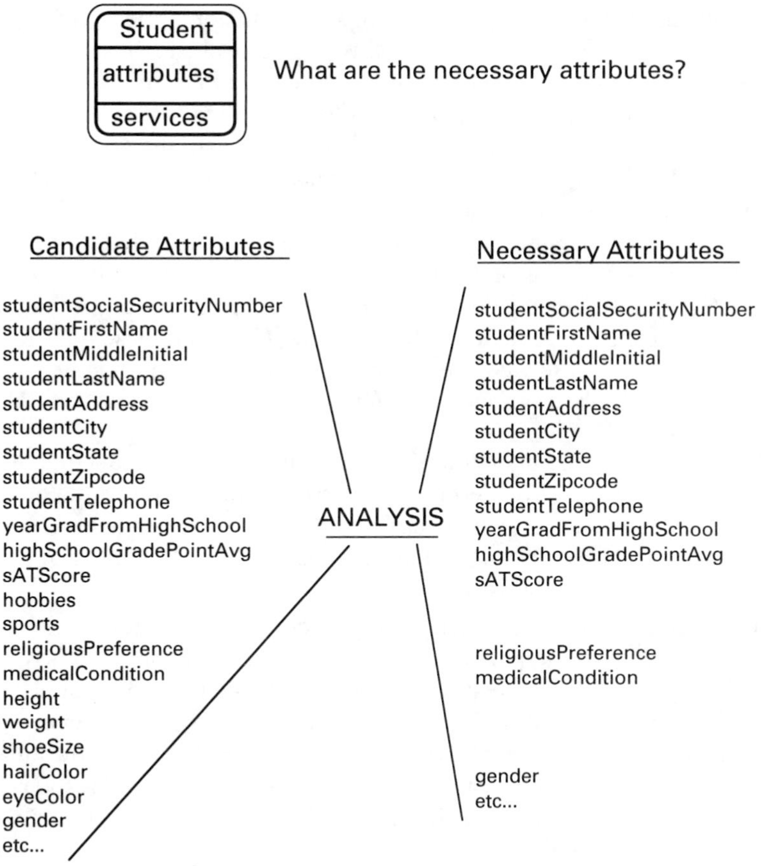

```
┌─────────────┐
│  Student    │
├─────────────┤
│ attributes  │        What are the necessary attributes?
├─────────────┤
│  services   │
└─────────────┘
```

Candidate Attributes

studentSocialSecurityNumber
studentFirstName
studentMiddleInitial
studentLastName
studentAddress
studentCity
studentState
studentZipcode
studentTelephone
yearGradFromHighSchool
highSchoolGradePointAvg
sATScore
hobbies
sports
religiousPreference
medicalCondition
height
weight
shoeSize
hairColor
eyeColor
gender
etc...

ANALYSIS

Necessary Attributes

studentSocialSecurityNumber
studentFirstName
studentMiddleInitial
studentLastName
studentAddress
studentCity
studentState
studentZipcode
studentTelephone
yearGradFromHighSchool
highSchoolGradePointAvg
sATScore

religiousPreference
medicalCondition

gender
etc...

Figure 5.3 Doing Analysis to Determine Necessary Attributes

whether or not this is a valid eye color data value. There are other ways of handling attribute value validations beyond the one just presented here, and several are discussed in a later chapter dealing with input and output design.

Determining Attributes

Determining attributes, like finding objects, is still a highly cognitive activity involving the analyst and the user. Many traditional information systems have predetermined attribute templates from which to begin, but even these must be investigated as to their significance within a user's specific problem domain. Just because every new car sold in the United States has a cigarette lighter in it doesn't mean that every buyer will use it. The same is true for using attribute templates for common information systems. However, unlike the car's unused cigarette lighter, which is there and never needs any attention beyond an occasional dusting, an attribute that is not needed but still part of the information system can add confusion and overhead in the form of increased processing and storage requirements.

Omitting an attribute can also be problematic if it is determined later that a specific output from the information system is not possible without that attribute. Retrofitting it is costly, just as retrofitting anything in a construction project would be. As the systems analyst and the user discuss the inclusion and exclusion of attributes, the systems analyst should ask the user to think of the future need for an attribute that is now not necessary and is tentatively going to be excluded. If the user says there is a good chance that the attribute might be needed within a year or two, the systems analyst might be wise to include it in the information system's requirements now in order to avoid the retrofitting later on.

There are endless questions that could be asked of users to help determine the required attributes for the many classes in the information system. In the following list are a few of the more generic-type questions that can be used in almost any situation and often can start the investigation moving with the users. Note that these questions are in the first-person mode because we have found that this technique often helps to "put yourself in the class-with-object's shoes for a few minutes" to better relate to it.

1. How am "I" described in general? For example, if you were a personal computer (just for a few minutes), how would you describe yourself generally? You would probably mention something about the brand, type, and speed of microprocessor you have, type of floppy disk drive you have, size of hard disk, amount of RAM, number of parallel and serial ports, type of monitor, and so on.

2. How am "I" described in this specific problem domain? For example, how would you describe yourself if you were a specific brand of personal computer, say an IBM model XYZ? You would tell us the specific components that make up the IBM model XYZ. Some of your responses would be identical to your responses to question 1, while others would be slightly different because we are focusing on a specific situation. In addition, some of your responses might be contrary to the "general" responses you gave in question 1, again because the IBM model XYZ may have something different than the generic personal computer.

3. What do "I" (as this class or object) need to know? In other words, what data are important to the success of this information system? Sometimes looking at the desired outputs will help determine the necessary inputs. The necessary inputs often tend to become the attributes.

4. What *state information* do "I" need to remember over time? This question can help identify additional attributes that are not obvious nor identified by the prior questions. Often this question is probing to identify attributes that may be needed in order to provide historical values over time for one or more attributes. If, for example, it is important to retain all temperature readings from a freezer's thermometer, and the readings are done every 15 minutes to prevent food thawing caused by an air-conditioner failure within the freezer, then one or more attributes and possibly a new class need to be created.

5. What *states* can "I" be in? Once attributes are identified, you can then ask the user this question for each attribute to determine the policy for each attribute and its accepted values.

In additon to these questions, you will no doubt ask questions that start with the words "what," "why," "when," "who," and "how."

Attribute Types

As you discuss attributes and their states, at least three types of attributes will be discovered in many information systems. These types are: (1) single-value attributes, (2) mutually exclusive value attributes, and (3) multivalue attributes, each of which is discussed next. Following the discussion of these attribute types, there will be a discussion of the strategy used in object-oriented models to accommodate these attributes.

The **single-value attribute** is perhaps the most frequently encountered attribute type. A **single-value attribute** is one that has only one value or state for itself at any moment in time. Referring to Figure 5.4, examples of this type of attribute are name,

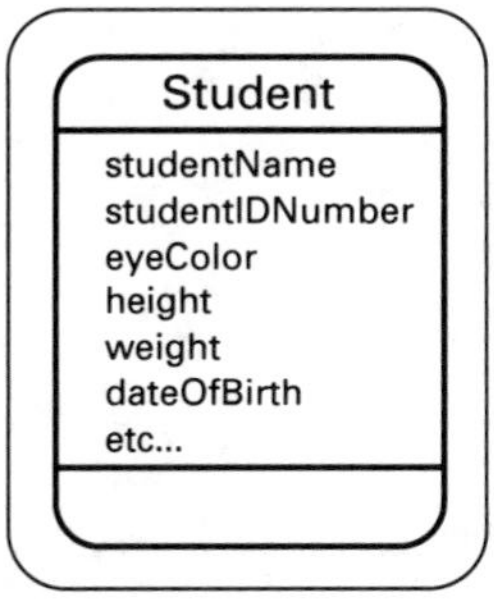

studentName	studentIDNumber	eyeColor	height	weight	dateOfBirth
MIchael W Smith	559-46-0912	Blue	5ft 9in	155	4-12-74
Amy Grant	371-38-7640	Brown	5ft 6in	118	12-2-73

Figure 5.4 Single-Value Attributes Example

studentIdentificationNumber, eyeColor, height, weight, and dateOfBirth. Even though a student's height and weight will change over time, at any single moment in time there is only one height and only one weight for a specific student. Many single-valued attributes have a descriptive nature to them, as do these examples.

The mutually exclusive value attribute is most often problem domain dependent. What is meant by this is that the identification of this type of attribute can only be determined by discussing it in its role as part of the problem domain and more specifically how it relates to the other attributes within a specific class in the problem domain.

An attribute is said to be a **mutually exclusive value attribute** if the presence or absence of its value is dependent upon the presence or absence of one or more other attribute values. The business policy decisions that this information system is representing play a vital role in determining whether an attribute is of the mutually exclusive value type. Figure 5.5 illustrates mutually exclusive value attributes. The business policy for this information system states that an employee may be either hourly or salaried, but not both. With this in mind then, either the hourlyRate attribute or the weeklySalary attribute would have a value for each Employee object instance.

The third attribute type is the **multivalue attribute.** This attribute is the opposite of the single-value attribute because it can have, as its name implies, multiple values at any moment in time. Figure 5.6 illustrates this, again using the Student class. For this illustration the Student class has name, collegeAttended, and collegeGrade-PointAverage attributes. A specific student, you for example, may be attending or have already attended one or more other colleges, community colleges, or vocational training schools. The business policy decision that is being enforced or supported by these attributes says that a student is to list all current and prior colleges, community colleges, and vocational training schools that he or she attended.

employeeName	employeeNumber	hourlyRate	weeklySalary
MIchael W Smith	559-46-0912	$9.75	
Amy Grant	371-38-7640		$475.00

Mutually Exclusive Attributes with Values

Figure 5.5 Mutually Exclusive Attribute Values Example

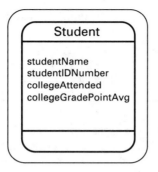

studentName	studentIDNumber	collegeAttended	collegeGradePointAvg
MIchael W Smith	559-46-0912	Grossmont	2.9
		Point Loma	2.7
Amy Grant	371-38-7640	NY Univ.	3.2
M.C. Hammer	270-73-9815	Golden Gate	2.2
		Georgia State	2.9
		U of San Diego	3.1

multi-value attributes with values

Figure 5.6   Multi-value Attributes Example

Object-Oriented Methodology Strategy for Different Attribute Types

One of the goals of developing information systems using an object-oriented methodology is to simplify the decision logic that is embedded within the programming code for the information system. For example, using Figure 5.5, the programming code embedded in one of the class's services would have to be able to distinguish an hourly employee from a salaried employee, since both types of employee objects are associated with this class. The reason the programming code would need to know about this distinction between hourly and salaried employees is that different processing actions would be performed on each employee type. An object-oriented methodology strategy for handling mutually exclusive attributes would be to create new classes, one for each employee type—hourly and salaried—as shown in Figure 5.7. With this structure all Employee objects below HourlyEmployee are hourly employees and all Employee objects below SalariedEmployee are salaried employees. There is still only one class—HourlySalariedEmployee—containing employee objects, but hourly employees are associated with the HourlyEmployee class and salaried employees are associated with the SalariedEmployee class.

In object-oriented methodology theory this works well. But just how practical is it in every situation? Well, it often depends on the number of mutually exclusive attributes belonging to a class. For a class having two mutually exclusive attributes, as in Figure 5.7, the theory works well. The class having two sets of two attributes that are mutually exclusive can probably also be structured similarly to Figure 5.7. But what about three, four, five, six, or more sets of pairs of mutually exclusive attributes in one

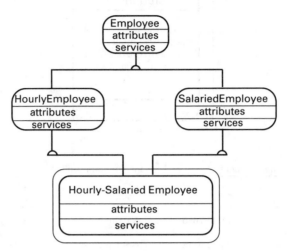

Figure 5.7 Object-Oriented Strategy for Mutually Exclusive Attributes

class? Structures to support these would get quite cumbersome. Also the example so far only has considered pairs of mutually exclusive attributes. What about the situation where there are three attributes (or four, five, six, and so on) that are mutually exclusive with each other? And what about sets of these groups of attributes? Again, elaborate object-oriented structures to accommodate these situations may be more problematic than they are worth, even though in theory mutually exclusive attributes should be split apart and represented by class gen-spec patterns, which are discussed in the next chapter.

Object-Oriented Strategy for Multivalue Attributes

Multivalue attributes also lend themselves to the creation of gen-spec patterns to more effectively represent them. Looking at Figure 5.6, there are three student names. Each would be an object. However, notice that because of the multivalue attributes, the first name has two colleges attended and the last name has three colleges attended. Because of these multiple values, there would actually be six Student objects—two for the first name, one for the second name, and three for the last name. Each Student object would have data values for each attribute, causing redundancy as shown in Figure 5.8.

  Multivalue attributes can be demoted to become the lower level in a whole-part object connection pattern (discussed in the next chapter) as illustrated in Figure 5.9. By doing this, redundancy is avoided. The object-oriented model now contains two classes—Student and CollegeAttended. Note that the name and studentIdNumber attribute data values are no longer duplicated, as was the case in Figure 5.8. Each Student object in Figure 5.9 is associated with the appropriate number of College-Attended objects, as illustrated to the right side of the whole-part pattern in the figure. The whole-part pattern object connection constraints are also noted, which allows the

studentName	studentIDNumber	collegeAttended	collegeGradePointAvg
MIchael W Smith	559-46-0912	Grossmont	2.9
mIchael W Smith	559-46-0912	Point Loma	2.7
Amy Grant	371-38-7640	NY Univ.	3.2
M.C. Hammer	270-73-9815	Golden Gate	2.2
M.C. Hammer	270-73-9815	Georgia State	2.9
M.C. Hammer	270-73-9815	U of San Diego	3.1

* Notice the redundancy of data values for these two attributes

Figure 5.8   Student Class Attribute Data Values

software to administer the number of "part" objects associated with each "whole" object and vice versa. In the figure, a Student object may have zero or more Colleges attended, and a CollegeAttended object must be associated with one and only one Student.

Theoretically speaking, whole-part object connection patterns are the most appropriate way to model multivalue attributes using object-oriented models. However, the practical side of this is about the same as was discussed earlier for the mutually exclusive attributes.

Finally, the foregoing discussion is more focused on "how" rather than "what," making it more appropriate for consideration during the design phase of information systems engineering. What this means is that during the user requirements determination and analysis phase, the object-oriented model may contain mutually exclusive and multivalue attributes in the same class. However, during the design, mutually exclusive value and multivalue attributes will be looked at from a "how to design and implement" perspective and adjusted via the gen-spec and whole-part patterns as dis-

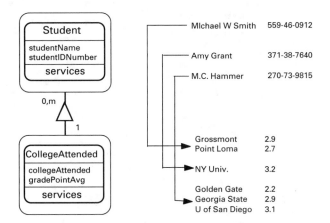

Figure 5.9   Student and CollegeAttended Class Object Connection

cussed earlier. As always, if during user requirements determination and analysis it is beneficial for the user's understanding of the problem domain's object-oriented model, then creating new class patterns to accommodate both of these attribute types is acceptable. Both gen-spec and whole-part patterns are discussed in more detail in the next chapter.

The Video Store Example—Identifying Attributes

The video store information system's user requirements model now includes classes. A few more brainstorming sessions yielded the attributes shown in Figure 5.10. Trying to show the object-oriented model with all of the attributes assigned to the various classes becomes cumbersome on paper since the model will overlap several pages. So the approach taken in Figure 5.10 is to just list the classes in the left column and their associated attributes in the right column. Almost all of the attributes were suggested by the students with little prodding from the instructor, which indicated that they had some familiarity with video stores.

An additional piece of documentation is extremely useful to maintain with object-oriented models and is called an attribute data dictionary. An **attribute data dictionary** is an alphabetized list of attribute names, which class each belongs to, and details regarding each attribute's definition and editing rules. CASE software that sup-

Class	Attributes
Member	memberNumber memberName memberAddress memberCity memberState memberZipCode memberPhone creditCardNumber creditCardExpireDate depositAmount
Video	barCodeNumber description qtyOnOrder price cost taxCode timesRented dueDate memberNumber quantitySold qtyOnHand
Game	(same as Video)
ConcessionItem	barCodeNumber description qtyOnOrder price cost taxCode quantitySold qtyOnHand
VCR	barCodeNumber description qtyOnOrder price cost taxCode timesRented dueDate memberNumber

Figure 5.10   Video Store Classes with Attributes—Part 1 of 2

Class	Attributes
SalesTransaction	transactionNumber employeeNumber transactionDate transactionTime barCodeNumber price salesTax quantitySold
RentalTransaction	transactionNumber employeeNumber transactionDate transactionTime barCodeNumber price salesTax memberNumber
Employee	employeeNumber employeeName employeePhone positionCode
StoreLocation	storeNumber address city state zipcode telephone
Vendor	vendorNumber vendorName vendorAddress vendorCity vendorState vendorZipCode vendorPhone vendorFaxNumber
PurchaseOrder	purchaseOrderNumber purchaseOrderDate purchaseOrderDueDate purchaseOrderCancelDate barCodeNumber quantityOrdered vendorNumber itemCost

Figure 5.10   Video Store Classes with Attributes—Part 2 of 2

ports object-oriented modeling automatically supports an attribute data dictionary, but the systems analyst must still tell it about each attribute's definition and editing rules. A sample of an attribute data dictionary for the video store example is shown in Figure 5.11.

SUMMARY

This chapter focused on "what 'I' know" about myself object responsibility. This responsibility addresses the identification of attributes as an important part of the object-oriented methodology for creating an object-oriented user requirements model of an information system. Attributes were defined, described, and illustrated with examples. Several questions were presented that can help a systems analyst discover attributes. Three types of attributes were discussed and illustrated—single-value, mutually exclusive value, and multivalue attributes. The object-oriented methodology modeling strategy to handle mutually exclusive and multivalue attributes was discussed. Finally, the video store information system example was further expanded to include attributes.

Attribute	Class	Definition/Rules
barCodeNumber	Inventory Transaction	Up to 12 characters Up to 12 characters
creditCardExpireDate	Member	MM/YY (month, year)
creditCardNumber	Member	Up to 19 characters
cost	Inventory	range: 0 to 999.99
depositAmount	Member	min: $25; max: $200
description	Inventory	40 characters
employeeNumber	Transaction	6 digits
memberAddress	Member	30 characters (no P.O. Box)
memberCity	Member	30 characters
memberName	Member	30 characters
memberNumber	Member	6 digits
memberPhone	Member	999-999-9999
memberState	Member	2 char; validate using State Obj.
memberZipCode	Member	99999-9999
price	Inventory Transaction	range: 0 to 999.99 range: 0 to 999.99
qtyOnOrder	Inventory	4 digits
salesTax	Transaction	range: 0 to 999.99
etc.....		

Figure 5.11   Video Store Attribute Data Dictionary (partial)

QUESTIONS

5.1 What is an attribute, and what is an attribute's purpose in an object model?

5.2 What is the most important method for identifying and defining attributes? Discuss why this is so important.

5.3 What are some of the specific questions that can be asked of a user in an attempt to best identify attributes?

5.4 What is a single-value attribute?

5.5 Define and give a few examples of a mutually exclusive attribute.

5.6 What is the purpose of demoting multivalue attributes to the lower level in a whole-part object connection pattern?

5.7 What aspect of using multivalue attributes in a whole-part object connection pattern makes this task better suited for the design phase of systems analysis and design?

5.8 What is an attribute data dictionary and what is its purpose in an object-oriented methodology?

REFERENCES

COAD, P., D. NORTH, and M. MAYFIELD, *Object Models: Strategies, Patterns, and Applications.* Englewood Cliffs, NJ: Prentice Hall, 1995.

COAD, P., and E. YOURDON, *Object-Oriented Analysis* (2nd ed.). New York: Yourdon Press/Prentice Hall, 1991.

SHLAER, S., and S.J. MELLOR, *Object-Oriented Systems Analysis: Modeling the World in Data.* New York: Yourdon Press/Prentice Hall, 1988.

6

Object Responsibilities: Class and Object Connections

CHAPTER OBJECTIVES (YOU SHOULD BE ABLE TO)

1. Define an object pattern and discuss its purpose in systems analysis and design.
2. Define and use the generalization-specialization pattern.
3. Define and use the whole-part: assembly-part, container-content and group-member patterns.
4. Define and use the participant-transaction pattern.
5. Define and use the place-transaction pattern.
6. Define and use the participant-place pattern.
7. Define and use the transaction-transaction line item pattern.
8. Define and use the item-line item pattern.
9. Define and use the peer-peer pattern.

Objects! Objects! Objects! They are everywhere, so it seems. They are certainly the focal point in an object-oriented systems analysis and design methodology. If you look at Figure 6.1, you will notice that there are quite a few classes. With all of these candidate classes floating around, you might be thinking that the object model is uncoordinated, cumbersome, hard to comprehend, unmanageable, overwhelming, and so on. Object models for average size information systems can have several dozen classes of objects. The video store information system has purposely been scaled down and simplified for textbook illustration purposes. Even so, a dozen classes can be overwhelming to look at and make sense out of for the user.

Objects (and classes) have responsibilities that they are expected to exhibit in an

116

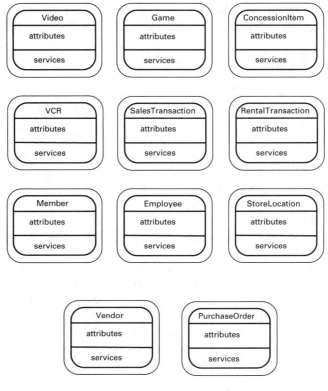

Figure 6.1   Video Store Candidate Class Symbols

object-oriented information system. There are three basic types of object responsibilities: (1) what "I" know, (2) who "I" know, and (3) what "I" do. Each object knows things about itself, called attributes; each object knows about other objects, called object connections; and each object knows what functions it has to perform, called services. This chapter addresses the "who 'I' know" responsibility of an object. The preceding chapter covered the first of the two basic responsibilities, and the next chapter covers the third one.

"WHO 'I' KNOW" RESPONSIBILITY OF AN OBJECT

Given many classes, there needs to be a way to organize and relate classes for two purposes: (1) to connect classes or objects one with another to complete part of a class's or object's responsibilities, and (2) to simplify the object model and make it more comprehensible for the user. In so doing, the information system model will become more coordinated, less cumbersome, easier to comprehend, more manageable, and less overwhelming. Obviously, this is extremely important for communication with the user and the others that are part of the project team. Research and empirical results have shown over and over again that the less understood an information system user requirements model is to the user, the less likely the user is to identify problems, oversights, and ambiguities early in the systems development life cycle.

OBJECT PATTERNS

One commonly accepted way to organize and relate classes is to use the notion of patterns. A **pattern** in object-oriented systems development is a template of objects (classes) with stereotypical responsibilities and interactions. The template may be applied again and again by analogy. Pattern instances are building blocks used to assemble effective object models. Patterns are commonly used in music, literature, art, architecture, linguistics, dressmaking, manufacturing, and many other disciplines. Your life has a general pattern to it which can change over time. The pattern for your life right now could be something like wake up, exercise, eat, go to classes, eat, go to work, eat, talk on phone, do homework, go to sleep.

An object model pattern can consist of a doublet, triplet, or other small grouping of classes that is likely to be helpful again and again in object-oriented development situations. These pattern types are likely to be repeatable multiple times within a specific problem domain and transferrable across many problem domains. Identifying patterns is much like finding objects. However, we start with the classes we have found so far and work from there in our identification activity.

Dozens of patterns have been documented in the chapter reference materials. The following patterns have been selected as representative and being most appropriate for this introductory book. Each pattern is shown with an example from a university student registration system. Following the discussion of all of the patterns, the video store example is revisited and the candidate classes shown in Figure 6.1 are grouped into patterns and finally assembled into the first pass of an object model. The eight patterns described and illustrated here are:

1. generalization-specialization
2. whole-part: assembly-part, container-content, and group-member
3. participant-transaction
4. place-transaction
5. participant-place
6. transaction-transaction line item
7. item-line item
8. peer-peer

Generalization-Specialization Pattern

Generalization-specialization (gen-spec) is a hierarchical parent-child pattern. It is not unique to systems development in general or object-oriented systems development specifically. In fact, everyone has probably seen the value of this ageless concept in other disciplines and areas of life. The gen-spec template is illustrated in Figure 6.2, and gen-spec examples are shown in Figure 6.3.

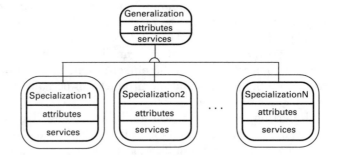

Figure 6.2   The Generalization-Specialization Pattern

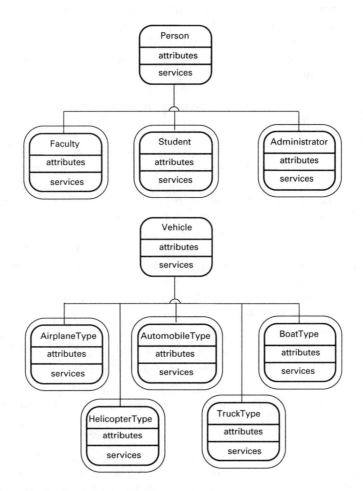

Figure 6.3   Generalization-Specialization Examples

Object-oriented programming's metaphor of metaclasses, superclasses, classes, and subclasses exploits this hierarchical pattern, and now object-oriented analysis and design does so as well. Using ideas and patterns that are already familiar to the user, such as the gen-spec pattern, can significantly contribute to the user's acceptance and comprehension of the object-oriented methodology and resulting object model. Both the process and the product can perhaps be better understood by the user.

Referring to the gen-spec figures, notice the arc or semicircle on the line that connects the generalization on top to the specializations on the bottom. This notation denotes a gen-spec pattern in the object model. Notice also that the lines always connect and touch the class portion (inner) of the class symbol. In other words, the gen-spec pattern does not connect one object to another object. Instead, it connects one class to another. Keep in mind also that each symbol in the pattern is often referred to as a node. Nodes are often referred to in a familial manner such as parent, child, grandchild, great-grandchild, and so on, going from top to bottom or down the hierarchy. Going bottom to top or up the hierarchy the nodes would be referred to as child, parent, grandparent, great-grandparent, and so on. Even though most of the gen-spec pattern illustrations in this chapter are only two tiers or levels, like both examples in Figure 6.3, there is no restriction on the number of levels that can make up a gen-spec pattern. You are free to use however many levels you need, keeping in mind that the two most important things are that (1) the model be as close to representing the real-world information system as possible, and (2) the model facilitates correct communication between the user and the systems analyst.

Please look at the top example of the gen-spec pattern in Figure 6.3. It was taken from part of a university information system. There are a few aspects of this gen-spec pattern that can be determined just by looking at it. First, Person is a class that contains no objects. In other words, all the objects that relate to Person are either part of Faculty, Student, or Administrator. There are no other choices given the pattern that is being shown. How was this gen-spec pattern determined? By discussing the application with the user. This is not something that the systems analyst does on his or her own initiative. This particular gen-spec pattern can be presented to the user for verification of correctness. The following represents a possible dialog with the user regarding this gen-spec pattern:

Systems analyst: "Bob [our user], I would like to review this part of your information system's user requirements model with you to see if we understand your requirements correctly."

Bob: "Okay, go ahead."

Systems analyst: "The requirements model document that we are working from indicates that all of the people in this system are either faculty, students, or administrators? Is this correct?"

Bob: "Yes." [If he would have said no, then we need to discuss the variation and determine how to deal with it in light of the requirements model.]

Systems analyst:	"Okay, so in the model of your system we are planning to group all of these people as hierarchically belonging to a group called Person in order to keep like things together and help to simplify the model."
Bob:	"Okay." [Bob nods okay.]
Systems analyst:	"The model uses the singular form of all words even though plural can be implied."
Bob:	"I understand that. Thanks."

If Bob were familiar with the terminology of class and even objects, then we would use those terms in the preceding scenario, but it is important not to overwhelm the user with methodology-specific jargon when it is possible to communicate with words that are already meaningful to him and will suffice just as well.

Why was Figure 6.3's gen-spec hierarchical pattern created? It was created for at least three reasons: (1) simplification of the problem domain by relating similar classes, (2) clarification and communication of the systems analyst's understanding of the problem domain, and (3) for relating the model effectively to the real-world problem that is being worked on. Keep these same three ideas in mind as you create other gen-spec patterns. The goal is not to see who can make the "neatest, most elaborate, or coolest" patterns, but to see who can create patterns that meet the foregoing objectives.

Chances are that during the "finding objects" discovery process for the university example, only Faculty, Student, and Administrator classes were identified. No Person class was identified. As we discuss these three classes' relationships with each other coupled with the fact that they may have some common characteristics (attributes) and functions (services), we create a class called Person to help organize them in a hierarchical gen-spec pattern. So the "identifying patterns" activity has actually created a new class, adding to the total number of classes, but simplifying the model as well.

Just two more things about this particular example before moving on. First, during your discussion with the user, if the user says that there are other people in the system beyond Faculty, Student, and Administrator who are important to the system, then approval to change the requirements model document must be obtained. Then the object-oriented user requirements model is modified to reflect the change. For example, if the discussion with the user discovered that the Trustees were an important group of people for this information system, then a new class would be created and called Trustee, along with changing the requirements model document. For verification, validation, and consistency, the user requirements problem domain object model of the information system must always tie directly back to the requirements model document or other requirements document that we are working from.

Finally, referring to Figure 6.3, during the discussion with the user, it was determined that the Administrator class consisted of a heterogeneous group of objects, such as managers, staff, custodians, attorneys, accountants, and so on. Because the Administrator grouping of objects is so diverse, calling the group "Administrator" may or may not meet the needs of the user. So there are a few options for drawing the gen-spec that contains this Administrator group of objects. For example, Figure 6.4a illus-

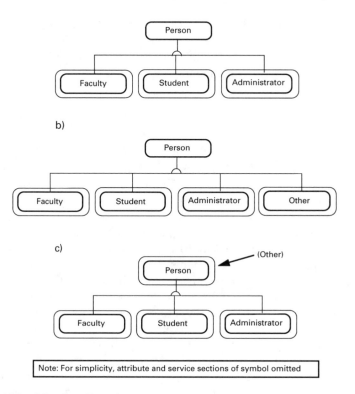

Figure 6.4   Additional Gen-Spec Examples

trates one way, which is the same as in Figure 6.3. Because the Administrator class is not homogeneous, that is, not able to be grouped easily into a common class like Faculty and Student, it might be more acceptable to the user to have the systems analyst create a class called "Other" to cover this specialization group of people, as in Figure 6.4b. The Administrator class would contain individuals that the user believed belonged in such a grouping, and those that did not would be moved to the Other class. Or it may be more acceptable to the user to make Person a class-with-objects in order to allow all the "other" persons to be objects in the Person class-with-objects as in Figure 6.4c. Either of these three gen-specs is technically correct. The best way is determined by the needs of the problem domain, and the user being most able to understand it. More than likely, the user will tell the systems analyst to choose whatever gen-spec pattern the systems analyst thinks would be best.

Class names that are vague, like "Other" in Figure 6.4b, are usually an indication that not enough analysis has been done to fully explore this part of the problem domain. The recommendation and guideline for thorough analysis would be to not allow such names in the user requirements object model even though classes can technically be called that.

Referring to the examples in Figure 6.3, another aspect of interpreting this pattern is how to read or say it. The rules of thumb or heuristics for reading or saying gen-spec patterns is to: (1) read or say them from the bottom up, going from child node to parent node, and (2) use the words "is a" or "is a kind of" in between the specialization and the generalization words. For example, we read the top example in Figure 6.3 as "Faculty is a Person" (we really are!), "Student is a Person," and "Administrator is a Person." Alternatively, it also could be read from the bottom to top, "Faculty is a kind of Person," "Student is a kind of Person," and "Administrator is a kind of Person." Either way is correct and sometimes just reading or speaking the gen-spec pattern helps to determine if what you think is a gen-spec pattern really is.

The gen-spec pattern in Figure 6.5 can be determined to be incorrect simply by reading it. "Arm is a Person" and "Leg is a Person." This just does not make any sense. Currently, this cognitive understanding of correctness or incorrectness of a gen-spec can only be made by a human. You see, the pattern is technically drawn correctly (syntax), but its content (understanding) is incorrect.

Some additional discussion about gen-spec patterns should be helpful as you begin the learning process for identifying them. Refer to the generic gen-spec hierarchical pattern in Figure 6.6 to illustrate the points discussed here. In Figure 6.6a, the top level of this gen-spec pattern has a generalization class symbol, G, and three class specializations—S1, S2, S3—below it. Depending on your familiarity with hierarchies, there is a tendency to think that an object associated with S1, S2, or S3 must relate or connect to an object behind class G. This is incorrect thinking for the gen-spec pattern only. Look again at Figure 6.4c. The Faculty objects do not connect to the Person objects, the Student objects do not connect to the Person objects, nor do the Administrator objects connect to the Person objects. The Person objects in that figure are unique and separate from Faculty, Student, and Administrator objects, and Figure 6.6b illustrates this concept. Notice that the gen-spec connection line always connects the class squircle (bold line) with another class squircle (bold line). Determining the gen-spec patterns in your problem domain still remains a highly cognitive human activity. However, as said before, the resulting patterns in the object model should represent the user's view of the real world as closely as possible.

Any combination of class and class-with-objects symbols is technically allowed in the gen-spec pattern with the important exception that the lowest-level node in every branch of a gen-spec pattern hierarchy must always be a class-with-objects. Why must the lowest-level node in each branch always be a class-with-objects? Sim-

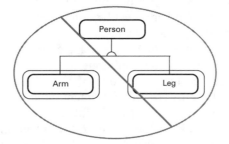

Figure 6.5   Incorrect Gen-Spec Pattern According To "is a" Heuristic

a) Generalization Class G with Specialization Classes S1, S2, and S3

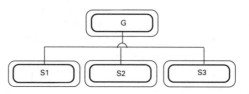

b) how to think about a) above. Show gen G's objects
and spec S1's (S2 or S3) objects with the gen-spec
pattern connection touching class G's class squircle
and class S1's (S2 or S3) class squircle.

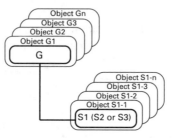

Figure 6.6   Gen-Spec Illustration Showing Pattern Connection

ply because it would not make any sense or be of any value to have a class symbol located as the lowest node of a branch with no objects associated with it and nothing below it in the hierarchy.

As you discuss the details of the problem domain with your user, you may find that things don't always fit neatly into one class symbol. For example, referring to Figure 6.3 again, you might discover that your user says that it is possible for a Faculty Person to also be a Student Person (faculty sometimes take classes), and that on a few occasions a Student Person is also a Faculty Person (a Ph.D. student may also be on the Faculty). How is such a situation modeled? Figure 6.7a illustrates a gen-spec pattern that represents this situation. Any Faculty Person who is not a Student would have his or her object behind the Faculty class; any Student who is not a Faculty would have his or her object behind the Student class; any Faculty Person who is also a Student or any Student who is also a Faculty would have his or her specific object behind the Faculty-Student class. Note then that there is only one object instance for each Faculty Person and only one object instance for each Student Person, and that objects can be located in one of three places—Faculty, Student, or Faculty-Student class. If you were only a Student, your object instance would be behind the Student class. If you were only a Faculty, your object instance would be behind the Faculty class. Finally, if you were both Student and Faculty, your object instance would be behind the Faculty-Student class. Figure 6.7b shows one way to mentally think about this pattern situation.

a) Object-Oriented Structure Model

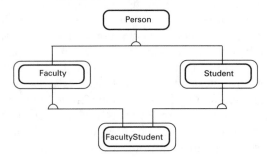

b) How to think about a) above. Note that FS 1 is a
completely different Object from F1 and S1, and
so forth.

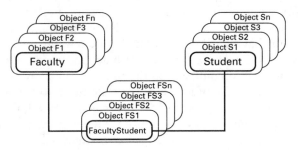

Figure 6.7   Gen-Spec Illustration Showing Multiple Generalizations Connecting to a Specialization

Generalization-Specialization Inheritance

The final topic to discuss about the gen-spec pattern is the notion of inheritance. As discussed in an earlier chapter, inheritance is a principle used for handling complexity in systems development. **Inheritance,** as used in systems development, is a principle in which all lower-level or children nodes in an inheritance hierarchy inherit the characteristics of the parent node. As mentioned earlier, grouping classes together in patterns helps to simplify the model. In addition to this, using a gen-spec pattern introduces the notion of inheritance in the sense of a parent-child pattern into the object model, also contributing to the model's simplicity. All inheritance in gen-spec patterns is one way—from parent node to child node. The child node inherits all of the characteristics (attributes) and the functions (services) of the parent node. Therefore, there is no need to list the attributes and services again in the children nodes. The same holds true for grandchildren nodes, great-grandchildren nodes, and so on, all of which also inherit everything from connected nodes above them in the gen-spec pattern hierarchy. In Figure 6.3a, Faculty, Student, and Administrator inherit all attributes and services associated with Person. In Figure 6.3b, AirplaneType, HelicopterType, AutomobileType, TruckType, and BoatType inherit all attributes and services associated with Vehicle. In addition to the "is a" or "is a kind of" heuristic used to help determine correct gen-spec patterns, reviewing attributes and services of a parent node can also be a heuristic to help determine correct children nodes. If the attributes or services of the parent node do not seem to make sense or be of any use when inherited down into the child node, then chances are that the pattern is not meant to be a gen-spec. The reverse of this is helpful also and will be discussed in more detail in the chapters on attributes and services. Attributes and services, independent of each other, can be promoted to a higher-level node in the hierarchy or demoted to a lower-level node in the hierarchy as a result of discussing whether each attribute or service applies to all object instances at the node level where the attribute or service has been placed.

In discussions about attribute and service inheritance, the concepts of overriding and extending inherited definitions arise. A detailed discussion of this topic is outside the scope of this book but is acknowledged here for completeness. As their names imply, these topics deal with (1) **overriding** (canceling) inherited attributes and services by using *same-named* attributes and services at the lower level, and (2) **extending** (expanding or enhancing) inherited attributes and services by using same-named attributes and services at the lower level with some indicator to extend the inherited definition having the same name. Figure 6.8 illustrates these topics. These two concepts, overriding and extending attributes and services, should be used with caution as they can lead to confusion in the user's mind, as well as possibly not being implementable when the time comes to create the software for the system, since different programming strategies (e.g., C++, Smalltalk, Eiffel, COBOL, and so on) may not be able to accommodate overriding and extending attribute and service concepts.

Finally, a heartily debated object-oriented analysis, design, and programming subject is the notion of multiple inheritance. As with the prior topic on overriding and extending attributes and services, this topic is outside the scope of this book but is

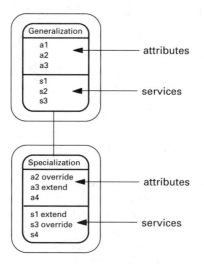

Figure 6.8   Inheritance—Extending and Overriding Concepts

acknowledged here for completeness. **Multiple inheritance** deals with the situation of a child node inheriting attributes and services from more than one parent node. Figure 6.7 illustrates a multiple inheritance gen-spec pattern, and Figure 6.9 illustrates the inherited attributes and services concept of multiple inheritance. Multiple inheritance, say its advocates, is very powerful. Its opponents caution against its use by saying that its use tends to make the model lose its simplistic nature as well as making it difficult to deal with the rules when inheriting the same-named attribute or service from more than one parent or inheriting incompatible attributes or services from more than one parent. Caution is suggested when considering the use of multiple inheritance for the same reason as mentioned earlier regarding overriding and extending attributes and services.

Inheritance is a powerful concept in object-oriented information systems. It is also a model simplification concept, which allows the systems analyst not to have to copy or repeat all of the inherited attributes and services to lower-level classes within a gen-spec pattern. At times the need to see all of these inherited attributes and services in the lower-level classes becomes helpful. Most CASE tools that support object-oriented models automatically display the inherited attributes along with the class-specific *local* attributes, indicating which ones are inherited and which ones are local to the class as shown in Figure 6.10.

As the discussion of gen-spec patterns ends, Figure 6.11 presents a summary of gen-spec pattern rules and guidelines. Sometimes the appropriate pattern, whether gen-spec or other patterns, which are described next, can only be determined after some attributes and/or some services are identified and discussed with the user. These additional parts of a class help give more clarity to the most appropriate pattern to apply.

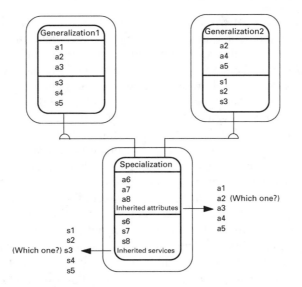

Figure 6.9   Gen-Spec Pattern of Multiple Inheritance

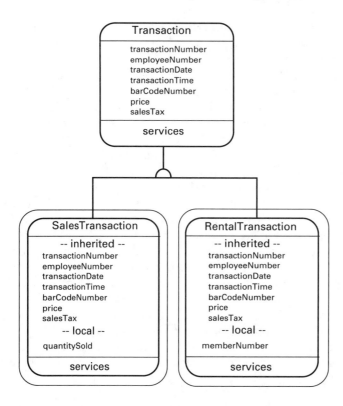

Figure 6.10   CASE Tool Display of Inherited Attributes in Gen-Spec Pattern

Whole-Part Object Connection Pattern

The whole-part pattern, like the gen-spec pattern, is part of everyday life for you and the user. It, too, is not unique to systems development. In fact everyone has probably seen the value of this ageless classification theory concept in other disciplines and areas of life. Fathers, mothers, and others who have had the privilege of assembling children's toys are well aware of the whole-part concept. All too often these people wish they would have purchased the assembled (whole) toy rather than the unassembled (parts) toy. Anyone who has put batteries in a portable radio, CD, cassette, or flashlight knows that they are an essential part of the whole item. Can you think of anything in its whole state that is not made up of one or more parts? Probably not. Again, using ideas and patterns that are already familiar to the user, such as the whole-part pattern, can significantly contribute to the user's acceptance and comprehension of the object-oriented methodology and resulting object model. Both the process and the product can perhaps be better understood by the user.

The whole-part object connection pattern template is shown in Figure 6.12. When discovered in problem domains, the whole-part pattern usually appears in one of three configurations—assembly-part, container-content, or group-member. When speaking or thinking about these three configurations, one says, "assembly and its parts," "container and its contents," and "group and its members."

1. A gen-spec follows the "is a" or "is a kind of" heuristic.

2. The gen-spec notation symbol is an ARC or semicircle.

3. The line connecting a generalization with a specialization starts with the generalization's class squircle (bold line symbol) and ends with the specialization's class squircle (bold line symbol).

4. Specialization objects are not connected to any generalization objects as they are separate and distinct.

5. A specializations have their own attributes and/or services as well as inheriting additional attributes and/or services from the generalization node above them.

6. Inherited attributes and services may be overriden or enhanced at the specialization node level.

7. Multiple inheritance is allowed but can complicate

Figure 6.11   Generalization-Specialization Pattern Guidelines

As you look at Figure 6.12, notice the triangle on each of the lines between whole and part. The triangle denotes a whole-part pattern in this methodology's object modeling. The triangle always points to the whole class. Notice also that both ends of the whole-part line always connect and touch the object squircle portion of the class symbol. In other words, one parent object instance is most often connected to some number of children object instances. Finally, notice the letter "x" alongside the lines between the whole and the part classes. The "x" is called an object connection constraint. An **object connection constraint** is an expression of "who the object knows." In other words, it indicates how many other objects an object is aware of in the whole-part object connection. Object connection constraints are a common characteristic of all whole-part patterns as well as all other patterns discussed here with the exception of the gen-spec. The gen-spec pattern has no object connection constraints because that pattern is a connection between classes, not objects.

For those systems analysts and programmers who are familiar with the cardinality concept associated with an entity-relationship diagram, object connection constraints are quite similar. The object connection constraint expresses how many other objects an object knows about. For example, the letter "x" in the generic template in Figure 6.12 represents many possible types of object connection constraints, such as 0-n objects, 1-n objects, 1-5 objects, 1 object, 25 objects, and so on. The letter "n" means "many" or "unlimited."

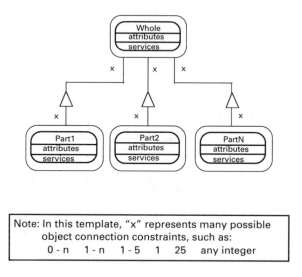

Figure 6.12   Whole-Part Pattern Template

The object connection constraint expresses the size of a mathematical set of numbers. Each number or number pair expresses a lower and upper numeric boundary range for how many other objects an object knows when viewed from the whole to the part or the part to the whole perspective. The number or number pairs closest to the whole object indicate the whole-to-part constraint or how many part objects a whole object knows about. The number or number pairs closest to the part object indicate the part-to-whole constraint or how many whole objects a part object knows about. The six object connection constraints in Figure 6.12 are interpreted as, "one Whole object knows about 'x' Part1 objects," "one Whole object knows about 'x' Part2 objects," "one Whole object knows about 'x' PartN objects," "one Part1 object knows about 'x' Whole objects," "one Part2 object knows about 'x' Whole objects," and "one PartN object knows about 'x' Whole objects."

Sometimes the object connection constraint is indicated with just one number instead of a number pair as in Figure 6.13a. The interpretation of these constraints is, "one Whole object knows about 5 Part1 objects," "one Part1 object knows about 1 Whole object," "one Whole object knows about 2 PartN objects," and "one PartN object knows about 3 Whole objects." When the lower bound is zero (0) and the upper bound is unlimited, the notation uses "0-n" to communicate this, as shown in Figure 6.13b. The interpretation of "0-n" is "one Whole knows about zero or more PartN objects." The upper bound can be infinite or many, and this is indicated with the letter "n" as in Figure 6.13b. The lower bound need not be zero or one, as also illustrated in the same figure.

As with the gen-spec pattern, each symbol in the whole-part pattern is often referred to as a node. Nodes are also often referred to in a familial way, such as parent, child, grandchild, great-grandchild, and so on. Even though the whole-part illustrations

a) Single Boundary Object Connection Constraint

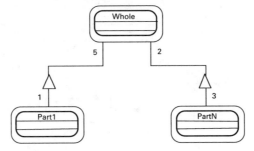

b) Additional Object Connection Constraint Alternatives

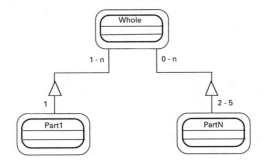

Figure 6.13   Object Connection Constraint Alternatives

in this chapter are only two levels, there is no restriction on the number of levels that can make up a whole-part pattern. You are free to use however many levels you need, keeping in mind that the most important thing is that the model be an effective communication tool and represent the real-world information system as closely as possible.

Figure 6.14 shows an example of a whole-part pattern taken from the same university information system as the earlier gen-spec pattern example. There are a few aspects of this whole-part pattern that can be determined just by looking at it. First, Faculty and Student are classes that each contain instances of Faculty objects—me, your instructor, other instructors, and so on—and Student objects—you, your roommate, your other classmates, and so on—respectively. According to the interpretation of the figure, each Faculty object (1) has one or more (1-n) degrees such as B.S., M.B.A., Ph.D., (2) teaches or has taught zero or more (0-n) courses (a new faculty person may not have taught a course yet), and (3) has zero or more (0-n) committee assignments, such as personnel, curriculum, or computer equipment, within the university. Each Student object (1) belongs to zero or more (0-n) campus clubs such as DPMA, ACM, Ski Club, and (2) has taken zero or more (0-n) courses at this university (a new student may be taking courses now but not completed any courses yet). How are these object connection constraints determined? By discussing the application with the user. This is not something that the systems analyst does on his or her

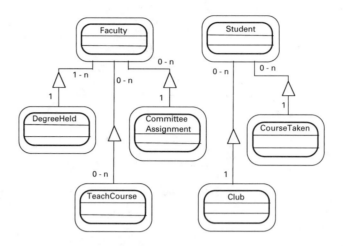

Figure 6.14   Whole-Part Pattern Example

own initiative. This particular whole-part pattern can be presented to the user for verification of correctness. The systems analyst would review Figure 6.14's interpretation with the user, being sure to get his or her approval of it.

The object connection constraints expressed by the whole-part pattern (or for any of the other patterns discussed later in the chapter) are very important. Discussing these with the user is critical to getting the application requirements correct. A constraint lower limit of zero is different than a lower limit constraint of one. Conversely, a higher limit of many (n) is different than a fixed upper limit such as 5, 10, 50, and so on. Getting these cardinalities correct during the analysis activity will avoid problems being discovered at a later time causing rework to be done.

Why create a whole-part hierarchy pattern such as the one shown in Figure 6.14? This is done basically for the same reasons that we create gen-spec patterns. Whole-part patterns also assist in simplifying the problem domain by grouping part classes together with their respective whole classes. In addition, they are created for clarification and communication of our understanding of the problem domain, and for relating the model effectively to the real-world problem being worked on. Keep these ideas in mind as you create whole-part patterns. Again, the goal is not to see who can make the "neatest, most elaborate, or coolest" patterns but to see who can create patterns that meet the foregoing objectives.

Just as the gen-spec pattern uses the "is a" heuristic, the whole-part pattern uses the "has a" heuristic. The gen-spec pattern is read bottom to top with the "is a" heuristic, while the whole-part pattern is read top to bottom with the "has a" heuristic. As you group whole-part classes, use this heuristic to help see if the pattern makes sense. Applying this to Figure 6.14, you would say, "Faculty has a Degree," "Faculty has a Course Taught," "Faculty has a Committee Assignment," "Student has a Club," and "Student has Taken a Course." Remember, like the "is a" heuristic, the "has a" heuristic is just a guideline and sometimes the specific words "has a" do not fit well (e.g.,

"Student has a Club" may bring up a mental image of a student walking around campus with a club in hand) so other similar words could be used such as "Student belongs to a Club."

Using Figure 6.14's object connection constraints, a good translation of the pattern when communicating it to the user would be to say, "Faculty can have one or more degrees," "Faculty has taught between zero and many unique courses," "Faculty has had zero or more committee assignments," "A Student belongs to zero or more Clubs," "A Student has taken zero or more Courses." Stating the same idea but using the constraint perspective of the part rather than the whole would sound like this, "A Degree is held by one faculty," "A Course is taught by zero or more Faculty," "A Committee Assignment is for Faculty," "A Club is for one Student," and "A Course taken is for one Student."

You may be asking "how do you know that 'Club is for one Student'?" or "how do you know that 'Course Taken is for one Student'?" or "how do you know . . .?" for any of the other relationships in Figure 6.14's whole-part pattern. The answer is that the systems analyst has talked to the user to determine this. Sometimes the attributes and services of a particular class within the pattern must be discussed in order to really determine this. On the surface, it seems that "Club is for one or more Students" would be more correct, but when we find out that the Club object has attributes whose values are specific to a certain student, such as "date joined the club" or "current office held in club," then we know that the Club object belongs to only one Student. If the user says something like "we only need to keep the name of the club as an attribute and nothing else," then we may change the constraint to "Club is for one or more Students" or remove this portion of the whole-part pattern since it now better fits another pattern template. Once again, discussion with the user is crucial for determining the correct whole-part patterns and object connection constraints. As with gen-spec patterns, the current cognitive understanding of correctness or incorrectness of whole-part patterns can only be made by a human. The whole-part pattern may be drawn correctly (syntax), but its content (understanding) may still be incorrect.

Some additional discussion about whole-part patterns should be helpful as you begin the learning process for finding them and assigning object connection constraints to them. Unless the part-to-whole constraint has a lower boundary of zero, all part objects must connect with one or more whole objects, as is illustrated in Figure 6.15.

Once again, remember that the notions of inheritance and multiple inheritance that exist with gen-spec patterns do not apply at all to whole-part patterns or any other patterns discussed in this chapter. There simply is no inheritance in the whole-part pattern or any other. It is unlikely that a service in the whole symbol would be useful or make sense in its part symbol. And it is just as unlikely that attributes in the whole symbol would be useful or make any sense in the part symbol.

The possibility exists to have combination patterns consisting of both gen-spec and whole-part patterns. Figure 6.16 illustrates this. You are free to combine both of these patterns (or any others) in any necessary manner to accomplish the task of creating a model that closely represents the real world it is modeling. Keep in mind that when combining gen-spec and whole-part patterns, the inheritance feature of gen-spec

a) Zero-to-Many Object Connection Constraint Example

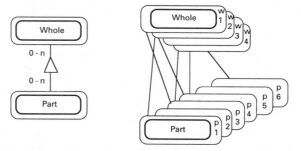

b) One Part-to-Whole Object Connection Constraint Example

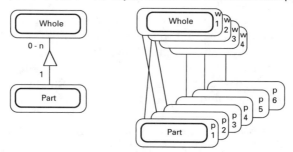

c) One-to-Many Object Connection Constraint Example

This example would look just like Figure 4.23a except both constraints would be 1 - n instead of 0 - n; and Whole w3 would be connected to one or more parts and Parts p2 and p5 would be connected to one or more whole.

Figure 6.15   Whole-Part Pattern Constraint Illustration

patterns is discontinued or broken at the point in the pattern where a whole-part pattern enters the compound pattern.

Heuristics for finding whole-part patterns. When attempting to identify whole-part patterns, three heuristics or guidelines are available to assist with their identification: (1) assembly and parts, (2) container and contents, and (3) collection and members. All three of these heuristics are variations of the whole-part theme.

Assembly and parts is probably the easiest to identify. This variation is often used to identify a manufactured item (assembly) and the different subassemblies and parts (part) that it is comprised of. For example, a personal computer is made up of power supply, mother board, chips, disk drives, screws, and so on. A car is made up of doors, windows, wheels, axles, trunks, engine, seats, and so on. Figure 6.17 illustrates the assembly-part type of the whole-part pattern.

Container and contents is a variation that can help with situations where the objects aren't really the traditional assembly and parts that you may be used to thinking about. For example, an airplane cockpit (container) consists of gages, dials, levers,

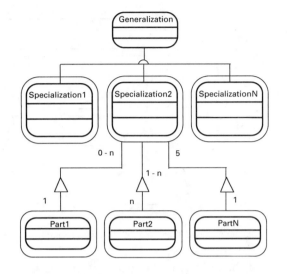

Figure 6.16   Combination Gen-Spec and Whole-Part Pattern

lights, and so on (contents). An office (container) consists of desks, phones, bookshelves, file cabinets, windows, and so on (contents). Figure 6.18 illustrates the container-contents type of the whole-part pattern.

The third variation of the whole-part pattern is the **group and members,** which helps with some situations the other variations do not address well. For example, Professional Organization may be a class that has the Association for Computing Machinery (ACM) professional group as an object along with the Data Processing Management Association (DPMA) and others. The people who are members of these professional associations are considered the members of the group. The Class of '97

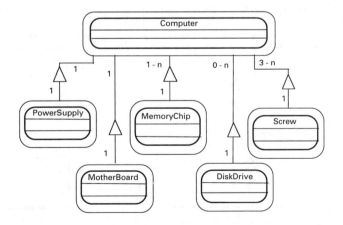

Figure 6.17   Assembly-Part Type Whole-Part Pattern Example

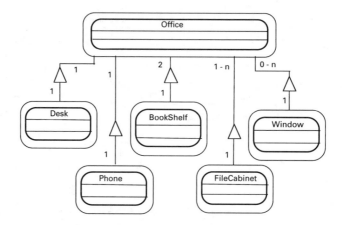

Figure 6.18   Container-Contents Type Whole-Part Pattern Example

may be one of the objects in the group of graduating classes from your university. The students who graduate in 1997 are the members. Figure 6.19 illustrates the group-members type of the whole-part pattern.

Object Connection Patterns

The remaining patterns belong to a classification called object connection patterns. The object connection pattern models the *association* principle for managing complexity in systems analysis and design, which was discussed in a prior chapter. All object-oriented methodologies have the association principle embodied within them in one way or another.

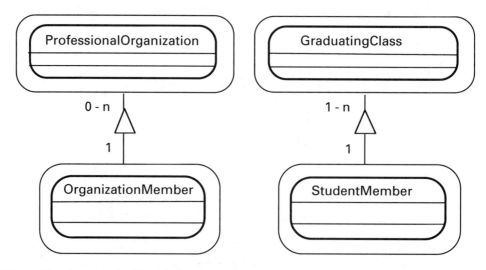

Figure 6.19   Group-Member Type Whole-Part Pattern Example

One important concept and tool that emerged with the structured and information modeling methodologies and the database management systems is the entity-relationship diagram (ERD). Entity-relationship diagrams are quite common to business systems analysts and database administration professionals. Much of the ERD concept can be translated and transferred very easily into "object think" terminology's object connections.

As discussed in the whole-part pattern section previously, an object connection connects one object to one or more other objects. It represents a relationship between two objects. An object connection constraint is an expression of "who the object knows" by indicating how many other objects an object is aware of in the object connection. In other words, the object connection and its constraint can be thought of as "every time you have one of these objects you have to have so many of those objects." The object connection really reflects the couplings within the object model that are necessary to complete its assigned work.

The object connection is a line with dual object connection constraints, as in Figure 6.20a. This is similar to the whole-part object connection with the exception that there is no triangle drawn on the line. As with whole-part object connection constraints, the object connection constraint closest to the class symbol indicates how many of the objects at the other end of the connection are known to this object. Also

a) Object Connection Template

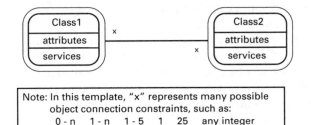

b) Object Connection Example

Figure 6.20   Object Connection Template and Example

note that the instance connection line starts at the object squircle (thin outer one) of the class symbol and ends at the object squircle of the other class symbol, as shown in Figure 6.20. The interpretation of the student-to-course object connection shown in Figure 6.20b is illustrated in detail in Figure 6.21. In this example, the policy relationship that exists between the Student objects and the Course objects was determined to be that a student is currently taking zero or more courses during the current semester and this object connection communicates this policy. From this perspective, the information system should be able to produce a list of students and the courses that each is taking this semester. All students could be listed but some would not have any courses this semester, such as S3 in the figure. Conversely, the university has a list of hundreds of courses, each an object instance behind Course. All courses are listed but not all are offered every semester, or a course that was offered did not get sufficient enrollment so it is dropped from the current semester. From the Course perspective then, a course is (currently) taken by zero or more students. From this perspective, we should be able to produce a list of students who are taking a specific course, say this one for example. Some courses would be listed but have no students, such as C2 and C5 in the figure.

Remember, an object connection is determined by the user and the policy that governs the aspects of the problem domain that you are addressing. Also object connections may occur in the other three components of an object model—human interaction, data management, and system interaction. Using Figure 6.20 or Figure 6.21, if a Course is optional (e.g., you can still be a Student object even if you drop out for two semesters), then the lower boundary of the object connection constraint between Student and Course would be zero. If Course were not optional but mandatory (e.g., your Student object would be removed if you dropped out of school for one semester), then the lower boundary of the constraint between Student and Course would be one (1). Hypothetically, if a Course could only be taken by zero or one student at a time (e.g., just you and the instructor in the room), then the upper boundary of the constraint between Student and Course would be one (1). A constraint of 0-1 is different than a constraint of 1-1 (always written as just 1). The first case is optional and, if chosen, only one; the second case is mandatory and limited to only one.

Many-to-many object connection constraints (0-n, 1-n, and so on at both ends of an object connection) as in Figure 6.22a reveal an additional class when the discussion

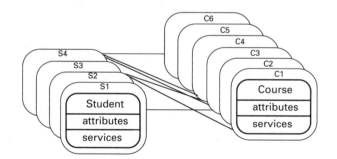

Figure 6.21   Object Connection Example Detail Illustration

turns to attributes for each class. Initially, Figure 6.22a is perfectly valid and legal during the early iterations of finding objects and identifying patterns. As you learn more about the problem domain, often as part of a discussion about class attributes, you discover that a new class is needed as in Figure 6.22b. The new class surfaces when you discover one or more attributes that do not belong with either of the classes per se. In fact, they belong to the union or relationship that exists between the two classes.

Keep in mind that object connection patterns are not as strong or obvious as the gen-spec and whole-part patterns, yet the pattern's relationship is necessary for implementing the policies established for the information system to perform its intended work. Keep in mind also that you do not need to struggle over identifying the "correct or proper" pattern type that you discover when looking at the classes and objects in the object model. In other words, it is most important to identify object connection relationships that are needed between classes in order for the information system to function correctly. The fact that the identified object connection can be called "such and such" a pattern is secondary. Simply put, please remember that object connection patterns are available to help you with the cognitive activity of identifying object connections that are necessary. If patterns help you with that process, great! Each of the object connection patterns is discussed next.

Participant-transaction pattern.   The **participant-transaction** object connection pattern connects a participant object, such as a person or organization, with a

a) Original Many-to-Many Object Connection

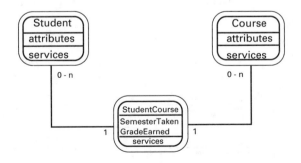

b) Many-to-Many Object Connection With Associative Class

Figure 6.22   Many-to-Many Object Connection

transaction object, such as a sale or payment. People and organizations perform trans-actions. Students pay registration fees to take courses at a university. Figure 6.23 shows the participant-transaction pattern template and a student-course example of the pattern.

Place-transaction pattern.   The **place-transaction** object connection pattern connects a place object, such as a store, office, or ticket window, with a transaction object, such as a sale or payment. Transactions are performed at locations. Students

a) Participant-Transaction Pattern Template

b) Participant-Transaction Pattern Example

Figure 6.23   Participant-Transaction Pattern and Example

a) Place-Transaction Pattern Template

b) Place-Transaction Pattern Example

Figure 6.24   Place-Transaction Pattern and Example

pay registration fees at a university cashier's window. Figure 6.24 shows the place-transaction pattern template and a registration fee being paid at a cashier's window example of the pattern. Figure 6.25 combines the participant-transaction pattern with the place-transaction pattern because these two patterns often exist in an object model.

a) Combination Pattern Template

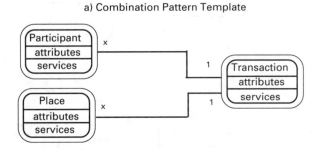

b) Combination Pattern Example

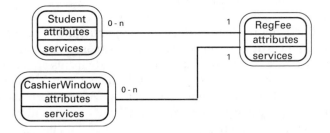

Figure 6.25   Combining Participant-Transaction Pattern with Place-Transaction Pattern

a) Participant-Place Pattern Template

b) Participant-Place Pattern Example

Figure 6.26   Participant-Place Pattern and Example

a) Transaction-Transaction Line Item Pattern Template

b) Transaction-Transaction Line Item Pattern Example

Figure 6.27   Transaction-Transaction Line Item Pattern and Example

Participant-place pattern.   The **participant-place** object connection pattern connects a participant object, such as a person or organization, with a place object, such as a store, office, or ticket window. Participants are organized by place. At some point in time each university student declares his or her major, which is housed in a particular college on the university campus. Figure 6.26 shows the participant-place pattern template and a student belonging to a college within the university example of the pattern. This pattern may also be combined with the participant-transaction pattern because these two patterns often exist in an object model.

Transaction-transaction line item pattern.   The **transaction-transaction line item** object connection pattern connects a transaction object, such as a sale or order, with a transaction line item object, such as sale line item or order line item. Transactions usually have one or more transaction line items. Students paying full university registration fees have several items that they are paying for, such as tuition, health, and student body fees. Figure 6.27 shows the transaction-transaction line item pattern template and the student registration fees to registration fees line items example of the pattern.

Item-transaction line item pattern.   The **item-transaction line item** object connection pattern connects an item object, such as a product, with a transaction line item object, such as sale line item or order line item. Each transaction line item has one item. The items that a university registration fees system is aware of are things like tuition, health, and student body fees. Each student's registration fee transaction will have one or more of these fees associated with it. Figure 6.28 shows the item-transaction line item pattern template and a master list of registration fees along with the student's registration fees transaction example of the pattern.

This pattern is worthy of further discussion and illustration since it is a very common one in many information systems. Transaction line items usually come from

a) Item-Transaction Line Item Pattern Template

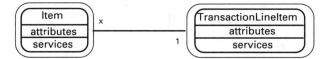

b) Item-Transaction Line Item Pattern Example

Figure 6.28   Item-Transaction Line Item Pattern and Example

a master list of items that are available to be included as a transaction line item. A master class-with-objects of student registration fees might look something like the one in Figure 6.29. A transaction line item class-with-objects that gets its registration fee items from the master student registration fees class might look something like the one in Figure 6.29. Master lists of items include customers, students, courses, products, faculty, rooms, and so on.

Peer-peer pattern.   There is a significant amount of robustness in the use of object connections even though the ones already discussed may address a large majority of your problem domain needs. The **peer-peer** pattern is an object connection between objects within the same class, as shown in Figure 6.30. In this example the user-determined policy was that a student could have zero, one, two, or three room-

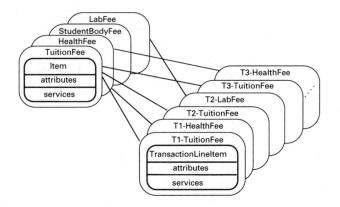

Figure 6.29   Item-Transaction Line Item Pattern Illustration

a) Peer-Peer Pattern Template

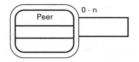

b) Peer-Peer Pattern Examples

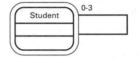

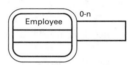

Figure 6.30   Peer-Peer Pattern Template and Examples

mates; therefore, it is possible to look at your Student object instance and find out the names of your roommates, assuming you had one, two, or three as in this example.

VIDEO STORE EXAMPLE

The video store candidate classes from Figure 6.1 can now be grouped according to patterns. Several were identified by the student/instructor analysis team. First, two gen-spec patterns were discovered, as shown in Figure 6.31. In this figure, the top gen-spec has three levels—the top and middle being classes and the bottom level being class-with-objects. The four types of products—videos, games, concession items, and VCRs—are all types of inventory that are rented or sold. The middle-layer generalizations—SaleItem and RentalItem—were created to simplify the distribution of attributes between classes that had objects for sale—Video, Game, and ConcessionItem—and classes that had objects for rent—Video, Game, and VCR. Attributes associated with the sale of an inventory object could be promoted to the SaleItem class. Attributes associated with the rental of an inventory object could be promoted to the RentalItem class. Note that three new classes with no objects were created to become the generalization for this gen-spec pattern.

The second gen-spec pattern in Figure 6.31 combines sales and rental transactions. Like the prior gen-spec, attributes that are common to both sales and rental transactions are promoted to the Transaction class.

Next several object connection patterns were discovered, as shown in Figure 6.32. For example, the PurchaseOrder class had multivalued attributes that led to the

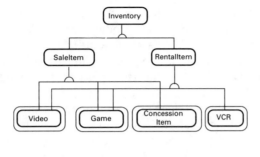

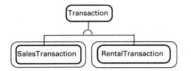

Figure 6.31   Video Store Gen-Spec Patterns

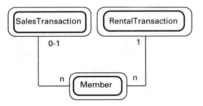

Figure 6.32   Video Store Object Connection Patterns

creation of the POLineItem class. Also the discussion centered on the fact that not all products are sold and not all products are rented. To accommodate this, two new classes were introduced—SaleItem and RentalItem. All of the new classes were organized into the first pass of the problem domain component's object model and are now shown in Figure 6.33 as the final-pass object model. Finally, the list of all of the classes is shown in Figure 6.34.

At this point in the analysis of the video store information system, the instructor would present new material concerning services and scenarios. These topics are found in the next chapter of this book. After reviewing the material, the instructor would engage the students in another series of brainstorming sessions about the video store information system.

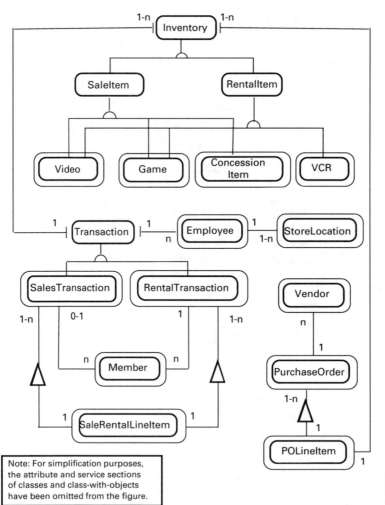

Note: For simplification purposes, the attribute and service sections of classes and class-with-objects have been omitted from the figure.

Figure 6.33   Video Store Problem Domain Component's Object Model—Final Pass

- Video

- Game

- ConcessionItem

- VCR

- SalesTransaction

- RentalTransaction

- SaleRentalLineItem

- Member

- Employee

- StoreLocation

- Vendor

- PurchaseOrder

- POLineItem

- Inventory

- Transaction

- SaleItem

Figure 6.34   Video Store Information System Candidate List of Classes—Pass 3

SUMMARY

This chapter addressed object responsibilities—patterns. There are dozens of patterns documented in various reference books. Eight representative patterns were presented in the chapter beginning with the gen-spec and followed by the whole-part object connection. The last section of the chapter presented six additional object connections. Patterns can be useful to help identify object relationships. They help support the "who I know" object responsibility.

QUESTIONS

6.1 Define and give an example of a gen-spec pattern.

6.2 Define and give an example of a whole-part object connection pattern for each of assembly-part, container-content, and group-member patterns.

6.3 Define and give an example of the participant-transaction pattern.

6.4 Define and give an example of the place-transaction pattern.

6.5 Define and give an example of the participant-place pattern.

6.6 Define and give an example of the transaction-transaction line item pattern.

6.7 Define and give an example of the item-line item pattern.

6.8 Define and give an example of the peer-peer pattern.

REFERENCES

ANDERSON, B., "Patterns: Building Blocks for Object-Oriented Architectures," *Software Engineering Notes.* (January 1994).

COAD, Peter, "Object-Oriented Patterns," *Communications of the ACM,* 35, no. 9 (September 1992), 152–159.

COAD, P., D. NORTH, and M. MAYFIELD, *Object Models: Strategies, Patterns, and Applications.* Englewood Cliffs, NJ: Prentice Hall, 1995.

COAD, P., and E. YOURDON, *Object-Oriented Analysis* (2nd ed.). New York: Yourdon Press/Prentice Hall, 1991.

GAMMA, E., R. HELM, R. JOHNSON, and J. VLISSIDES, *Design Patterns: Elements of Reusable Object-Oriented Software.* Reading, MA: Addison-Wesley Publishing, 1995.

SHLAER, S., and S.J. MELLOR, *Object-Oriented Systems Analysis: Modeling the World in Data.* New York: Yourdon Press/Prentice Hall, 1988.

WINBLAD, A.L., S.D. EDWARDS, and D.R. KING, *Object-Oriented Software.* Reading, MA: Addison-Wesley, 1990.

WIRFS-BROCK, R.I., B. WILKERSON, and L. WIENER, *Designing Object-Oriented Software.* New York: Prentice Hall, 1990.

7

Object Responsibilities: Services and Scenarios

CHAPTER OBJECTIVES (YOU SHOULD BE ABLE TO)

1. Define and give examples of business policies and procedures.
2. Define the two types of services.
3. Describe the methods for finding services.
4. Describe some of the typical problems associated with narrative specification documents.
5. Define and use service scenarios.
6. Define Structured English and use it to document a business procedure.
7. Define decision table and decision tree and use them both to describe a policy that has complex decision logic.
8. Describe a state-transition diagram.
9. Describe ways to identify *home* classes for services.

What do the following statements have in common? "Print a report showing the courses each student is taking next semester," "Display my savings account balance on the ATM display screen," "Scan the barcode on the groceries as I go through the checkout line at the supermarket," "Turn the heater on," and "Change the traffic signal light from yellow to red." What's common among them? Well, first of all, these statements are all action oriented, and second, each is an example of a potential service in some appropriate information system.

Objects and classes have responsibilities that they are expected to exhibit in an object-oriented information system. There are three basic types of object responsibilities: (1) what "I" know, (2) who "I" know, and (3) what "I" do. Each object knows things about itself, called attributes; each object knows about other objects, called object connections; and each object knows what functions it has to perform, called services. This chapter addresses the "what 'I' do" responsibility of an object. The preceding two chapters covered the first two object responsibilities.

"WHAT 'I' DO" RESPONSIBILITY OF AN OBJECT

Services, often called **methods** in other object-oriented methodologies and object-oriented programming, are the **actions** that the information system must perform in order for it to fulfill its intended purpose and meet the information system needs of the user. Another way to think of services is that they are the information system's response to an event. An **event** is something that happens at a point in time, such as a user input or request, money being dispensed from an ATM machine, a traffic light turning red, an automobile insurance policy expiring, flight 119 from San Diego to San Francisco, or an air conditioner being turned "on." To summarize then, services are actions in response to events. In business situations, events often have a transactionlike quality to them, such as withdrawing cash from an ATM or adding a course to your class schedule. Most of these types of events are supported by business policies and business procedures, which are discussed in the next section.

Business Objectives and Tactics, Information System Objectives and Tactics, and Policies and Procedures

Do you recall the discussion in Chapter 2 about business objectives, business tactics, information system objectives, and information system tactics? Well, another way to think about services is that they are the information system tactics. As such, they support the information system objectives, which in turn support the business tactics, which in turn support the business objectives. Finally, business objectives support the overall goals and mission of the business.

Business tactics are the methods and ways the business believes are necessary to meet the business objectives. Another way of saying this is that business tactics are usually the things people do within the business to support the business objectives. For example, a counter worker at McDonald's takes customers' orders for food and drink, collects money, dispenses change, assembles the food and drink that were ordered onto a tray or into a bag, and gives the order to the customer. Business tactics include events—things that happen, such as a customer orders food and drink, a customer gives the counter worker the exact amount of money for the order or more money than the total cost of the order. Because things happen, business tactics are often supported by business policies and business procedures. The absence of business policies and procedures could create business inconsistencies, chaos, conflicts, and other problems.

A business **policy** is a set of rules that governs some activity within a business. Policies often are the basis for decision making within the business. You encounter policies every day, such as "No smoking allowed," "Parking allowed between 6 p.m.

and 6 a.m. only," "Ride the Double Twister at your own risk," and "You need 128 university credits to graduate." Businesses also have policies, such as "Two weeks vacation after three years of service," "Credit card purchases allowed for amounts over $50," "1.5 percent interest will be charged each month on your outstanding balance," and "Check cashing allowed with two picture identifications." Sometimes policies can be directly represented by an information system, but most often policies need to be translated into operational procedures.

Procedures are the step-by-step instructions for carrying out a policy or accomplishing some task(s). As stated previously, not all policies need to have procedures to carry them out. However, a policy such as "You need 128 university credits to graduate" must be translated into detailed procedures for the "graduation check" staff to use to be consistent in enforcing such a policy. Procedures can also be thought of as the steps taken in response to an event that has occurred, such as a student is requesting a "graduation check," so the "graduation check" clerk follows a procedure—set of steps—to determine how many graduation units the student has. Once determined, the clerk can then compare the student's number of units to the university's policy—"128 units to graduate"—to determine the response to give to the "graduation check" event. Of course, it is understood that a real university's graduation policy is much more rigorous than just saying "128 units to graduate." More than likely, the entire graduation policy or a significant amount can be translated into an information system's services, which can be called upon in response to a "graduation check" event.

As stated earlier, once the policies and procedures are determined, they become the basis for the information system objectives. For example, an information system objective related to the "graduation check" example would be "to support and enforce the 'graduation check' policy." The required information system tactics to support this information system objective would be an automated version of the manual procedures performed by the "graduation check" clerk. Problems tend to arise in businesses that either have undocumented policies and procedures, or have documented policies but do not have well-documented procedures for carrying out those policies.

Types of Services

In any information system's object model, there can be dozens and dozens of services. Every service is associated with a class and is listed in the lower portion of the symbol as in Figure 7.1. At the highest level of abstraction of the object model, just the word *service* is displayed, as in the template in the figure. At the next lower level of abstraction, as the example in the figure, just the name of each of the services is listed in the symbol.

There are two general types of services in Coad's object-oriented methodology: (1) those that are basic to all classes, and (2) those that are dependent on the specific functional requirements of the problem domain. All services fall into one of these two broad categories.

Basic Services

Basic services are the most prevalent and common services in Coad's object-oriented methodology. In fact, basic services are so basic that they are often not even listed in

a) Template

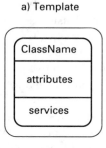

b) Example

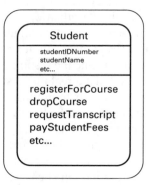

Figure 7.1   Class with Services

the service section of a class symbol in order to keep the symbols more simplistic. Basic services are implicit and automatically exist for every class. However, depending on implementation platforms, operating systems, and languages, some or all of these basic services may have to be explicitly addressed at implementation time. The basic services are: create object, search, get and set attribute values, add and remove object connections, and delete object. Basic services are the actual services that manipulate objects, their connections, and object attribute values. Each is discussed next.

The **create object service** creates new objects within a class. This is a basic class service, which allows the class to create a new object instance, as illustrated in Figure 7.2. Note in the figure that a new student object is created; however, any attributes associated with the Student class will initially have null or no data values. Once created, the set attribute value service would be used to populate the object instance's attributes with data values. For example, the values for each of the following attributes could be established—student's name, identification number, address—with the set attribute value service.

The **search service** locates a specific object within the class and is illustrated in Figure 7.3. For data-oriented classes, which are common in business information systems, this is much like a database table lookup feature in traditional programming lan-

a) Template

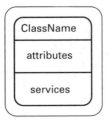

b) CREATE Object Service Example

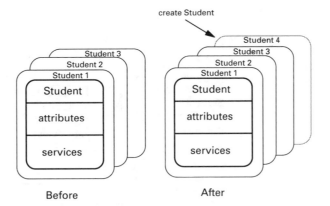

Before   After

Figure 7.2   CREATE Object Service Example

guages. The search service will attempt to find an object or a group of objects that meets certain criteria. For example, "search Student, StudentIDNumber = 123-45-6789" would be a request to find the student object that has this student identification number; "Search Student, StudentName = Smith" would be a request to find the student object or objects that have this student name. Once found, other services do their specified actions on the object(s).

The get and set services are the opposite of each other. The **get service** retrieves the value of an attribute or group of attributes. The **set service** updates the attribute's value. Both are illustrated in Figure 7.4. The get service is invoked anytime a problem domain specific attribute value is required for processing. For example, to display the name of a student, a get service would be issued to retrieve the value of the student name attribute. The set service simply replaces current attribute values with new ones when requested to do so. Anytime there is a need to update attribute values with new ones, this service is invoked.

The add and remove services are also the opposite of each other. The **add connection service** connects objects one to another, while the **remove connection service** disconnects objects from one another. Both of these services are used to couple or uncouple object connection patterns as discussed in the preceding chapter. Objects

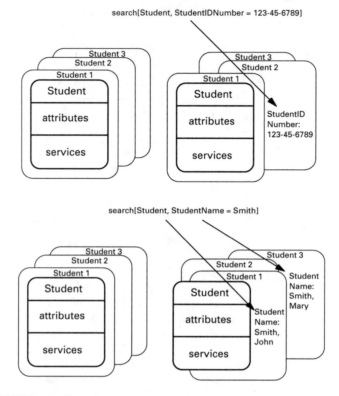

Figure 7.3   SEARCH Service Example

must be connected before a delete object service can perform its job properly. Figure 7.5 illustrates the function of these services. In the ADD connection portion of the figure, a student (Student N) is becoming a member of the DPMA club. In the REMOVE connection portion of the figure, that same student (Student N) is being removed as a member of the DPMA club.

The **delete service** deletes or removes objects from a class when requested to do so and is illustrated in Figure 7.6. This service deletes the whole object so that there is no longer any way to retrieve the object again without some type of "restore," "recover," or "undelete" feature within an object-oriented language. Deletion of an object needs to be preceded by an object disconnection if it is part of an object connection pattern.

Problem Domain Specific Services

As the name implies, problem domain specific services are unique to the problem domain and the specific class defined within the problem domain. An example of a problem domain specific service, using a car rental problem domain, might entail subtracting the value of the endingMileage attribute from the beginningMileage attribute

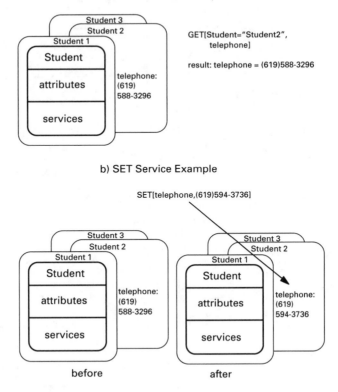

Figure 7.4   GET and SET Attribute Value Service Example

in the Automobile class in order to find the total miles driven. Another example could be multiplying the rental charge by the current tax rate to find the tax owed, and so on for additional services. Figure 7.7 illustrates both of these services.

Problem domain services come "in all sizes and shapes." What this means is that you are free to create services that fit the needs of the problem domain being automated, keeping in mind that services should accomplish something specific. For instance, the Car Rental System mentioned previously may be set up to produce a report each morning listing all rented vehicles that were due to be returned as of yesterday or earlier but have not yet been returned by the renter. A Rental Agent person would then use this report for various administrative tasks. The vehiclesOverdueReport service in the Report Class produces this report. As part of its normal processing, this monitor service sends a message to checkReturnDate in the Automobile class, which checks the scheduledReturnDate attribute of all rented vehicles (hence, a monitor service). The vehicles whose scheduledReturnDate is earlier than today would be reported back to the vehiclesOverdueReport service by the checkReturnDate service. Figure 7.8 illustrates this service.

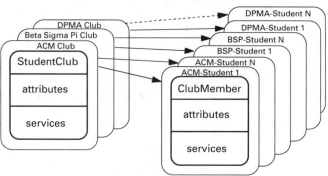

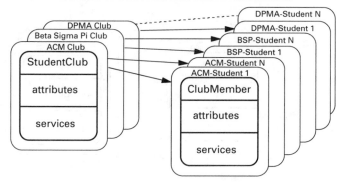

Figure 7.5   ADD and REMOVE Object Connection Service Example

Services should conform to the information-hiding (encapsulation) principle for managing complexity discussed in an earlier chapter. In other words, a specific service should perform one or, at most, a few more cohesive functions, such as dispensing cash, scanning a product, adding a course to your course schedule, dropping a course from your course schedule, turning on a heater, changing a signal light from one color to another, and so on. Each of these examples performs a single function in the mind of the user, but in reality there may be several actions required by the information system "behind the scenes" in order to perform what looks to the user to be just one function.

Often one service needs to request the assistance of another service either within the same class or within another class in the information system. This is analogous to you asking someone to assist you with an activity, for example, asking someone at the lunch table to pass you the ketchup. You, as an object instance, send a message, "Please pass the ketchup," to another object instance, the person sitting across the table from you. In order for you to accomplish your task, which is to put ketchup on your hamburger, you had to ask for help from another object instance to make it happen. You also had to receive the results of your request—take the ketchup from the ob-

a) Template

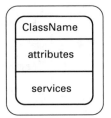

b) DELETE Service Example

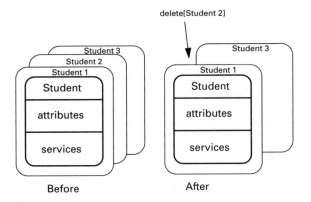

Figure 7.6   DELETE Service Example

ject instance passing it to you. In an information system of any consequence, services communicate with other services on a regular basis. This is normal and similar to the division of labor concept in business. The division of labor in business means that certain people do certain jobs hoping to increase efficiency and effectiveness due to that person's experience and expertise with the job.

When one service requests assistance from another service, the requesting service is considered the *sender* service and the service receiving the request is considered the *receiver* service. Generally, the receiver service completes the action(s) it was created to do, then returns a result of its action(s) to the sender service. For example, in Figure 7.9, a sender service called addClass from the user's perspective looks like it performs a single function in the information system, which is to allow a student to add a class to his or her schedule. The addClass service has lots to do in order for the student to actually add a class, such as (1) have student using the addClass service enter his or her student identification number, (2) validate that the student is really a student using the student identification number, (3) have the student enter the class schedule number of the class he or she wants to add, (4) validate that the class schedule number entered is a valid class schedule number for this semester, (5) determine if there is an available seat in the class for the student, (6) see if the student meets any re-

Sender = consumer
receiver = provider

Example 1

calculateMilesDriven[VIN = 12345, totalMiles]

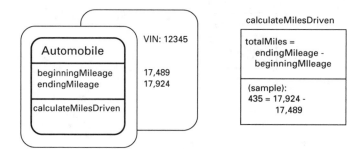

Example 2

salesTaxOwed[VIN = 12345, taxRate, totalTax]

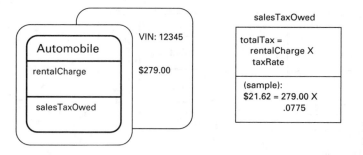

Figure 7.7   Problem Domain Service Examples 1 and 2

strictions that might be associated with registering for this class (there could be many of these), (7) reserve a seat in the class for this student, and so on.

Each of these (and other) actions could be separate services in order to take advantage of the reuse principle for managing complexity. The more cohesive the function is, the more potential it has to be reused. For example, a service that did all seven of the foregoing actions would have very little reusability, whereas a service for each of these seven actions could be very reusable within an information system.

Using object-oriented methodologies and programming languages, it is possible for a receiver service to return a result as discussed earlier, but continue to take ongoing action. A scheduler service, Figure 7.10, illustrates this. For example, we could have a scheduler service activate itself at the close of business each business day, say 6:00 p.m., and terminate itself at the start of the next business day, say 8:00 a.m. During the time the scheduler is active (6:00 p.m. to 8:00 a.m. the next business day), it is checking the current time with times associated with batch-processing jobs waiting to be run during these hours. When a particular batch job, say payroll, has its scheduled time, say 9:00 p.m., equal to the current time, the scheduler service sends a message to the payroll service to activate it for running payroll. The scheduler continues to monitor the time even while the payroll is running, and may even activate other batch jobs while payroll is running.

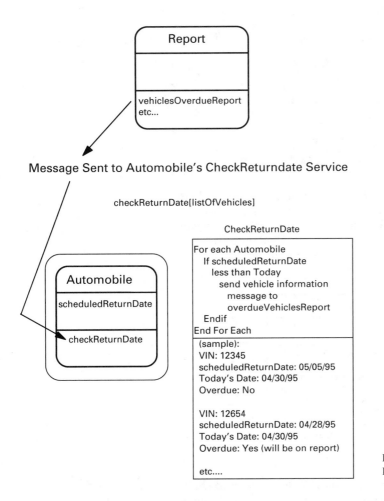

Figure 7.8 Problem Domain Service Example 3

Receiver services do not necessarily have an obligation to send anything back to the sender service unless the sender service is expecting some kind of notification or data. This feature allows object-oriented systems to perform in a concurrent fashion if that kind of processing makes sense for the business problem being automated. For example, the scheduler service shown in Figure 7.10 sends a message to start certain services but may not be expecting any kind of response back from those services.

Finding and Identifying Services

How do you go about finding services that are required in the user requirements information system model? Service identification is not yet a science; therefore, several heuristics are possible and encouraged. Not all of them are explored in this book; however, several of the documented techniques for finding services are. Services can be found and identified using one or more of the following techniques: (1) identify events and associated services, (2) identify class states, (3) identify needed messages, and (4) specify the details of the services. Each technique is described here.

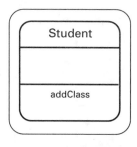

The addClass service requests the assistance of each of the following services (which are either part of the Student class or some other class within the information system) in order to carry out its intended purpose:

```
ClassName

enterStudentIDNumber
validateIDNumber
enterClassScheduleNumber
validateClassScheduleNumber
checkSeatAvailability
checkStudentRestrictions
reserveSeatInClass
```

Figure 7.9   Encapsulation and Reuse Services Example

The first technique, **identifying events and their associated services,** is probably the most appropriate starting place for identifying services. After all, events are things that happen in the problem domain that can be readily identified by users. This technique fits very well into the object-oriented methodology being presented in this book. The really interesting thing about identifying events is that they can be identified before, during, or after all of the other methodology tasks that have preceded this task in the methodology and already discussed in this book—finding classes, identifying patterns, and identifying attributes.

Identifying events is compatible with most techniques and strategies used to determine user requirements, such as those presented in Chapter 2—PIECES, requirements model, reviews, research, site visits, surveys, and interviews. Brainstorming with the user is a very effective technique for identifying events. To find services, a systems analyst could start out by basically asking the user, "What are the objects in your problem domain?" Some users can respond to this question well, while others may not. It might be helpful to ask this group of users an alternative question, such as "What are the events (things that happen) in your problem domain?" Or the two questions could be intermixed, or, as in the case of this book, the events question could come much later in the methodology. Again, the flexibility of the object-oriented methodology allows this.

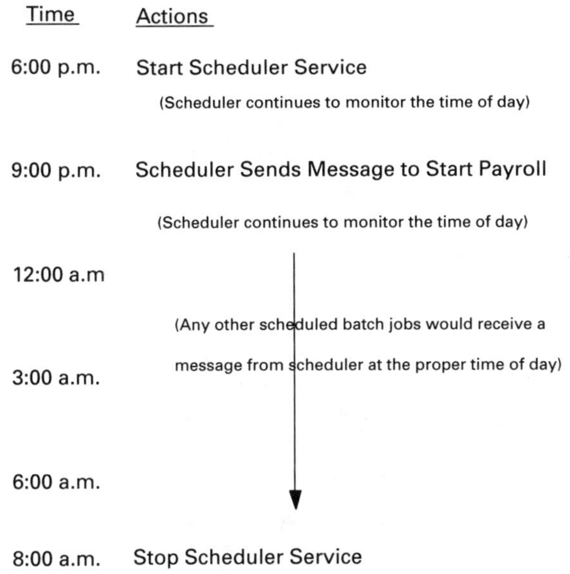

Time   Actions

6:00 p.m.   Start Scheduler Service

            (Scheduler continues to monitor the time of day)

9:00 p.m.   Scheduler Sends Message to Start Payroll

            (Scheduler continues to monitor the time of day)

12:00 a.m

            (Any other scheduled batch jobs would receive a
3:00 a.m.   message from scheduler at the proper time of day)

6:00 a.m.

8:00 a.m.   Stop Scheduler Service

Figure 7.10   Service Example Showing Messages and Ongoing Actions

The Video Store Example—Identifying Services

Perhaps this is a good time to bring in the video store example again. The video store information system user requirements model was presented in its most current form at the end of the last chapter. At that point, the model had already identified classes, patterns, and attributes. Now is the time to identify services. After discussing the notion of events and services as in this chapter, the student team again held a brainstorming session. Figure 7.11 lists the events and/or services identified during this session. As with previous brainstorming sessions, no attempt was made to critique the identified events and services. Nor was there an attempt to really distinguish events from services at this time. Doing so may have restricted the brainstorming activity.

Looking at the list in Figure 7.11, you will probably notice that most of the items listed are services, not events. That's okay; however, some other team leader may have documented the brainstormed items so that they more closely represented events. That's fine also. Perhaps the most important thing about the list is that it looks pretty good from a user's perspective. Every item in the list is a service from the user's point of view. In other words, the task was to identify events and/or services from the user's perspective and that is exactly what the students did.

After a few additional user requirements centered discussion sessions with the students, the initial list of services was modified to more closely represent the real and true user requirements, as shown in Figure 7.12. Actually, there were two other intermediate passes in the list of services prior to finalizing the one in the figure. In reality, these were the user requirements that the instructor wanted to pursue for the remain-

Acquire Membership
Scan a Product
Verify Membership
Re-Stock Shelfs
Print Receipt
Print Catalog of Movies
Renting a Movie
Inquire about Available Movies
Pay for Movies and Delinquency
Determine (During Sale) If Delinquent
Purchase "For Sale" Items/Inventory
Update Credit Card Information
Update Membership Information
Cancel Membership
Credit the Membership Account
Print Membership Labels
Update Inventory Quantities
Provide Display of Customer Information
Print Over-Due Movie List
Print Daily Sales/Rentals
Update Employee Information
Add New Inventory Items
Change Inventory Item Information
Delete/Remove Inventory Items
Security

Figure 7.11   List of Video Store Information System Events and/or Services—Pass 1

der of the video store project in order to keep its boundaries contained and allow the students sufficient time to complete the project before the end of the term.

The next task in the video store project would be to identify the *home* class for each of the services listed in Figure 7.12; however, this task will be deferred until later in this chapter in order to return to the discussion of other techniques for identifying services. Be assured that the user-centered discussion sessions for identifying services for the video store example with the students included several of the techniques discussed here, demonstrating that these techniques can be integrated to meet the needs of the project.

Other Techniques for Identifying Services

The second technique for identifying services is to **identify class states.** While discussing attributes and their states (data values) with the user, the systems analyst often discovers services that are needed to manipulate the attribute values. For example, if you have a Student class having a telephone attribute among many others, there must be at least one service within the class that allows the telephone number to be changed (updated) in the event that an individual student object instance gets a new telephone number.

1. Acquire Membership
2. Scan a Product
3. Verify Membership
4. Order Inventory
5. Print Receipt
6. Print Catalog of Movies
7. Rent an Item
8. Inquire About Available Inventory
9. Pay for Transaction
10. Determine If Delinquent
11. Purchase "For Sale" Items
12. Update Credit Card Information
13. Update Membership Information
14. Cancel Membership
15. Update Overdue Amount
16. Print Membership Labels
17. Update Inventory Quantity-On-Hand
18. Display Member Information
19. Print Overdue Movie List
20. Print Daily Sales/Rentals
21. Update Employee Information
22. Add New Inventory Item
23. Change Inventory Item Information
24. Delete/Remove Inventory Item
25. Security
26. CheckIn a Rental Item
27. Provide Store Information
28. Update Rental Informaton
29. Update Quantity Sold
30. Update Quantity-On-Order
31. Add New Vendor Information
32. Change Existing Vendor Information
33. Delete Vendor
34. Provide Vendor Information
35. Create New Purchase Order
36. Delete Existing Purchase Order
37. Compute Item Total Cost
38. Compute Purchase Order Total Cost

Figure 7.12   List of Video Store Information System Events and/or Services—Final Pass

The third technique involves the discussion of messages. A **message** is a request for a service to be performed, sent from a sender service to a receiver service. As messages are discussed, additional services may be identified. A message is the technique by which services communicate with one another within the same or different class in order to complete their assigned responsibility. The name of the service that is needed is the name of the message. Messages are discussed in more detail later in this chapter.

The final technique to discuss is **identifying the details of the identified services.** As you begin discussing the details of a specific service, you may discover that it needs assistance from another service in the same or other class in order to accomplish the needed action. In so doing, you may identify additional support services, that is, services expressly added to the model to support some other service(s). These support services are quite necessary in the object-oriented user requirements model in order for the information system to perform properly. Support services are functionally similar to all of the support personnel surrounding a play, movie, or television show. These support people are never seen by the audience but are integral to the overall success of the event. The same holds true for the support services. They are the behind-the-scenes services that allow the user-defined services to accomplish their tasks. Now let's look closer at service details followed by some techniques used to document service details.

SERVICE DETAILS

Up to this point, services have mostly been identified by name only, such as those in Figure 7.12. Very little has been said about the detail actions within the named services. As the user-defined problem domain related services are investigated in more detail (e.g., the services that are familiar to the user as discussed before), eventually the systems analyst has to discuss the step-by-step action details of the service in user terminology. The step-by-step action details are those that must take place in order for the service to accomplish its responsibility, such as "addClass" in Figure 7.9.

When the systems analyst gets to this point in the information systems analysis and layout of the problem domain component of the object model that represents user requirements for the information system, the following question virtually always comes up, "Just exactly how much detail do we need to specify in the details for a service?" This is difficult to answer specifically because there are a number of factors that play a role in what the real answer is. One suggestion based on many years of information systems analysis and design experience is that the amount of detail specified during the analysis activity should be sufficient enough that you could "walk" the user through each user-defined service via the steps you have detailed, and your user would be able to get a sense that the service actually does accomplish the intended work. This walk-through activity is a good verification and validation step and represents the requirements assurance portion of the requirements determination activity discussed in an earlier chapter.

There is considerable debate whether service details are considered as part of systems analysis or part of systems design. As far as I am concerned, service details could fit either place. Some service details seem very appropriate to document during analysis while others, more platform specific perhaps, may not even be able to be documented until systems design. So if your instructor has a different view of when service details should enter the object model, that's no problem from my perspective.

Techniques for Documenting and Describing Service Details

Historically, requirements specification documents have been large volumes of the written word. Pages and pages of written specifications, sometimes in the thousands of pages, were used to communicate the system requirements. These large documents were "affectionately" referred to as "Victorian novels." The three major problems encountered with such specification documents are:

1. Too many words; the documents are simply overwhelming for a user to validate and verify their correctness.
2. The words were often confusing, so more words were used to try to clarify, often leading to additional confusion.
3. These wordy documents were often attempting to create a visual image of the information system in the user's mind and simply did not do it well.

There are several generally accepted techniques for specifying the details of a service during analysis. Here one object-oriented technique and three traditional techniques are presented. All four of them are generally understood by practicing systems analysts and programmers. As these systems analysts and programmers begin their transition from some other methodology to an object-oriented one, they are already familiar with these techniques, which is a psychologically positive contribution to their emigration or migration process. The four techniques are:

1. scenarios
2. Structured English or pseudocode
3. decision tables and decision trees
4. state-transition diagrams

Scenarios

The first technique for documenting and specifying service details is called scenarios. Scenarios are specific to Coad's methodology and notation although other object-oriented methodologies utilize similar concepts, such as Jacobson's *use case* scenarios. A **scenario** is a specific time-ordered sequence of object interactions, one that exists to fulfill a specific need within the information system. Scenarios are developed with the help of a scenario view that is part of the object model. More than likely, the most interesting service scenarios are ones that begin or end with human interaction making them easier to begin with. Remember, the scenario view is similar in concept to you asking for assistance from someone else, such as "What time is it?" or "Can you give me a ride to school today?", or "Here is the money, buy me one too." In all of these examples you wanted or needed assistance from someone else in order to accomplish some action, idea, or task you have.

Since scenarios show the interrelationship between one service of one class and all other services that it needs whether in the same or different class, it is important to remember that any given scenario view for an object model will be quite involved and large. With this in mind, Figure 7.13 illustrates a generic service scenario template.

A service scenario has been equated to a subroutine, procedure, or paragraph call within different programming languages, and it is a common phenomenon in all object-oriented methodologies. A service scenario conceptually consists of three parts:

1. the sender class and service
2. the receiver class and service
3. parameters and number of invocations of the service

There are certain service scenarios in which one or both of the sender information and the parameter information may be optional; however, the service scenario must always know the receiver class and the specific service that it wants to have perform some action(s). When the sender class expects some kind of a response from the

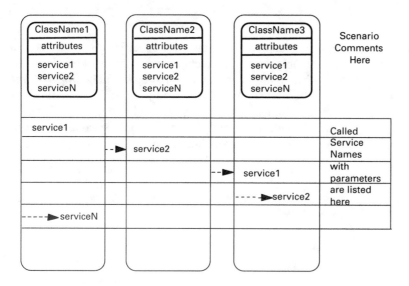

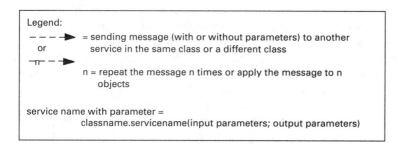

Figure 7.13   Service Scenario Template

receiver class and service, it must let the receiver service know the sender class and service name.

Much of the time a receiver service sends results to the sender service. Sometimes the receiver service needs some type of data to be given to it when it is asked to do its work. These data are sent as part of the invocation message sent to the receiver service and are called **parameters.** For example, in Figure 7.14, there is a service called calculatePay, which is designed to multiply hours worked times hourly wage. The sender service sends a message to calculatePay (receiver service). As part of the message, the sender service must also send it two parameters, which would be actual values for some employee object's hours worked, say 40 hours, and hourly wage, say $15.00 per hour. CalculatePay can then compute the pay as $40 x $15 = $600. The $600 is then passed back to the sender service as the result or answer.

Structured English or pseudocode

The first of the more traditional detail documentation techniques is Structured English, sometimes referred to as pseudocode. **Structured English** (pseudocode) is an abbreviated and action-oriented version of the English language. There is no standard way of doing either of these, but emphasis should be placed on being concise.

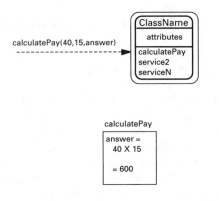

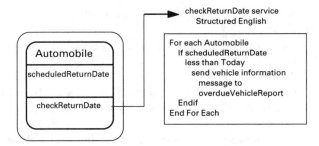

Figure 7.14   Sender Service to Receiver Service Message Communication

Figure 7.15 illustrates one possible example of Structured English, taking advantage of an outline form of it. Note how this example concentrates on "what" the information system must do from the user perspective with little or no technical jargon. This is done for two reasons: (1) Structured English, like the rest of the object-oriented user requirements model, is to be a strong and clear communication tool between the systems analyst and the user, and (2) the details embodied in Structured English are to be as nonimplementation specific as possible this early in the project. Later, during design, there may be a need to add implementation-specific details depending on implementation hardware and software platforms.

Quite often at this early point in the project, error handling is omitted. For example, if a student entered an incorrect student identification number in the preceding figure, an error service with appropriate message should be activated. If, however, the user insists on addressing error handling this early in the project, by all means add it to the service's details, creating any new services necessary. Also if you want to assure that you have the correct error-handling procedures at this point in time, then put them in the service's details or create new error-handling services that you can review with your user at some later time.

The details of most services may be more elaborate than the prior example. Therefore, you are allowed to use additional Structured English constructs such as in-

Figure 7.15   Structured English Example

dentation and decisions, as illustrated in Figure 7.15. Remember, the important point is to communicate the essence of the service with the user and others on the project team.

Decision Tables and Decision Trees

Sometimes decision situations within a service get complicated to the point that it is difficult to comprehend them in the written Structured English form alone. When such a situation occurs, two additional documentation tools are available to use along with the Structured English or stand alone—the decision table and the decision tree. Both of these tools are illustrated by the templates in Figure 7.16. Both the decision table and the decision tree accomplish the same thing but visually represent it differently. Often it is a matter of personal preference regarding which one to use in a given situation.

Both of these tools can help identify and resolve three common problems with words:

1. **Missed conditions,** such as "If the student's grade point average is less than 2.5, do not admit to the university; if greater than 2.5, admit." What about the student who has exactly 2.5?

2. **Conflicting conditions,** such as "If the student has a 2.5 or greater grade point average, admit to the university; if the student's SAT score is not in the upper 10 percent, do not admit." What do you do with a student who has a 2.9 grade point average but has an SAT score below the 10 percent level?

3. **Missed actions;** sometimes new actions are discovered as a result of identifying missed and conflicting conditions. In addition to this, new actions can be identified as you combine the various conditions and discover that no action has been identified for this particular combination of conditions.

Both tools have the advantage of being exhaustive, which means that they can represent all of the possible combinations of conditions as well as determine all of the appropriate actions.

The **decision table** is partitioned into three sections:

1. **Conditions** that describe the factors or conditions that affect the policy or situation.
2. **Actions** describe via action statements the possible actions or decisions that are to occur.
3. **Rules** describe what actions are to be taken for a specific combination of conditions.

To illustrate, we will use a story. The story is about three university students, named William, Winifred, and Wendy, who stopped by a mobile radio broadcast booth at a professional football game one afternoon. The disc jockey was running a contest for three all-expense-paid trips to the upcoming Super Bowl and was offering one to each of the three students. This is what she told the students:

a) Decision Table Template

Conditions	Rules 1. 2. 3. 4. 5. 6. 7. 8. etc..
1...	
2...	
3...	
4...	
5...	
etc...	
Actions	
1...	
2...	
3...	
etc...	

b) Decision Tree Template

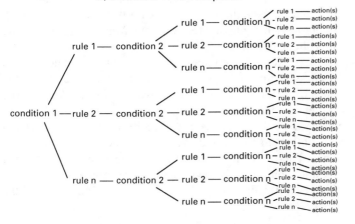

Figure 7.16 Decision Table and Decision Tree Templates

I have five playing cards—three aces and two kings. After blindfolding each of you, I will give each of you one card to place in your pocket. One at a time, I will remove each blindfold. As I do so, you may look at the other two students' cards but not at your own card. At that time you have the choice of guessing or not guessing the card you have. If any one of you can tell me the card you have—ace or king, then all of you will win the all-expense-paid trips to the Super Bowl. If any one of you guess wrong, then none of you win the trips. It's all or nothing for all three of you."

The students agree to the rules. The five cards are shown to the students. Then they are blindfolded and each given one of the five cards to put in his or her pocket.

William's blindfold is removed first. He looks at Winifred's and Wendy's cards, but decides it is too risky to guess so he does not guess. Winifred's blindfold is removed next. He looks at William's and Wendy's cards and also decides to not guess due to the risk.

Finally, just as Wendy's blindfold is to be removed, she says to the disc jockey, "If I leave my blindfold on and guess correctly, will you give each of us two all-expense-paid trips to the Super Bowl so that we can each take a friend?" The disc jockey said to Wendy, "Okay, you have a deal; six all-expense-paid trips if you guess correctly without removing your blindfold." Wendy then said, "Because my fellow students did not guess, I know that my card is an ace." The disc jockey almost fainted for Wendy had guessed correctly.

How did Wendy do it? Well, let's set up a decision table to show you how. Figure 7.17 is a decision table for solving the playing card problem. All the possible combinations of playing cards are listed as rules—A = ace and K = king—opposite the names of the three students, who are considered the conditions in this example. Looking at rule 1's column, three kings are impossible since we only had two to give out; therefore, the first action—Impossible, only two kings—is marked with an "X."

Both rule 3 and rule 5 were eliminated by Wendy as possibilities because if either William or Winifred had seen two kings, both of them would have known that their card was an ace. Rule 2 cannot be eliminated because Wendy never did see the other two students' cards.

Look at each of the remaining rules—4, 6, 7, and 8—and focus on the third condition, Wendy's card. Only rule 7 results in Wendy having a king, the others all show her having an ace. She concluded that she did not have a king from the fact that neither of the other two intelligent students guessed. Here's how Wendy eliminated rule 7, thus guessing correctly that she had an ace.

If rule 7 were true, Winifred would have known that his card was an ace. Why? Well, since William did not guess, Winifred knew that William did not see two kings; he had to either have seen an ace and a king or two aces. If we assume rule 7 to be true as Winifred's blindfold is removed, then Winifred would have seen a king in Wendy's pocket. Therefore, his own card could not possibly have been a king because, if William had seen a king in both Winifred's and Wendy's pockets, he would have known that his card was an ace. Therefore, Wendy eliminated rule 7 because she knew that Winifred was intelligent and would have known that his card was an ace. Having eliminated rule 7 as a possible (valid) rule, Wendy knows her card is an ace in all the remaining rules.

Conditions	Rules	1	2	3	4	5	6	7	8
William		K	K	K	K	A	A	A	A
Winifred		K	K	A	A	K	K	A	A
Wendy		K	A	K	A	K	A	K	A
Actions									
Impossible—only two kings		X							
William would have guessed					X				
Winifred would have guessed				X					
Eliminate due to William's and Winifred's non-responses								X	
Wendy knows her card is an Ace			X		X		X		X

A = Ace   K = King   X = Action to Take

Figure 7.17 Decision Table Solution to the Ace and King Contest

Creating a decision table to illustrate and clarify a complex policy or procedure in decision logic can be as simple as the following steps:

1. Identify all the conditions in the specific policy or procedure you are dealing with, such as "[if the] student's grade point average is greater than 2.5."

2. Identify all of the possible values that each condition can take on, such as binary values like yes, no, true, false. Often values can be more than binary, such as "student undergraduate grade level" condition can result in four values: freshman, sophomore, junior, or senior.

3. Calculate the number of rules (columns) in the decision table by multiplying the number of possible values for each condition together. For example, in the preceding playing card example, there were three conditions (William, Winifred, and Wendy) each having two possible values (ace or king), so the number of rules is 2 x 2 x 2 = 8 rules (columns). If there were a situation that had a total of five conditions, three having two values possible, one having three values possible, and one condition having four values possible, then the number of rules would be 2 x 2 x 2 x 3 x 4 = 96 rules (columns).

4. Identify all the possible actions to take using the policy or procedure.

5. Create a decision table and fill in the conditions and the actions (e.g., the rows of the table).

6. Create the rule columns, numbering them consecutively. The total number of rule columns was determined in step 3.

7. List all the combinations of condition values below each of the rules.

8. Examine each rule column's condition values and place an "X" in the appropriate action(s) row(s) for this combination of conditions and associated values.

9. Identify any missed conditions, missed actions, and conflicting or ambiguous conditions.

10. Discuss what was discovered in step 9 with the user and modify the decision table appropriately.

A Decision Table Example

Using the following university tuition policy statement, you will be walked through the creation of a decision table to represent the policy. Note that there is more than one way to lay out most decision tables; therefore, your version may not look like this one, but you should arrive at the same actions for the combination of conditions and rules that you come up with.

"Tuition fees for the university are as follows: resident students taking six or less units, $200/unit; nonresident students taking six or less units, $350/unit; resident students taking more than six units, $2,000; nonresident students taking more than six units, $3,800; nonresident students having teaching assistantships pay the resident student tuition fees; students who are employees of the university or in the employee's immediate family (son, daughter, spouse) receive a 50 percent discount on their tuition."

Step 1:   Identify the conditions:
           Resident student
           Six or less units
           Teaching assistantship
           Employee or family

Discussion: There is no need to list "nonresident student" or "more than six units" as conditions in this policy because each is the opposite of "resident student" and "six or less units," respectively.

Step 2:   Identify the possible condition values:
           Resident student............................yes (Y), no (N)
           Six or less units..............................yes (Y), no (N)
           Teaching assistantship ....yes (Y), no (N)
           Employee or family ....yes (Y), no (N)

Discussion: A "no" value for "resident student" implies that the student is a "nonresident"; similarly, a "no" value for "six or less units" implies that the student takes "more than six units."

Step 3:   Determine the number of rules (columns):
           2 x 2 x 2 x 2 = 16 rules

Step 4:   Identify possible actions:
           $200/unit
           $350/unit
           $2,000
           $3,800
           50 percent discount

Step 5:   Create decision table and fill in conditions and actions. See Figure 7.18.

Step 6:   Create rules (columns) based on result of step 2. See Figure 7.18, which shows 16 rules (columns).

Conditions	1	2	3	4	5	6	7	8	9	10	11	12	13	14	15	16
Resident student																
6 or less units																
Teaching assistantship																
Employee/family																
Actions																
$200/unit																
$350/unit																
$2,000																
$3,800																
50% discount																

Figure 7.18   Decision Table with 4 Conditions, 5 Actions, and 16 Rules

Step 7: List possible condition values below the rules. See Figure 7.19 for this.

Step 8: Examine rules (columns) and place an "X" in the appropriate action row(s). See Figure 7.20 for this.

Step 9: Identify missed conditions, missed actions, conflicting or ambiguous conditions, and duplicate actions that could be consolidated with another rule. See Figure 7.21 for this.

Step 10: Discuss with user and modify decision table appropriately by repeating steps 1–9. See Figure 7.22 for this, and notice the consolidated columns due to duplicated actions. As you discuss the decision table conditions, rules,

Rules

Conditions	1	2	3	4	5	6	7	8	9	10	11	12	13	14	15	16
Resident student	Y	Y	Y	Y	Y	Y	Y	Y	N	N	N	N	N	N	N	N
6 or less units	Y	Y	Y	Y	N	N	N	N	Y	Y	Y	Y	N	N	N	N
Teaching assistantship	Y	Y	N	N	Y	Y	N	N	Y	Y	N	N	Y	Y	N	N
Employee/family	Y	N	Y	N	Y	N	Y	N	Y	N	Y	N	Y	N	Y	N
Actions																
$200/unit																
$350/unit																
$2,000																
$3,800																
50% discount																

Figure 7.19 Decision Table with 4 Conditions, 5 Actions, 16 Rules, and Rule Values

Rules

Conditions	1	2	3	4	5	6	7	8	9	10	11	12	13	14	15	16
Resident student	Y	Y	Y	Y	Y	Y	Y	Y	N	N	N	N	N	N	N	N
6 or less units	Y	Y	Y	Y	N	N	N	N	Y	Y	Y	Y	N	N	N	N
Teaching assistantship	Y	Y	N	N	Y	Y	N	N	Y	Y	N	N	Y	Y	N	N
Employee/family	Y	N	Y	N	Y	N	Y	N	Y	N	Y	N	Y	N	Y	N
Actions																
$200/unit	X	X	X	X				X	X							
$350/unit											X	X				
$2,000					X	X	X	X			X	X				
$3,800														X	X	
50% discount	X		X		X		X		X		X		X		X	

Figure 7.20 Decision Table with 4 Conditions, 5 Actions, 16 Rules, Rule Values, and Action Values

Rules

Conditions	1	2	3	4	5	6	7	8	9	10	11	12	13	14	15	16
Resident student	Y	Y	Y	Y	Y	Y	Y	Y	N	N	N	N	N	N	N	N
6 or less units	Y	Y	Y	Y	N	N	N	N	Y	Y	Y	Y	N	N	N	N
Teaching assistantship	Y	Y	N	N	Y	Y	N	N	Y	Y	N	N	Y	Y	N	N
Employee/family	Y	N	Y	N	Y	N	Y	N	Y	N	Y	N	Y	N	Y	N
Actions																
$200/unit	X	X	X	X					X	X						
$350/unit											X	X				
$2,000					X	X	X	X					X	X		
$3,800															X	X
50% discount	X		X		X		X		X		X		X		X	

Duplicates* ⟶ 3 4 1 2 7 8 5 6

* Based on:
  1. Same actions.
  2. All but one condition have the same values.
  3. The condition with different values represents all possible values
     for that condition.

Figure 7.21 Decision Table Showing Duplicated Rule Actions

and actions with your user, it might point out situations that have not been fully covered. For example, in our example, a full-time, resident student with an assistantship pays $2,000 while a full-time, nonresident student with an assistantship pays the same, $2,000. It appears that there is no tuition benefit for a resident student to be a teaching assistant. Note also that the absence of a value (e.g., a blank space) under a rule, such as rules 1 and 2 for resident student, means that condition and its values (e.g., Yes or No in this example) are irrelevant for satisfying the conditions of the rule.

A Decision Tree Example

The **decision tree** is another tool used to visually represent complex policy or procedure decision logic. Often it becomes a matter of personal preference between using the decision table or the decision tree to represent the decision logic. The decision tree is perhaps easier to create for beginners than is the decision table. Figure 7.23 illustrates the foregoing university tuition policy using a decision tree. As with the decision table, there are multiple ways to create the decision tree; therefore, your own version of illustrating the decision logic for the university tuition policy may differ from this one. Notice with the decision tree that nodes are used to represent decision points and the lines branch out from the nodes to represent either further conditions or, eventually, the action(s) to be taken based on the conditions expressed by following a particular set of nodes and branches.

Rules

| | (1 | 2 | 5 | 6 | 9 | 10 | 11 | 12 | 13 | 14 | 15 | 16)* |
| | (3 | 4 | 7 | 8) | | | | | | | | |

Conditions	1	2	3	4	5	6	7	8	9	10	11	12
Resident student	Y	Y	Y	Y	N	N	N	N	N	N	N	N
6 or less units	Y	Y	N	N	Y	Y	Y	Y	Y	N	N	N
Teaching assistantship					Y	Y	N	N	Y	Y	N	N
Employee/family	Y	N	Y	N	Y	N	Y	N	Y	N	Y	N
Actions												
$200/unit	X	X			X	X						
$350/unit							X	X				
$2,000			X	X					X	X		
$3,800											X	X
50% discount	X		X		X		X		X		X	

* Rule numbers in parentheses above the table are the original rule numbers and would <u>not</u> normally be shown on the decision table.

Figure 7.22   Final Decision Table

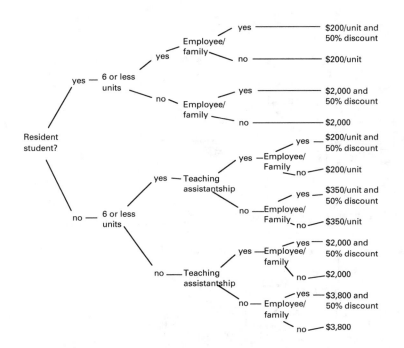

Figure 7.23   Decision Tree Solution

State-Transition Diagrams

The last of the documentation techniques to discuss is the state-transition diagram. Certain problem domains and services within them lend themselves to the use of state-transition diagrams to more effectively communicate the details of a service, a portion of a service, or the interaction between services. A **state-transition diagram** illustrates: (1) the states (data values) of a class, and (2) the operations and exceptions that cause the transitions (changes) between the states of the class. Figure 7.24a illustrates a state-transition diagram template. Note that the template provides for any state (the rectangles) to have the potential to be switched to any other state based on the conditions and actions (arrows) within the diagram. Figure 7.24b illustrates a state-transition diagram example for a telephone answering machine device.

Historically, state-transition diagrams have frequently been used to illustrate the parts of an information system having a time-dependent behavior component to them. Such systems, often referred to as **real-time systems,** include, for example, the monitoring of the ambient temperature in order to turn on or off a furnace or air conditioner, receiving satellite (or other) signals that become pictures, military command and control systems, filling liquid into containers, monitoring a robotics operation on an assembly line, and so on. These types of systems typically monitor a steady stream of input data and respond in certain ways to the incoming data. Years ago we created a prototype stock and commodity market information system that received radio frequency signals of the stock and commodity activity as it occurred. Certain actions were to occur based on fluctuations in stock and commodity prices.

a) Template

(Note: Each arrow should be labeled "condition – action".)

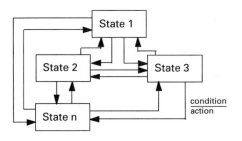

b) A Telephone Answering Machine Example

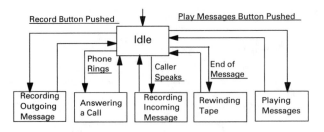

Figure 7.24   State-Transition Diagram

Today's business applications are often created with some real-time dependent component to them such as those described previously. A heuristic to help you identify such a component is to look for classes that represent or model some sort of hardware device. For example, an ATM machine class is basically waiting in an idle state for a person to insert his or her ATM card; then, certain actions occur. A supermarket barcode scanner device class at the checkout counter is waiting idly until a product is moved across the scanning device, at which time it attempts to read the barcode. Upon doing so, the display portion of the device can display the product being purchased and its purchase price, the calculating portion of the device can accumulate the total of the prices, and another part of the device can update inventory quantities on hand for the products being purchased.

Often it is difficult for first-time users of state-transition diagrams to know when to use them. The use of a state-transition diagram is not a technique that fits every situation; in fact in business applications, it often only fits a small percentage of situations within the problem domain. As was mentioned earlier, look for classes that represent some sort of hardware device as a clue. In addition, do not confuse the "state" portion of a state-transition diagram with the "state" or value of an attribute within a class. For example, a Student class has a telephone attribute. The fact that there exists a need to change some object instance's (e.g., a student's) telephone number (its state or value) does not necessitate the use of a state-transition diagram.

The Video Store Example—Assigning Services to Classes and Message Connections

The video store's user requirements problem domain object model is nearing completion. The list of user-defined services has been created (see Figure 7.12). The next step is to identify which classes should be regarded as the *home* for each service. Service names within the object model should be unique in order to avoid conflict and miscommunication. If two service names start off the same, every attempt should be made to make the names different no matter how slightly. If the names are exactly the same, then theoretically they should be the same reusable service with the same program logic within it. One of the two should be assigned a home class and reused in order to avoid duplication.

How do you assign services to classes? A few guidelines are helpful:

1. Look at the service names. Sometimes the service name includes a class name or a variation of it. For example, the Verify Membership service in Figure 7.12 has the Membership class name as part of its name. So consideration is first given to locate it with the Member class. In most instances this assignment will be most consistent with how the user perceives the real-world problem domain. In other words, things associated with membership belong in the Member class.

2. Check the real-world physical persons, places, and things in the problem domain. Talk about these and what things happen to them. You might discover that some services naturally belong to a certain class because that is how they best represent the real world. For example, the Display Membership Information service would initially be assigned to the Member class according to the preceding guideline. However, when discussing the point-of-sale (P.O.S.) physical device used to scan the barcode on products and to display information, it seems apparent that the actual display of member information happens on this device. Therefore, locate the service to do so—Display Membership Information—in the P.O.S. Terminal class instead of the Member class. In doing so, keep in mind that a support service will be needed in the Member class to provide the P.O.S. Terminal's Display Membership Information service with the member data in order to display it. The support service needed may just be the Member class's search and get basic services. The P.O.S. Terminal class is not shown on the problem domain model because it is part of the human interaction component of the model.

3. Consider which services involve class attributes. Often services will add, change, display, or delete one or more attribute data values within a class. Chances are that these services belong in the class owning the attributes.

Assigning services to classes is not yet a science. So your assignment of services for the video store application may be somewhat different than the ones shown in Figure 7.25. As systems analysts gain more experience with the object-oriented methodology, their assignment of services will improve. Remember, services can be relocated

Class	Attributes	Services	page 1 of 2
Member	memberNumber memberName memberAddress memberCity memberState memberZipCode memberPhone creditCardNumber creditCardExpireDate depositAmount	acquireMembership verifyMembership updateCreditCardInformation cancelMembership updateOverdueAmount determineIfDelinquent	
Inventory	barCodeNumber description qtyOnOrder price cost taxCode	orderInventory inquireAboutAvailableInventory addNewInventoryItem changeInventoryItemInformation delete/RemoveInventoryItem updateQuantity-On-Order	
SaleItem	quantitySold qtyOnHand	updateQuantitySold updateInventoryQty-On-Hand	
ConcessionItem	(inherit from Inventory)	(inherit from Inventory)	
RentalItem	timesRented dueDate memberNumber	updateRentalInformation	
Video	(inherit from Inventory)	(inherit from Inventory)	
Game	(inherit from Inventory)	(inherit from Inventory)	
VCR	(inherit from Inventory)	(inherit from Inventory)	
Employee	employeeNumber employeeName employeePhone positionCode	updateEmployeeInformation	

Figure 7.25a   Video Store Classes, Attributes and Services.

Class	Attributes	Services
Transaction	transactionNumber employeeNumber transactionDate transactionTime barCodeNumber price salesTax	payForTransaction
SalesTransaction	quantitySold	purchase"For Sale"Items
RentalTransaction	memberNumber	rentAnItem checkingInARentalItem
SaleRentalLineItem		transactionNumber barCodeNumber price SalesTax
StoreLocation	storeNumber address city state zipCode telephone	provideStoreInformation
Vendor	vendorNumber vendorName vendorAddress vendorCity vendorState vendorZipCode vendorPhone vendorFaxNumber	addNewVendorInformation changeExistingVendorInformation deleteVendor provideVendorInformation
PurchaseOrder	purchaseOrderNumber purchaseOrderDate purchaseOrderDueDate purchaseOrderCancelDate vendorNumber	createNewPurchaseOrder deleteExistingPurchaseOrder computePurchaseOrderTotalCost
POLineItem	barCodeNumber quantityOrdered itemCost	computeItemTotalCost

Figure 7.25b Video Store Classes, Attributes and Services.

to a different class later during analysis or design if a good reason comes up to do so. Again, flexibility is available with the object-oriented methodology to do so.

The video store user requirements model now contains classes, gen-spec patterns, whole-part object connection patterns, object patterns, attributes, and services. Service details and support services are missing. If this were a real project, the service details would definitely be done at this time, most likely using service scenarios as discussed earlier. There would be service details for all services, 38 of them in the case of the video store project, documented similarly to Figures 7.13 through 7.24. Remember, some of the services shown in Figure 7.12 will be located with human interaction component classes; therefore, they are not shown in Figure 7.26. Figure 7.26 shows a conceptual portion of an object-oriented model down to the level of service details. Note that each class could be decomposed similarly giving a large number of service detail pages.

Figure 7.14 illustrated the message communication concept. Again, the concept is that a message is a sender service requesting help from a receiver service. As each of the many service details in the video store's model are worked on, other existing (and new) services would be identified that need to be used to help the service perform its task. When this condition arises, more message communication would be required to communicate the need for assistance between the two services. For example,

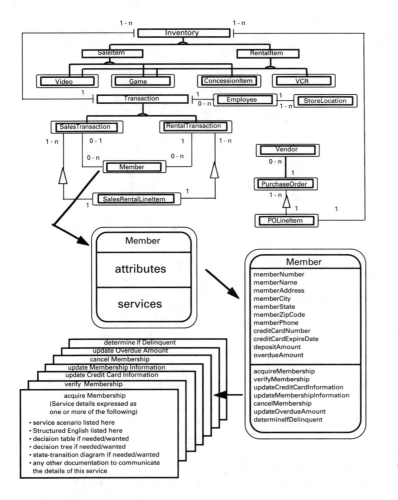

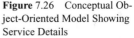

Figure 7.26 Conceptual Object-Oriented Model Showing Service Details

the human interaction component's P.O.S.Terminal class has a service called displayMembershipInformation. This service is responsible for displaying membership information on the P.O.S. terminal. It would definitely need help from the Member class's getMemberData, search and get basic services in order to retrieve the member's information to display on the P.O.S. terminal. Figure 7.27 conceptually illustrates how this message communication would work between these three services. Note that getMemberData sends a message to both create and get basic services. Consider also that getMemberData does more than is shown in the example, hence, the need to send a message to it rather than just sending a message directly from displayMembershipInformation to the search and get services.

Transition from Systems Analysis to Systems Design

Traditional information systems development is partitioned into two major activities—analysis and design. Design in this context includes implementation as well. With the problem domain component of the object model developed, the shift takes place from analysis to design. The next chapter begins the discussion of the design ac-

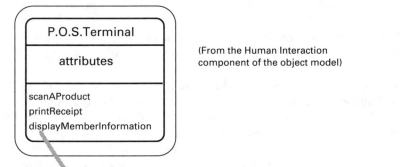

P.O.S.Terminal

attributes

scanAProduct
printReceipt
displayMemberInformation

(From the Human Interaction
component of the object model)

getMemberData[memberNumber = 123456, memberInfo]

(Conceptually, "display Member
Information" sends a message to
"get Member Data" along with the
member number it wants. "get
Member Data" sends a message to
its search basic service requesting a
lookup of member number
123456. If found, the search would
locate the Member object with
member Number = 123456
otherwise it would indicate not
found. If found, "get Member Data"
would send a message to the get basic
service to get attribute values.
"get Member Data" would
then send back to "display Member
Information" the attribute values
for this object instance in the
conceptual parameter "memberInfo"
if found, otherwise it would send
back null values in memberInfo)

Search[memberNumber = 123456]
Get[memberInfo]

Note: memberInfo = list of attributes
getMemberData does more than is shown in this example

Member

memberNumber
memberName
memberAddress
memberCity
memberState
memberZipCode
memberPhone
creditCardNumber
creditCardExpireDate
depositAmount

-- Basic Services --
create
search
add, remove, set, get
Delete
-- User-Defined Services --
getMemberData
acquireMembership
verifyMembership
updateCreditCardInformation
cancelMembership
updateOverdueAmount
determineIfDelinquent

Figure 7.27   Video Store Conceptual
Example of the Message Communication

tivity. However, with an object-oriented methodology and object model, there is no
radical transition as was discussed in an earlier chapter. Rather, a progressive expansion of the user requirements problem domain component analysis model includes the
additional three object model components—human interaction, data management,
and system interaction. All three of these are implementation platform specific requirements for the information system. Remember, the same model will be passed
along to the design activities and enhanced and expanded. Additional classes, patterns, attributes, and services will be added as the project moves through design to
implementation.

SUMMARY

This chapter defined business policies and business procedures and gave an example
of each. Services were defined and both basic services and problem domain specific
services were discussed along with examples of each. There are several strategies or
methods for finding services. Each was presented and illustrated. The traditional user

requirements document is a written narrative, which has been plagued with problems since its inception. Several problems and issues with the narrative-type documents were discussed.

Service scenarios were presented as the most common way to document the time-dependent sequence of tasks carried out by services as they communicate via messages to each other. For documenting additional details of services, Structured English was presented as a method to communicate them to the user and programming staff. Decision tables and decision trees were also presented and illustrated as additional methods to help document complex and conditional business procedures. A state-transition diagram was defined and illustrated as another method for documenting the details of some business procedures.

The message communication concept was discussed and illustrated as the technique for communicating between services. Finally, the video store information system user requirements model was revisited and expanded to include services.

QUESTIONS

7.1 Define services in the context of object-oriented programming.

7.2 How do services relate to business or information system objectives and tactics?

7.3 What are the two types of services in Coad's object-oriented methodology?

7.4 List and briefly define the different basic services.

7.5 Define and give an example of each of the basic services.

7.6 Define and give an example of a problem domain specific service.

7.7 Discuss the characteristics of services that are sometimes compared to a division of labor.

7.8 Briefly discuss a few of the techniques that can be used to identify services.

7.9 When attempting to identify services by specifying the details of the services, what is meant by a "walk-through" process? What can be accomplished by this activity?

7.10 What are some of the problems associated with using "written word" documentation when specifying the details of a service?

7.11 Define a service scenario and give an example of one.

7.12 What are some of the advantages of using Structured English (pseudocode) as a method of specifying service details?

7.13 Discuss the problems with using Structured English that can be alleviated by a decision table or decision tree.

7.14 When creating a decision table, how are duplicated actions treated?

7.15 In what situations should you use a state-transition diagram?

7.16 Define and illustrate a message.

7.17 What are the three parts of a message and how do they relate?

REFERENCES

COAD, P., D. NORTH, and M. MAYFIELD, *Object Models: Strategies, Patterns, and Applications.* Englewood Cliffs, NJ: Prentice Hall, 1995.

COAD, P., and E. YOURDON, *Object-Oriented Analysis* (2nd ed). New York: Yourdon Press/Prentice Hall, 1991.

RUMBAUGH, J., M. BLAHA, W. PREMERLANI, F. EDDY, and W. LORENSEN, *Object-Oriented Modeling and Design.* Englewood Cliffs, NJ: Prentice Hall, 1991.

WHITTEN, J.L., L.D. BENTLEY, and V.M. BARLOW, *Systems Analysis and Design Methods* (3rd ed.). Burr Ridge, IL: Irwin, Inc., 1994.

YOURDON, E. *Object-Oriented Systems Design: An Integrated Approach.* New York: Prentice Hall, 1993.

8

Systems Design

CHAPTER OBJECTIVES (YOU SHOULD BE ABLE TO)

1. Discuss historical versus object-oriented information systems design.
2. Discuss the information systems design activities and deliverables.
3. Discuss the difference between transformed design and refined design and identify advantages of a refined design.
4. Describe the different types of feasibility that are addressed during design.
5. Discuss the different kinds of information systems documentation.
6. Describe the *creeping commitment* concept and its significance during information systems development.
7. Describe the different kinds of tests that are performed during design.
8. Discuss conversion and its two types within information systems development.
9. Describe the three phases of information systems implementation.
10. Discuss the advantages and disadvantages of prototyping.
11. Discuss the human interaction component of the object model.
12. Discuss the data management component of the object model.
13. Discuss the system interaction component of the object model.
14. Discuss alternative object-oriented information systems development strategies.

A colleague once said, "When discussing systems analysis and design with a user, three information systems development characteristics come to mind—good, fast, and cheap. You can always offer the user the choice of any two of the three, but never all three!" What he was saying to me was that systems analysts and programmers can generally develop a quality information system in a reasonable amount of time and for a reasonable price. However, trying to develop a quality system (good), in a very short amount of time (fast), and inexpensively (cheap) will result in serious problems. Keeping this proverb-like thought in your mind as you study systems analysis and design and subsequently in your professional career will be a valuable principle. However, the hope is that someday in the future this principle will no longer be true.

The purpose of this chapter is to present and discuss the design activity of information systems development, giving you the big picture of this important activity. Design will be addressed from both the historical information systems design perspective as well as some of the object-oriented specifics as they relate to design. The first section of the chapter presents and discusses the design activity from its historical perspective. The last part of the chapter looks specifically at object-oriented design.

INFORMATION SYSTEMS DESIGN

Historical Information Systems Design

Historically, the **design** activity of systems analysis and design has been separate and distinct from that of the systems **analysis** activity. Figure 8.1, repeated from an earlier chapter, depicts this traditional view. The view itself is still relevant in most information systems analysis and design environments. However, there are some differences between historical information systems design and currently practiced information systems design. These differences are primarily buried in the details of the design symbol shown in the figure.

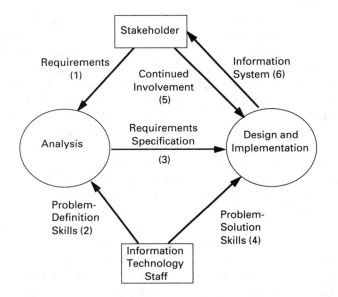

Figure 8.1   General Model of Systems Analysis, Design, and Implementation ("Partnership")

Note that the primary input to design is the **requirements specification** document, which is the output from analysis. Much of the contents of this document get *transformed* into design equivalents as the systems analysts begin to customize the proposed information system for the specific hardware and software platforms to be used for its creation and ultimate operation in some production environment within an organization. Historically, the further systems analysts progress into design, hence moving away from analysis, the less likely it will be for them to utilize the original requirements specification document to verify and validate their design work because they have transformed it into a design document.

To help picture the transformation from analysis to design, consider the following example. In the construction industry a general contractor starts with an artist's or architect's conceptual sketches of a building to be built. In information systems development this is analogous to the requirements specification document containing the requirements for the proposed information system. Next an architect creates the detailed blueprints of the proposed building. In information systems development this is analogous to the creation of the detailed design documents for the proposed information system. As the building's construction begins, the construction crew will probably focus almost entirely on the detailed blueprints for constructing the building. In information systems development this is analogous to the programmers creating the software using the detailed design as the guide for doing their work. In the building construction example, as the construction crew does its work, it is quite common to encounter situations that require construction changes not called for in the blueprints for a variety of reasons—this is just a construction fact of life. In information systems development this would be analogous to the programmers discovering that changes are required that were not documented in the detailed design for the information system.

Recently, a new university campus building was being constructed in the area next to my faculty office building. Over a year's time I was able to watch the construction of that building. At one point during the construction I noticed the construction crew taking several days to "modify" a section of the second-floor concrete flooring so that it would accommodate a set of stairs that needed to be attached to it from the ground floor. When I inquired, I was told that this adjustment was not anticipated in the detailed blueprints but was necessary for the stairs to be properly anchored to the second floor. The physical adjustment was made, and hopefully the construction supervisor reflected the adjustment to the detailed blueprints so that they would reflect the way the actual construction was done. In other words, I hope that the "map" (blueprints) matched the real "territory" (stairs attached to the second floor).

The requirements specification document for information systems development is critical to the success of the system because it represents the users' requirements for the system. As the contents of this document get transformed into design equivalents for the design, construction, testing, and implementation activities, changes will surface for a variety of reasons—a fact of life for systems development too. In almost all situations, the design documents get revised to reflect these changes, but rarely do the changes get reflected in the requirements specification document. This is partly due to

(1) time constraints, (2) little or no use of this document during or after design, and (3) difficulty keeping this document in agreement with the design documents.

You may recall from the analysis section of the book that the requirements specification document for a proposed information system was created with significant interaction between the software development team and the user. Therefore, the user should be familiar with the contents of this document. The user may not be familiar with the contents of all of the subsequent design documents for the information system, simply due to the technical nature of some of them. With this in mind, it is essential to validate and verify the completed information system with the original requirements specification document. However, historically this validation with the requirements document has been downplayed due to the reasons cited in the preceding paragraph.

An object-oriented model of information systems may go a long way to overcoming this deficiency in information systems development because the object-oriented model created during requirements determination is *not* transformed as with traditional systems development methods, but rather *refined* during the design activity. In many systems developers' minds, this represents a major contribution to systems development. The model created during analysis is enhanced with more detail during design *but* is still the same model—not a transformation to some other model(s) as with traditional and structured methods.

An Object-Oriented Analysis and Design Methodology

Using an object-oriented analysis and design methodology overcomes this significant and historical software development problem by progressively expanding the original requirements specification model rather than transforming it into design documents. Systems analysts continue to work with the original model all the way through implementation of the new information system. This should make it easier to validate and verify the completed information system with the user's original requirements specification document. Changes made as development progresses through design, construction, and testing are automatically reflected in the original document whenever real changes are made to the current design state of the information system under development.

Information Systems Design Strategy Choices

Information systems design is partitioned into several activities, each contributing one or more deliverables to the information system as depicted in Figure 8.2. From the general model, Figure 8.1, the single user deliverable from design is the information system, but recall that the information system includes hardware, software, data, people, and procedures; hence, only one all-encompassing deliverable—the information system. Each of these activities is briefly discussed next.

The **physical design** activity addresses all of the environment, platform, and human-computer interface specifics for the proposed information system. Issues to be discussed and decided here include all aspects of the five components of an information system such as:

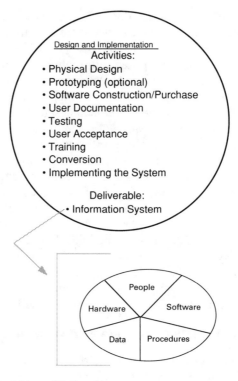

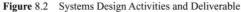

Figure 8.2   Systems Design Activities and Deliverable

1. **Hardware**—issues such as PCs, client-server, mainframe, networking and telecommunications, memory, disk capacity, response times, throughput, connectivity to other hardware, and so on.

2. **Software**—issues such as operating system, networking and telecommunications software, languages, backup and recovery software, productivity software such as word processors, spreadsheets, databases, work group software such as e-mail, calendars, schedulers, and so on.

3. **Data**—issues such as security, integrity, redundancy, physical structure, access paths, file managers, database managers, backup and recovery strategy, conversion from old system to new, disaster plan (includes hardware, software, data, people, and procedures), and so on.

4. **People**—issues such as training, organization culture, change, and readiness, staffing, and so on.

5. **Procedures**—training materials, user documentation and on-line help or "wizard"-type system, operational documentation for systems personnel, backup and recovery procedures, manual operation procedures in the event of a system failure, and so on.

All of these issues need to be addressed in the context of economic, operational, and technical feasibility along with security planning. **Economic feasibility** is a measure of the costs to create and operate a new or changed information system compared to the benefits it will offer the organization. This is often called **cost-benefit analysis.** **Operational feasibility** is a probability measure for the new or changed information system being successful in this organization's culture. Organizational culture covers a lot of issues, such as management philosophy, homogeneity of the employee work force, employee technology expertise, and management support of new or changing situations within the organization. **Technical feasibility** is a measure of the technical practicality of a new or changed information system given the availability of technical resources and schedule limitations. **Security planning** is often necessary for proposed information systems because unauthorized use or access and virus infiltration are becoming more and more important to organizations as computer connectivity options continue to expand. Users continue to expand their connectivity options, such as calling in to work from a home computer, and connecting to the Internet and on-line services and bulletin boards such as CompuServe, America On Line, Prodigy, and Microsoft Net.

The requirements specification document from systems analysis becomes the input to the physical design activity. Using this document, the preceding issues are discussed and decided in order to move ahead with the project. Often it is inevitable that many of these issues have been discussed or at least thought about at some cursory level earlier in the project. As the new or changed information system begins to take shape during analysis, many of these issues come into focus in the minds of users, systems analysts, and managers. In many situations, some of the decisions are arrived at simply on the basis of the incumbent hardware and software already existing in the environment.

Prototyping is once again available as an optional activity to further clarify, demonstrate, and test some portion of the impending information system. Its use was also an optional activity during systems analysis. Prototyping is discussed in more detail in a later section of the book. Its intent is to create a working model of all or part of the proposed information system. It can be done almost anytime deemed necessary. However, if used early in design it can often yield very insightful information regarding the proposed information system. Sometimes prototyping takes place even earlier than design. For example, during the requirements determination activity of information systems analysis, prototyping may be utilized in order to help clarify and crystallize the user's thoughts and ideas regarding the new or changed information system. Users tend to like prototypes because it gives them something tangible to look at, evaluate, and critique rather than just looking at diagrams and words on video displays or on paper. Prototyping offers the advantage of producing something visible for the users early in the project, but has the potential disadvantages of (1) elevating user expectations of system completeness, and (2) expansion of the scope of the system's boundaries beyond what is documented in the requirements specification document.

The **software construction or purchase** activity concentrates on the creation, generation, and purchase of some or all of the software necessary to complete the information system. The creation activity is the traditional programming effort found in most information systems development environments; the generation activity is the use of one or more programming code or application generators to automatically generate some or all of the program code in the chosen programming language; and the purchase activity is involved with the investigation and purchase of some or all of the software for the project.

Today's information systems are more complex than ever before due to the heterogeneous computing environments where the systems are implemented. What is meant by this is that many of today's computing environments in large organizations, such as MCI, General Motors, Chase Manhattan Bank, Merrill Lynch, and others, have a very sophisticated network of computers from different vendors utilizing a complex suite of software for their operating systems needs, telecommunications needs, networking needs, graphical user interface needs, and productivity software needs. The variations of computing environments are virtually infinite with no two being exactly alike.

Since programming from "scratch" every time is very costly and time-consuming, many organizations are turning to a variety of alternatives to totally developing the software in house. Some examples include the use of one or more of the following: program code and application generators, purchasable and shrink-wrap software, and reusable libraries of program code with the hope of reducing the cost and time to create the software portion of the information system. Many of today's systems development projects are time sensitive in order to compete in a global economy, hence the need to reduce the development time and get the system up and running quickly. Companies are counting on information systems to give them competitive advantage in the marketplace more often now than ever before. American Airlines with its SABRE computerized reservations system and American Hospital Supply with its interorganizational purchase ordering system were among the pioneers to use computer technology for competitive advantage. Over the last few years Hertz and Avis car rental companies have been observed utilizing computer technology to keep ahead of each other in the rental car industry. Do you remember seeing some of their television commercials? Did you notice the use of technology in them, such as hand-held devices and printers on Avis employees to quickly check you in when you return a car to the rental lot (no need to go into their office), or your name on Hertz's big board when you arrive at their airport parking lot to get your car?

Documentation as an activity is an ongoing activity throughout the entire systems development project. However, the **user documentation** activity is one that is focused specifically on developing documentation of all kinds for the user. For example, training and use documentation, training videos, computer-aided instruction, user manuals, on-line help assistance, and so on are all potential forms of user documentation. It is well known that very few users look at their user manuals. They tend to have the belief that when all else fails, then use the manuals. Therefore, today's systems demand significant on-line assistance available at the fingertips of the users in order to facilitate effective use of the information system. This is no small order; in fact, the project manager will need to seriously consider the time, people resources, and cost associated with documentation of all forms in order to avoid time and cost overruns on the project. Under no conditions should documentation as discussed here be considered a luxury or an afterthought! It takes skilled professionals to create quality documentation for information systems. Many of the PC software products have built-in "wizards" to assist the users, which certainly raise the standards for custom-designed information systems, such as the ones mainly addressed by this book.

The **testing** activity, discussed in more detail in a later chapter, addresses the different types of testing performed prior to implementation of the information systems. Although the testing types have different names in different environments, we refer to them as (1) module, (2) function, (3) subsystem, (4) system and integration, and (5) user acceptance testing. As Figure 8.3 illustrates, the different types of tests are designed to start with a small portion of the information system and, through successive tests, conclude with the testing of the entire information system under development. The testing activity culminates with the user approving and signing off for acceptance of the system.

A **module** can be as small as a few lines of program code that a programmer wants to test; a **function** is usually a number of modules put together to perform a specific user-identified function. A user-identified function is a function included in the requirements specification document, such as deposit cash, withdraw cash, print a report, display an invoice, or add a course to a course schedule. The user community may get involved in testing at this level and stay involved with testing straight through acceptance testing. In many projects this makes a great deal of sense, as it reflects the notion of a **creeping commitment** on the part of the user. As the user continues to be involved with the testing, the user is exposed to the system's functionality at an early stage and can suggest changes early rather than at the very late time of acceptance testing.

A **subsystem** is several functions put together as an integrated unit. The definition of a subsystem is specific to the information system being developed; in

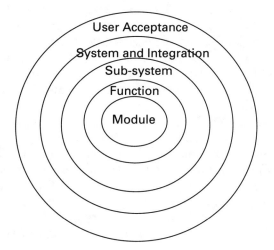

Figure 8.3   Information Systems Testing Layers

other words, we cannot predetermine a subsystem without first having the system and then partitioning it into one or more coherent subsystems. The **system and integration testing** is conducted for the entire information system being developed. Quite often development projects use the terminology of *alpha* and *beta testing* at this stage of testing. **Alpha testing** usually means the entire system is being tested internally by the developing organization, which could include users. **Beta testing** usually means having outsiders get involved in the tests such as (1) users who are not involved in the development project but would be users of the system when finally implemented, and (2) hand-picked outside organizations or customers who would eventually utilize the system and agree to test the system and provide feedback during the beta testing.

The final test prior to implementation, called user acceptance testing here, is the last opportunity for the user to make any recommendations for change prior to implementation or wide-scale distribution of the information system. **User acceptance** testing can be very rigorous and should in some way tie directly back to the original requirements specification document, since that document represents the user's requirements.

Some businesses have very rigorous user acceptance policies and procedures. The formality of such policies and procedures can lead to a very distinct and specific **user acceptance** activity that either is in place of the foregoing user acceptance testing part of testing or is in addition to that testing. Some information systems are so critical to the business's operations or are so huge that formal user acceptance is essential in order to increase the probability of success for the information system in its intended environment.

The **training** activity, as its name implies, refers to the preparation of the users to efficiently and effectively utilize the newly implemented information system. Training can begin very early during design, but not so early as to be of little use when the actual time comes to use the implemented system. Training has always been a critical success factor for user endorsement and utilization of a new system. Think about your own utilization of some productivity tool such as a word processor, database, spreadsheet, e-mail, or scheduler package. You (and I) probably only know a small fraction of the full capability of the software that you use. Part of the reason for this is more than likely because you do not need all of the included functionality, along with the fact that you may have had little, if any, training for full use of the software. What's more, you probably don't even care much about additional functionality beyond what you already know about the software.

The **conversion** activity focuses on all aspects of moving the user community's data from the old system to the new system. Sometimes this activity can be very involved and consume huge amounts of time and human and computing resources. For example, an old system has one piece of data called NAME; the new system has split the NAME piece of data into three pieces—FIRST NAME, MIDDLE INITIAL, and LAST NAME. A conversion program(s) would be required to look at all NAMEs in the old system and determine how to split those data into three parts. Now what about the other hundreds of data items that are also being immigrated from the old system to the new? Each data item has to be evaluated for its compatibility with the new system's requirements.

Conversion programs are often run in a test mode to identify problems prior to doing the actual conversion. In many situations the live conversion must be done during a time when the old system is not running, such as over a weekend. On Monday morning the new system is run instead of the old. This type of conversion is called an **abrupt cutover conversion**—at one moment in time you are using the old system and at the next moment in time you are using the new system. If the old system is run in conjunction with running the new system, this is referred to as a **parallel conversion.** At some point in time the old system is turned off, but not until the users are satisfied that the new system is working satisfactorily.

The **implementation** activity, discussed in more detail in a later chapter, has three subactivities—install, activate, and institutionalize. Implementation is the activity that places the newly developed information system into service, called **production** in many organizations. The **install** portion of implementation is involved with all of the work necessary to put the new system into production, such as hardware installation and testing, software installation and testing, procedure (documentation) availability, data availability, and people readiness. The **activate** portion of implementation deals with all aspects of getting people to use the new system. Organizational readiness, management support and commitment, and other factors need be addressed here. For example, if the new system has poor response time (e.g., you press a key and it takes several seconds to respond to your key press), then people are going to be reluctant to use the new system unless they have no other alternative. **Institutionalization** is making the new system the status quo in the organization. At this point the new system is operational and the old system is retired.

As we head toward the twenty-first century, users are becoming more computer literate and knowledgeable of a computer's capabilities. Seeing Apple Macintosh television commercials in the mid- to late 1980s sparked the user community to begin to ask for windowing capabilities on PCs similar to what they saw that the Macintosh already had. More recently, television commercials utilizing sophisticated graphics, voice recognition, and virtual reality in some clever way may spark future users of yours to ask you for these technologies to be a part of their proposed information system. That's just the way it is with technology. It keeps on changing the way things are done individually and collectively.

Object-Oriented Design

If you have already read and worked through the analysis section of this book, you will recall that it focused on capturing and documenting user requirements into a user requirements specification document. A significant part of this document is the problem domain (PD) object model that represents the user's requirements.

The problem domain object model is one of the four components that make up the entire object-oriented model of the proposed information system. There are three other components—human interaction (HI), data management (DM), and system interaction (SI)—as illustrated in Figure 8.4. If we were to "open" up the problem domain component for the video store information system analyzed earlier in the book, it would reveal the classes shown in Figure 8.5.

Using this textbook's object-oriented approach to systems development, only the problem component has been developed so far. Of the four components, this component is most representative of the user's requirements. However, it does need consider-

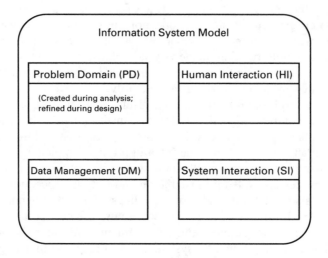

Figure 8.4   Coad's Object Model Components

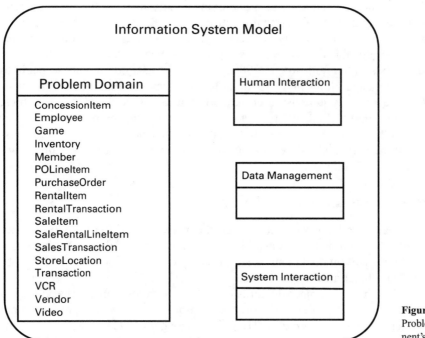

Figure 8.5   Video Store Problem Domain Component's Classes

able support in many information systems from all of the other three components. The human interaction component represents the part of the information system model that, as its name implies, carries out the human-computer interface responsibilities. The data management component represents the part of the information system model that, as its name implies, carries out the data storage, update, and retrieval responsibilities. Finally, the system interaction component represents the part of the information system model that is specific to the hardware, software, and communications platforms that are required to implement the information system for the user. Both the HI and DM components will be discussed in more detail here. The SI component will be omitted due primarily to the fact that this component is specific to platforms—hardware, software, and communications—and needs to be addressed according to the specific platforms you may have available to you as you implement an information system as a lab project or for a live user.

Even though this book only created the problem domain component of the object model so far, there is nothing methodologically prohibiting you from creating some or all of the human interaction and/or the data management components during analysis as well. From a pedagogical perspective, developing the problem domain component first seems easier to grasp, followed by the human interaction and data management components as is being done here.

Human interaction component.   In today's sophisticated information systems, the human interaction objects (classes) are usually windows or reports. What windows should you consider? At least three types:

1. **Security/logon window.** This type of window is the gateway for users to gain access to the information system.
2. **Setup window.** This type of window is for:
    a. Creating and initializing objects necessary to get the system ready to go, such as windows for creating, maintaining, and removing persistent objects. Persistent objects are similar to the data records commonly found in relational database information systems, such as customers, students, courses, inventory, and transactions.
    b. System administrator functions, such as adding and deleting authorized users and their privileges for the use of the information system.
    c. Activating or deactivating devices by a human, such as activating a printer, CD-ROM, or video camera.
3. **Business function windows.** This type of window is to accommodate the business interactions required to be carried out by the information system and its users. For example, a window for taking orders, scheduling classes, displaying status information from an elevator, or displaying daily business statistical information.

Reports are another common aspect of most information systems and are addressed in the human interaction component also. Report objects (classes) can contain almost any information requested by the user. Some examples are a daily sales report,

report list of classes scheduled to take next semester, sales receipt, income statement, balance sheet, accounts receivable report, payroll report, and overdue videos report.

Quite often during user requirements, reports are identified by the user as being a requirement for the system. Once identified, a report can be added to the human interaction component of the object model. By doing so, the human interaction component begins to take form as well as act as a "memory" bucket of requirements that are needed but not part of the problem domain component. Some of the human interaction windows and reports for the video store example are shown in Figure 8.6 and Figure 8.7.

Each of the objects (classes) identified as candidates for the human interaction component need to be expanded, just as the problem domain classes were, to include its responsibilities—"what 'I' know," "who 'I' know," and "what 'I' do." "What 'I' know" is the metaphor for identifying the attributes that are necessary for each class.

"Who 'I' know" is the metaphor for identifying any gen-spec connections or whole-part or other object connections between classes. One note here is that the object model with the four components—problem domain, human interaction, data management, and system interaction—tends to get a bit cluttered with object connections if connections are made across components, such as a class in the human interaction component connecting with a class in the problem domain, data management, or sys-

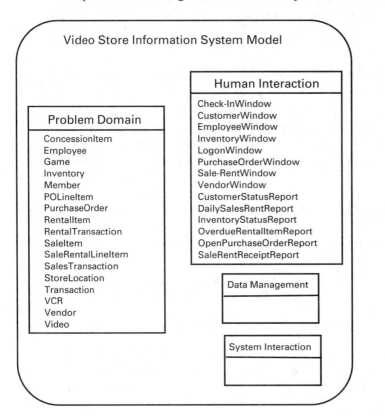

Figure 8.6   Video Store Problem Domain and Human Interaction Component's Classes

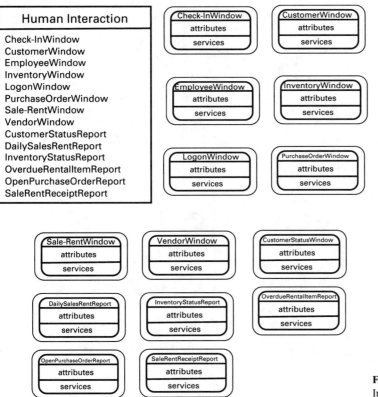

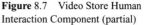

Figure 8.7 Video Store Human Interaction Component (partial)

tem interaction component. As a general guideline, Coad suggests that no object connections that cross components be made in the object model. Scenarios are the most appropriate place to show class connections across components. Figure 8.8 illustrates the suggested way to do this along with the way not to do it.

"What 'I' do" is the metaphor for identifying the services that are appropriate for the class to accomplish its intended purpose. Windows-type classes have certain characteristics such as size, scroll bars, and colors. These characteristics are usually inherent in the programming language chosen to implement a windowing graphical user interface so services to support such characteristics are implicit and do not need to be specified by the systems analyst. The services that do need to be identified are the ones that carry out the problem domain requirements which require human interaction with the computer. As was done with problem domain component services, human interaction component services should be further described using scenarios. Often these scenarios connect directly with other scenarios from the problem domain component as shown in Figure 8.8.

Data management component. The data management component of the object model serves the following purposes:

1. It stores problem domain objects (classes) that need persistence. That is, for objects that need to be stored between invocations of the information system, it provides the interface to the platform's data management storage system—flat

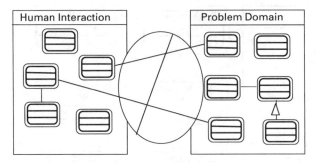

(Do not connect classes across components.)

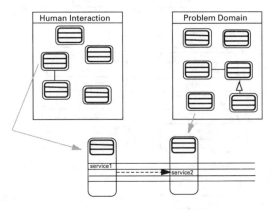

(Connect classes across components using scenarios.)

Figure 8.8   Suggested Connections Across Components

file, relational, indexed, object oriented, or other. By so doing, the data management component isolates the data storage, retrieval, and update aspects of the information system from the remaining parts of the system, thus improving its portability and maintainability.

2. It encapsulates search and storage mechanisms across all of the objects within a problem domain class that need persistence.

Using Coad's methodology and object model notation, each problem domain component class that needs persistence is associated with a data management component class of like name, as illustrated in Figure 8.9. Each of the data management component classes has one attribute, which is the plural name of the associated problem domain class, as shown in Figure 8.10 for the video store information system.

After thinking about the data management component a little, you may have the tendency to say something like, "The classes in the data management component seem identical and redundant to those in the problem domain component. Why do we need the data management component?" **The main reason is for maintainability of the object model across multiple hardware, software, and data management platforms.** Theoretically, Coad's object model components function much the same as the "plug and

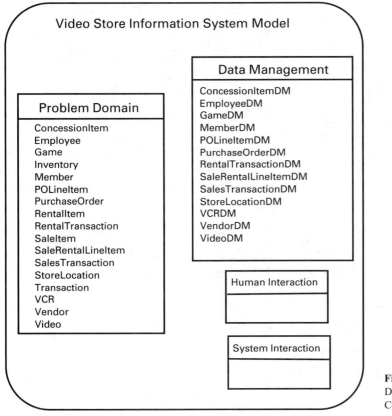

Figure 8.9 Video Store Problem Domain and Data Management Component's Classes

play" concept found in home stereo systems and as advertised by the Windows '95 operating system. Interoperability across multiple hardware, software, and data management platforms can be maximized by the object model's four-component notion of "separation of concerns" for the various aspects of the information system, as illustrated in Figure 8.11. The concerns are one of problem domain specifics, human interaction specifics, data management specifics, and system interaction specifics. Any of these components may be replaced by another "plug and play" compatible component as needed.

An example of a scenario involving at least one class from each of human interaction, problem domain, and data management is illustrated in Figure 8.12. As you view the figure, notice that the data management class—VideoDM—has an attribute called videos. This attribute can be mapped directly to all of the attributes associated with the video class in the problem domain, as shown in Figure 8.12 also.

The interpretation of the figure is as follows. The saleRent service in the SaleRentWindow of the human interaction component would allow the input of a specific video barcoded number. This number would be unique to one specific video in the video store. Videos having the same title (e.g., 36 copies of *Jurassic Park*) would each have a unique barcoded number so that one specific copy of the video could be distinguished from all other copies of the same video title. By doing so, the video store could keep track of how many times a specific video is rented, and after so many

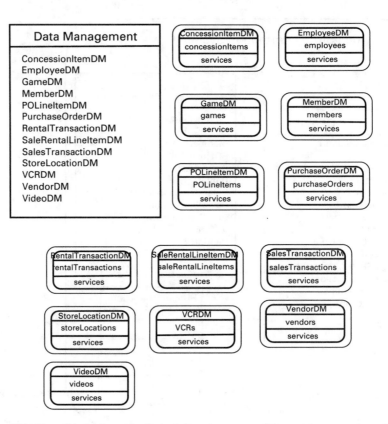

Figure 8.10   Video Store Data Management Component

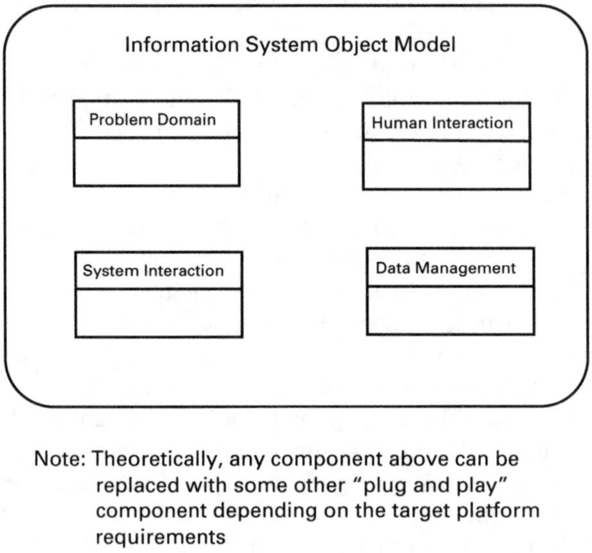

Information System Object Model

Problem Domain

Human Interaction

System Interaction

Data Management

Note: Theoretically, any component above can be
replaced with some other "plug and play"
component depending on the target platform
requirements

Figure 8.11   Separation of "Concerns" Via Object Model Components

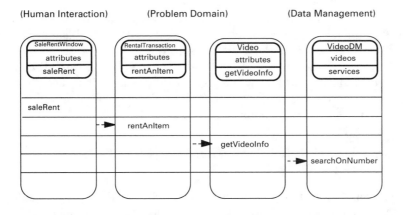

(Human Interaction)   (Problem Domain)   (Data Management)

Figure 8.12
Human Interaction, Problem Domain, and Data Management Scenario with Attribute Mapping

rentals of that videotape the store may decide to sell that video tape. As part of its processing, the saleRent service would send a message to the rentAnItem service belonging to the RentalTransaction class in the problem domain component. This service would contain the problem domain functionality associated with renting a video. As part of its processing, it would need to validate a barcoded videotape as input to the saleRent service. To do this validation, the rentAnItem service would send a message to the getVideoInfo service of the Video class in the problem domain component. The getVideoInfo service is expected to send certain attributes back to the rentAnItem service. But before it can do this it needs to find the actual video object in the VideoDM class of the data management component where persistent objects are stored. So the getVideoInfo service sends a message to the searchOnNumber service of the VideoDM class. The searchOnNumber service attempts to find the video object that matches the barcoded number being sent to it through the message. If it cannot find it, it sends an error message back to the sending service—getVideoInfo, which knows what to do with such a message. If it finds the object, it then sends the attribute information to getVideoInfo, which in turn passes the attribute information to the rentAnItem service, and the rental process continues as you can imagine it being done in video stores everywhere.

System interaction component.   The purpose of the system interaction component is to interface the proposed information system, such as the video store information system, with other systems and with physical devices that are controlled by the information system. Such devices could include sensors, gauges, elevators, heaters, air conditioners, bottle-filling machines, and so on. The kind of devices that are normally omitted from the system interaction component are the devices generally associated with a computer system, such as printers, plotters, keyboards, magnetic strip readers, barcode readers, display screens, and so on. Similar to the data management component, the system interaction component affords the object model to be modular with the "plug and play" concept for various devices and interfacing systems.

The system interaction component is developed in much the same way as the data management component. A class is established for each device that interacts with the problem domain classes. Scenarios are extended to include the system interaction classes as they interact with the appropriate classes. Figure 8.14 illustrates the system interaction component, but do not limit yourself to just what you see in the figure as there are many other types of devices and systems that could be candidates for inclusion as classes in the system interaction component of some other object model. Figure 8.15 shows a simple example of how a few device classes interface with the human interaction and problem domain classes.

ALTERNATIVE OBJECT-ORIENTED INFORMATION SYSTEMS DEVELOPMENT STRATEGIES

Over the years and especially prior to object-oriented analysis and object-oriented design, object-oriented programming has been preceded by various analysis and design strategies. This is not uncommon since structured programming in the 1960s and early 1970s was preceded by other than structured analysis and structured design until structured design surfaced in the mid-1970s and structured analysis surfaced in the late 1970s.

Today, many information systems development organizations are heavily invested in structured analysis and structured design by virtue of the fact that their systems analysts are well versed in these methodologies or strategies. So, it is not too unusual for an organization wishing to experiment with object-oriented programming to have a structured analysis and design project culminate with object-oriented programming. Similarly, it is not too unusual for an organization wishing to experiment with object-oriented analysis and design to have an object-oriented analysis and design project culminate with conventional, 3rd or 4th generation language programming. Figure 8.13 illustrates a few of the information systems development paths through analysis, design, and programming that may be selected to either start with object-oriented analysis or end with object-oriented programming. Research and industry practice continue to suggest that the optimal strategy in the long-run would be

ANALYSIS	DESIGN	PROGRAMMING
Structured	Structured	Object
Structured	Object	Object
Object	Structured	Conventional, 3rd or 4th generation languages
Object	Object	Conventional, 3rd or 4th generation languages
Object	Object	Mixture of Conventional and Object
Object	Object	Object

Notes:
1. There are many variations on these and other strategies.
2. Object Database Management Systems (ODMS), Relational Database, and Extended-Relational Database are options to consider for the database.

Figure 8.13   Alternative Object-Oriented Analysis, Design, and Programming Strategies

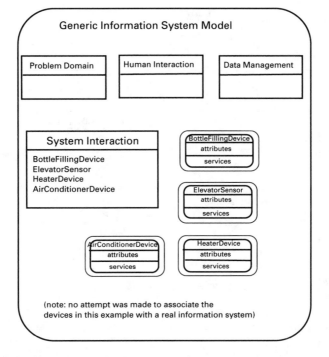

Figure 8.14   Example of System Interaction Component's Classes

to use object-oriented analysis, design, and programming. This makes sound technological sense, however the realities of business often dictate that less than optimal
technological paths be selected as an organization transitions from its incumbent information systems development strategy to a new one.

SUMMARY

This chapter has provided an overview of the design activity within information systems
engineering. An analogy to the construction industry was presented to help understand
the inner workings of the design activity. An object-oriented methodology was discussed
as an approach which allows the resulting information system to be validated back to the
original object-oriented model created during analysis, since this model does not get
transformed into something else during design, but rather gets refined with more detail.

All of the design activities, physical design, prototyping, software construction
and purchase, documenting, testing, training, conversion, and implementation were
discussed and examples given. The result of design is the implemented information
system which belongs to the user.

The three remaining components of the object model—human interaction, data
management, and system interaction—were presented with examples of classes for
each along with the general strategy for developing each of these components.

(Problem Domain)   (System Interaction)

Bottle		BottleSI	Scenario: fillBottle
attributes		attributes	Purpose: fill a bottle
fillBottle		services	
fillBottle			Bottle.fillBottle (vol; status)
	----------▶	startFill	startFill(vol; status)
		--▶monitorLevel	monitorLevel(vol; status)
		--▶stopFill	stopFill(; status)

(Problem Domain)   (System Interaction)

Elevator		ElevatorSI	Scenario: moveToFloor
attributes		attributes	Purpose: to move elevator
moveToFloor		services	car to a floor
moveToFloor			moveToFloor(floor; status)
	--▶	checkLocation	checkLocation(flooron)
		--▶computeDirection	computeDirection(floor,flooron; dir)
		--▶computeNumFloors	computeNumFloor(floor,flooron; num)
		--▶moveCar	moveCar(dir, num; status)
		--▶monitorCar	monitorCar(curfloor, num; status)
		--▶stopCar	stopCar(; status)

Figure 8.15 Scenarios Showing System Interaction Component's Classes

QUESTIONS

8.1 Traditionally, what happened to the requirements specification document as systems analysts moved from the analysis to design segments of information systems development?

8.2 What is one of the main drawbacks of traditional information systems development with regard to the requirements specification document?

8.3 How is an object-oriented model of information systems different from a historical model of an information system?

8.4 What are the advantages of a refined rather than a transformed requirements specification model?

8.5 List and give a brief description of the five components of an information system within the physical design activity.

8.6 List and discuss the different areas of feasibility that need to be addressed within the physical design activity.

8.7 Discuss some of the different kinds of documentation and why these particular kinds of documentation have become so prevalent in today's information systems.

8.8 What is meant by *creeping commitment* and what is its significance within the testing phase of information systems development?

8.9 Why is user acceptance testing such an important part of information systems testing?

8.10 What is meant by conversion as part of information systems development, and what are the two main types?

8.11 List and briefly describe the three phases within the implementation activity.

8.12 What are some of the advantages and disadvantages of prototyping?

8.13 What are the three types of windows most likely to be found in the human interaction component of the object model?

8.14 What other object type is most often found in the human interaction component of the object model?

8.15 How does the Coad notation show message communication across object model component boundaries?

8.16 Why does the Coad object model have a data management component?

8.17 What is the purpose of the system interaction component and why is it necessary in the object model?

8.18 Are devices such as printers and barcode readers candidate devices to be included as objects in the system interaction component of the object model? Why or why not?

8.19 Discuss each of the alternative object-oriented information systems development strategies.

8.20 Which object-oriented information systems development strategy is the best?

REFERENCES

COAD, P., D. NORTH, and M. MAYFIELD, *Object Models: Strategies, Patterns, and Applications.* Englewood Cliffs, NJ: Prentice Hall, 1995.

COAD, P., and E. YOURDON, *Object-Oriented Design.* New York: Yourdon Press/Prentice Hall, 1991.

COAD, P., and J. NICOLA, *Object-Oriented Programming.* New York: Yourdon Press/Prentice Hall, 1993.

KENDALL, K.E., and J.E. KENDALL, *Systems Analysis and Design* (3rd ed.). Englewood Cliffs, NJ: Prentice Hall, 1995.

STEVENS, W.P., G.J. MYERS, and L.L. CONSTANTINE, "Structured Design," *IBM Systems Journal,* 13, no. 2 (1974).

WHITTEN, J.L., L.D. BENTLEY, and V.M. BARLOW, *Systems Analysis & Design Methods* (3rd ed.). Boston: Irwin, 1994.

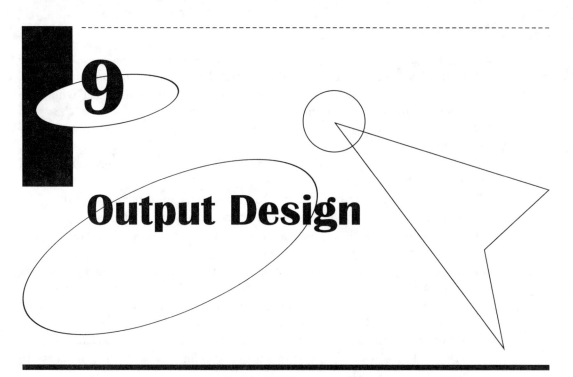

9
Output Design

CHAPTER OBJECTIVES (YOU SHOULD BE ABLE TO)

1. Define outputs and their value to people and the business.
2. Describe the characteristics of high-quality, usable information.
3. Describe internal, external, and turnaround outputs and distinguish their differences.
4. Describe static and dynamic outputs.
5. Identify the different types of output media.
6. Discuss the different types of output formats.
7. Identify and discuss the different report types.
8. Identify and discuss the different graph types.
9. Discuss some internal controls for output.

What makes arcade and video games such as Game Boy, Nintendo, and Sega so popular? Cost is one consideration, but not the overriding one. Those who play these games usually do so for the sport or challenge of it. The challenge comes in the general form of obstacles to overcome such as monsters, weapons, opponents, bombs, time, speed, and so on.

When thinking about a video game from a systems point of view, realize that the game is basically an information system. The game gets its **input** from the person playing the game as he or she moves the joystick, presses buttons and keys, and so on. The game then **processes** the player's input in the context of what is happening in the game and displays its response to this in the form of an **output** display complete with sound, score, and dazzling graphics. Getting down to the bottom line in video games, without the output, they would have little, if any, appeal to video game players. To repeat, without the output, video games would have little, if any, appeal to those that play these games. Someone could probably make the same argument about the input and processing done by video games as well, but for this chapter's illustration the focus is on the significance of output.

As was discussed in an earlier chapter, any manual or automated system that provides information to its user can be called an information system. Information is the **output** of the system; data are the **input,** and **processing** transforms the data into information. Information is what makes a system useful and valuable to its user—remember the video game example. Given erroneous, incomplete, or untimely information, a user may decide the system is not useful. So users must feel that they are winners when it comes to output—winners in the sense that they are getting the output they need in order to be more effective and efficient on the job, at school, at play, or at home.

The similarity between video games and business information systems may end with these three concepts—input, process, and output. Very few users approach their business information systems with the same enthusiasm they take to their video games. Perhaps the word *game* has something to do with it. Or perhaps the fact that with the video game you have the opportunity to control almost all that happens within the *system* or game. This is not the case with information systems because you may be sharing the information system with dozens of other users. Your job may be that of providing input by keying in phone or mail orders into the information system or scanning barcodes on a product, never really seeing the meaningful outputs of the system such as daily, weekly, or monthly sales, inventory levels, purchase orders, and so on. The point is, as the video game illustrates, the output of the business information system is what is most valuable to individuals and businesses.

OUTPUT: HIGH QUALITY, USABLE INFORMATION

How do we know when information is really high quality and really usable? Is this kind of like the notion of "beauty is in the eye of the beholder," or "one person's junk is another person's treasure"? It is true that one person's information is another person's data. What is meant by this is that things that you consider information may be simply considered data to another person and vice versa. You see, data are usually considered of little value in their raw state. Only when data become information do they become useful. For example, a 50-page report showing all of the sales transactions at a clothing store for a given week may be information to a supervisor, but may be considered data by a manager who is only interested in a single-page report showing summary sales information of each type of clothing sold, such as shirts, sweaters, ties, and shoes.

Another example that may help to distinguish data from information is the situation involving a university class schedule for the upcoming semester and your need to schedule classes. The master list of all of the courses being offered, their time of day, days of week they meet, and instructor is raw data. When you scan the schedule for a particular course, the raw data for that course become information to you. After seeing the time of day, days of week that the course meets, and instructor, you are able to make a decision about signing up for that course. You have turned the raw data into information to help you make a decision regarding your upcoming semester's class schedule.

Sometimes data are manually or electronically transformed into information, as in sales transaction data being aggregated and summarized to show sales information for groups of clothing. Other times data simply become information as a person focuses attention on it, as was the case with the class schedule example.

There are several characteristics that make information a candidate to be considered both high quality and usable:

1. Accessibility. This characteristic refers to how easy it is to use the information when a person gets it. For example, if the actual information the person needs is easily identified on a particular report, this characteristic would be rated high; conversely, if the person has to browse through dozens of pages of a report, then this characteristic may be rated low. Another example would be the situation in which you have a bank checking account and an associated bank savings account, but when you use an ATM machine you find that you can only interact with your checking account. You simply have no accessibility to your savings account via the ATM machine. You either accept this situation, ask the bank to make your savings account accessible through an ATM machine for you, or change banks.

2. Timeliness. This characteristic refers to the amount of elapsed time it takes to obtain specific information that is needed. The following three examples illustrate timeliness by showing untimely receipt of information. First, let's say that you took a test at the university today. After the test, you talked to another student and she told you that she went to a test review session the night before the test. That piece of information is untimely to you because you have no choice to attend or skip the review session. It may also be irrelevant (see number 3 following) to you if you study and test well.

Second, an example that can happen in a business setting is the situation in which a business manager needs raw data summarized in the form of a graphical report but the information system is not programmed to give such a report. So the manager submits a written request to the information systems staff to produce the custom graphical report. By the time the information systems staff fulfills the request, the need for the graphical report may have come and gone depending on how soon the manager needed the report and how much time it took to get the report from the information systems staff.

Finally, if another business manager is receiving expenditure information three months after money is spent, it may be too late for the manager to take any corrective action; hence, timeliness of this information is again poor. All three of these situations would receive a very low mark for timeliness.

3. Relevance. This characteristic refers to whether the information as presented is free from trivial or superfluous details or lacking in sufficient detail. The preceding test review example may represent irrelevant information to you because you have no need for test review sessions. This characteristic reminds me of the story of a first grader who asked his mother where he came from. His mother went into a long discourse trying to explain human sexuality and the birth process to him. When she was all finished, he said, "No Mommy, Johnny came from Pittsburgh, where did I come from?"

4. Accuracy. This characteristic addresses the issue of error-free information. Have you ever missed a final exam because you had the incorrect final exam time? Have you ever balanced your checking account's checkbook? In this example, are your calculations more accurate than the bank's? In the last quarter of 1994, Intel Corporation announced that its Pentium microprocessor had a bug in it that caused certain calculations to yield incorrect answers. You can imagine the thoughts of those who were using Pentium microprocessors in their PCs when they heard this announcement. In business, managers can have multiple reports with conflicting sets of numbers to explain a particular situation. Accurate information is simply essential for people to use it and place their trust in it.

5. Usability. This characteristic refers primarily to the presentation style of the information, in other words, its format. Some users prefer text and numbers, whereas others prefer graphics and graphs. Even though information may be accessible, timely, relevant, and accurate, if its presentability renders it unusable, then people simply will not readily use the information.

The goal for software engineers is to strive to have an information system's output exhibit high marks on all five of these characteristics. All of us as users of information systems' outputs continually evaluate our information based on these and possibly other characteristics.

OUTPUT TYPES

Internal, External, and Turnaround Outputs

Outputs can also be classified as either internal or external. **Internal** outputs are intended to be used by people or other information systems within a business. For example, professors use an output report, called a roster, that lists all of the students enrolled in one of their classes. Professors are persons within the university (business). Not all outputs of an information system are used by people, although people-oriented outputs are the ones most users and software engineers think of when discussing outputs. Sometimes output from one information system becomes the input to another information system.

External outputs are intended to be used by people or other information systems outside the business. The two distinctions between internal and external outputs are (1) internal outputs tend to be proprietary or confidential in nature, hence the need to keep them within the business, and (2) external, people-oriented outputs tend to have a more formal look to them, such as being printed by a high-quality laser or color printer, use of a business's logo, use of vertical and horizontal lines to create rows, columns, and boxes with shadows, and so on. External output that is electronically transmitted between an information system in one business and another information system within another business, such as the transmission of your IRS tax data electronically from your personal computer to an IRS tax processing center, only conforms to the foregoing nonproprietary distinction. Figures 9.1 and 9.2 illustrate internal and external output reports. Note the differences, which are mainly cosmetic or format.

The final classification of outputs is known as a **turnaround** output. This type of output is generally external and people oriented and has the dual purpose of serving as output to the user and input to the information system from the user. Typical examples of turnaround output include most bills you receive in the mail, such as cable television, telephone, and utility bills. A portion of the bill is usually perforated so that you can separate it from the rest of the bill and return it with your payment.

September 5, 1996 Student and Course Report Page 1 of 348
Fall Semester 1996

Last, First Name	Student ID	Course ID	Units	Section
Adams, Mary	387-33-8610	Bio101	3	2
		Eng100	3	1
		Soc105	3	1
		Phl108	3	1
		Eco104	3	3
		Total	15	
Aherns, Madhi	559-68-0348	Act102	3	1
		Bio101	3	2
		Chm109	3	2
		Phl108	3	1

September 5, 1996 Course Summary Report   Page 1 of 28
Fall Semester 1996

Course ID	Course Name	Total Units	Total Sections	Total Enrolment	Avgerage Enrollment/Section
Act102	Accounting Principles	3	4	300	75
...	...	...	..	...	...
...	...	...	..	...	...
Bio101	Introduction to Biology	3	6	600	100
...	...	...	..	...	...
...	...	...	..	...	...
Chm109	Organic Chemistry	3	2	90	45
...	...	...	..	...	...
...	...	...	..	...	...
Eco104	Macroeconomics	3	2	60	30
...	...	...	..	..	..
...	...	...	..	...	...
Eng100	Beginning English	3	8	208	26
...	...	...	..	...	...
...	...	...	..	...	...
MIS111	Introduction to Computers	3	3	330	110
...	...	...	..	...	...

Figure 9.1   Internal Reports

Invoice Number 204387

ABC Incorporated
INVOICE FORM

Invoice Date 06/02/96

Customer Number 82790

Customer Name Ticket-Ron,Inc.

Street Address 9347 Jackson Drive

City San Diego **State** CA **Zipcode** 92115-2611

Product Number	Product Name	Color	Unit Price	Quantity	Total Price
1 IC-PENT	Intel Pentium CPU	Bn	$675	1	$675
2 PS-220	220 V. Power Supply	Sl	$150	1	$150
3 KB-102	102-key Keyboard	Tn	$ 75	1	$ 75
4 MO-675	Mouse-Serial	Tn	$ 65	2	$130
5 HD-550	550 MB Hard Disk	Sl	$325	1	$325
6					
7					

Come to ABC Incorporated for all your technology needs.

Thank you for your patronage.

You are a valued customer.

ORDER TOTAL	$1,355
SALES TAX	$ 95
SHIPPING	$ 25
GRAND TOTAL	$1,475

Figure 9.2 External Report

Static and Dynamic Outputs

Outputs can be grouped into two broad categories—static and dynamic. **Static** outputs are those that can be predetermined and planned for. As the analysis portion of systems development progresses with the requirements specification document, these outputs are discussed, conceptually sketched, and included in this document. Examples of static output are hard-copy reports and soft-copy displays used within the business, such as weekly sales, inventory levels, accounts receivable and accounts payable; monthly bills such as utility, phone, and credit card bills; receipts such as ATM slips, sales slips from cash registers, credit card sales and returns receipts; production reports such as work-in-process reports, finished goods reports, bills-of-materials, and scrap reports. What makes all of these reports static? The format, not the content, of the reports does. Each of these reports has a static format. Consider your cable television bill, your telephone bill, and your utility bill. Each has a static format to it. Each time you receive the bill, the format of it looks the same as the last time although the content of the transaction information on the bill changes each time you receive it.

In traditional information systems development, outputs have been the center of attention. Regardless of whether the software engineers were using a structured or data modeling methodology, their discussion with the user usually starts with an investigation of outputs for the system because this is what the user needs from the system. In the object-oriented methodology the focus is on classes, attributes, and services. Included in an initial list of candidate classes or services were the output (report) classes or services that produced the outputs (reports). Ultimately, in an object-oriented methodology, outputs become the result of services.

As you might expect, **dynamic** outputs are those that cannot be easily predetermined or planned for during information systems development. In most business information systems these kinds of outputs have traditionally kept scores of software engineers and programmers busy after the information system is implemented. I cannot count how many "one-time" (dynamic) reports I have created during my professional career—hundreds no doubt! As I was writing this chapter, one of my clients (users) requested six (!) one-time or dynamic reports for an upcoming management meeting the following week. He didn't give me much time to create these reports, but that is reality! Fortunately, I have a good report generator software package for doing these types of reports. Some day I hope to train my client how to use it so he can create his own dynamic reports. Up to now he has refused to learn it, saying, "that's what I pay you for." Sometimes these one-time or dynamic reports are just that—one time; other times these reports become part of the information system and are run on a certain time schedule or whenever the user requests them. When this happens, they become part of the static outputs for the information system.

Businesses are dynamic, and so are their managers. Is it any wonder then that their information needs are also dynamic? As the proposed information system takes shape, you will get a better feel for the magnitude of the need to accommodate dynamic outputs. The question is not whether there will be dynamic outputs, but rather how many, how often, and how complex! Even though the dynamic outputs themselves cannot be planned for, the systems analyst can anticipate the need for them and address this along with the other needs covered in the general design discussion earlier in this chapter.

OUTPUT DEVICES AND MEDIA

The types of output devices and media described in this section are intended to produce output for people to use rather than output that is electronically transmitted to another information system. The output devices that most of us are familiar with are **printers** and **video display terminals** so they are not discussed here. **Plotters** are most often used to draw maps, architectural drawings and blueprints, graphs, and pictures. **Computer output on microfilm or microfiche (COM)** are devices that, as their name implies, produce microfilm or microfiche output. The actual microfilm and microfiche are considered output media and are discussed in the next section.

Output media is what the output devices record the output information on. The most common output media are paper, video, microfilm, microfiche, and sound.

Paper is still the leading medium for computer output due to its low cost, portability, and permanency characteristics. Its biggest drawback is that it represents a snapshot of the state of that information at the moment it was printed; seconds later the data that contributed to the information could change. For example, your telephone bill is printed at 12:00 a.m. Friday morning; at 12:05 a.m., while studying for an exam, you make a toll call across town to talk to a fellow student about the exam; a charge is assessed to your phone account, which will not show up on a printed bill for another 30 days. Another example of a paper-based output is a telephone book that has incorrect data in it even before it is delivered to your door. Regardless of its shortcomings, paper is still inexpensive per page, and it seems that every nonnetworked personal computer has a printer attached to it these days. In addition, being portable, paper can be sent through the mail, and it has a long life expectancy if stored properly.

Video output is quite popular today as virtually every personal computer has one attached to it. Video offers us convenient access to our information but only on a temporary basis until the image leaves the video screen. For a more permanent record of what you are viewing, many computers have a "print screen" feature, which allows you to print a copy of what you are looking at on the video screen.

The video display terminal is one of the most interesting output devices due to the variety of ways one can present information on this device. In the past video display terminals were essentially limited to 24 rows by 80 columns of white text on a black background coupled with character blinking and bolding, rows of characters scrolling up and down, and being able to reverse the video image (black text on a white background). Today a typical video display terminal can do all of this as well as display thousands of colors, dozens of different character fonts such as Courier, Prestige, Times Roman, and others in many different font sizes, an infinite number of graphical symbols or icons, and several overlapping windows of moveable, expandable, and shrinkable information—all of this with the quality of a high-definition, photographic picture.

Microfilm and microfiche are two media used in permanent storage situations where the bulkiness and handling of paper is difficult. For example, libraries usually keep copies of old newspapers and journals on microfilm or microfiche in order to save space and preserve the output on these longer-lasting media. **Microfilm** comes in a roll of photographic film, similar to film used in 35mm cameras, and is used to record dozens of pages of information in a reduced size onto the film. COM equipment, as described in the preceding section, can directly output information from the computer onto the microfilm. **Microfiche** is a single sheet of film about the size of a postcard or 4 x 6 note card and is capable of storing dozens of pages of output in reduced size on a single sheet. COM equipment is used to record output onto microfiche similarly to microfilm.

Many people are well aware that **sound** output is common in arcade and home video games. Some of the background music from home video games played over and over again may even be etched in people's brains forever! I may never forget the Mario Brother's Nintendo music! Like video games, business information systems have placed greater emphasis on sound in recent years. Actual human voices have become an integral part of some information systems. The U.S. Navy ship repair facility in San Diego has had a voice output inventory management system for over 20 years. Telephone companies have been using voice output for directory information for many years. Banks are now using voice output when you call to inquire about your account. Supermarkets tried unsuccessfully to implement voice output at its checkout stands in the mid-1980s. These voice output systems spoke the name of each item purchased and its price as it was passed across the scanner device by the checker. It only took a few weeks to find out that the consumer did not want everyone waiting in the checkout lines to hear that he or she purchased certain "embarrassing" items. So, in most supermarkets, the checkout voices have been turned off. Many companies have installed voice-actuated telephone systems to handle incoming calls more efficiently without having to hire additional people to answer the phone for the company.

OUTPUT FORMATS

Today's information systems present information in three basic formats—zoned, graphic, and narrative—as illustrated in Figure 9.3. **Format** refers to the look of the output on the medium; hence, sound is not included here. **Zoned** output is the traditional row- and column-oriented textual and numeric information. **Graphic** output represents drawings, symbols, icons, graphs, windows, and video images. Finally, **narrative** output is the kind of output most often associated with word processing software, which is used to produce letters, memos, contracts, documents, and books. The exciting part of output design is the creativity the user and software engineer can exhibit to combine all three of these output formats to create truly elegant output. **Desktop publishing** typically incorporates all three of these format types to produce newspapers and magazines. Video display terminals can also display all three of these in addition to the playing of selected digitized video segments with voice, music, and other sounds much like your home VCR equipment.

Budgets and other constraints often impact the use of output formats, and not every output is a candidate for all three of the preceding formats. Zoned, graphic, and narrative formats are not solutions looking for a problem to solve; they are, in fact, opportunities to present output information in as meaningful a way as the user would like. Most users are aware of all three output formats but may be reluctant to ask for something as exotic as video for their application. The normal progression for typical users is that as they see different information systems' output capabilities, they begin to ask for more features. So video may become quite pervasive in the future information systems that you develop.

OUTPUT: REPORT TYPES

The content variations of output reports are endless. However, we are able to categorize many types of reports, while realizing that these categories do not necessarily cover every conceivable report type. There is no universally accepted categorization of

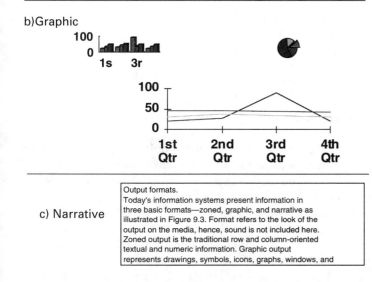

a)Zoned

September 5, 1996 Student and Course Report Page 1 of 348				
Fall Semester 1996				
Last, First Name	Student ID	Course ID	Units	Section
Adams, Mary	387-33-8610	Bio101	3	2
		Eng100	3	1
		Soc105	3	1
		Phl108	3	1
		Eco104	3	3
		Total	15	
Aherns, Madhi	559-68-0348	Act102	3	1
		Bio101	3	2
		Chm109	3	2
		Phl108	3	1
		Total	12	

b)Graphic

100
0
1s   3r

100
50
0
1st   2nd   3rd   4th
Qtr   Qtr   Qtr   Qtr

c) Narrative

Output formats.
Today's information systems present information in three basic formats—zoned, graphic, and narrative as illustrated in Figure 9.3. Format refers to the look of the output on the media, hence, sound is not included here. Zoned output is the traditional row and column-oriented textual and numeric information. Graphic output represents drawings, symbols, icons, graphs, windows, and

Figure 9.3   Output Format Examples

reports; therefore, you may come across other categorizations in other books. Nonetheless, the output report types discussed in this section are detail, summary, analytical, and historical.

Detail reports are just that—full of details, details, details. A detail report is equivalent to someone asking you, "What courses are you taking this semester?" and you responding with the details of each course, such as its course number, its name, what time it meets, what days of the week it meets, what building and room it meets in, and the name of the instructor. Another example, Figure 9.4, illustrates a detail report showing every course taken by every student in your university this semester. There would probably be several hundred pages in this report if it were real. A telephone book for your city is another example of a detail report. The day is coming when telephone books will be replaced with some other device and medium, perhaps video display of the phone book information recorded on a CD-ROM.

Typically, detail reports show all transactionlike activity within a business for a specific time period. A report showing every ATM transaction that occurred at a spe-

Last, First Name	Student ID	Course ID	Units	Section
Adams, Mary	387-33-8610	Bio101	3	2
		Eng100	3	1
		Soc105	3	1
		Phl108	3	1
		Eco104	3	3
		Total	15	
Aherns, Madhi	559-68-0348	Act102	3	1
		Bio101	3	2
		Chm109	3	2
		Phl108	3	1
		PEd118	1	5
		Total	13	
Banks, Jamal	371-49-3256	Chm109	3	1
		Eco104	3	2
		Eng100	3	1
		MIS111	3	2
		Mkt114	3	1
		Soc105	3	1
		Total	18	

vvvvvvvvvvvvvvvvvvvvvvvvvvvvvvvvvvvvvvvvvvvvvvvvvvvvvvvvvvv

Figure 9.4   Detail Report

cific on-campus ATM machine yesterday would be an example. Another detail report could show just the deposit transactions or just the withdrawal transactions at a specific ATM yesterday. Another variation on this example could be a detail report showing all the deposits made at all the ATMs in the city or county yesterday. Detail report possibilities for a given information system are endless.

Detail reports can include subtotal rows scattered throughout the report whenever appropriate. For example, a detail ATM report for a city might organize its detail transactions by ATM machine, and for each ATM machine the detail transactions may be further ordered by transaction type, such as deposits, transfers, and withdrawals. As all the detail data are being printed on the report, each time there is a change in the type of transaction or a change from one ATM to the next ATM, a subtotal row is printed summarizing the grouping of data above it. The key to remember here is that detail reports always include the details and optionally may include subtotal information.

Prior to businesses providing video display terminal availability to virtually everyone in the business, people made heavy use of detail reports in their daily jobs. Today these reports are most often used as a permanent record of the time period's detail activities. Also note that repetitious pieces of information, such as eliminating the repetition of the same Last Name, First Name, and student ID on successive rows, can be removed from the report in order to enhance readability, as illustrated in Figure 9.4.

Summary reports aggregate detail data into meaningful higher-level groupings and only print the higher-level grouping information on the report. A summary report is equivalent to someone asking you, "How many courses are you taking this semester?" or "How many units are you taking this semester?" Your response, in the form of a summary, would be something like, "4" or "5" to the first question, and

"12" or "15" in response to the second one. Regardless of how quickly or easily you responded to either of these two questions, you had to mentally summarize some of the details of your courses in order to respond. In other words, it required that you process detail data in order to allow you to respond with summary information.

Figure 9.5 is a summary report that shows the number of students enrolled in each course at your university this semester. This report needed all the rows of detail information as shown in the partial report in Figure 9.4 in order to produce this summary report. The difference is merely choosing not to print all the details.

Often the data stored in the information system's database or files are not in the order or sequence necessary to produce a summary report. In this situation, summary reports would require a reordering (sorting) or indexing of the data temporarily in order to produce the report in the desired order. For example, using Figure 9.4's information as the starting point, you can see that the sequence of the report is student Last Name, followed by student First Name, followed by course number (alphabetic and numeric), and finally, course section number. If the data are truly stored in this sequence in the database or files, then there is no need to order it differently in order to produce this report. However, if the data were stored in any other sequence in the database or files, it would be necessary to reorganize it temporarily (usually via a tem-

September 5, 1996 Course Summary Report			Page 1 of 28		
Fall Semester 1996					
			Total	Total	Avg.
Course ID	Course Name	Units	Sections	Enrolment	Enrollment/Section
Act102	Accounting Principles	3	4	300	75
...	...	...	..	...	...
...	...	...	..	...	...
Bio101	Introduction to Biology	3	6	600	100
...	...	...	..	...	...
...	...	...	..	...	...
Chm109	Organic Chemistry	3	2	90	45
...	...	...	..	...	...
...	...	...	..	...	...
Eco104	Macroeconomics	3	2	60	30
...	...	...	..	..	..
...	...	...	..	..	..
Eng100	Beginning English	3	8	208	26
...	...	...	..	...	...
...	...	...	..	...	...
MIS111	Introduction to Computers	3	3	330	110
...	...	...	..	...	...
...	...	...	..	...	...
Mkt114	Principles of Marketing	3	2	110	55
...	...	...	..	...	...
...	...	...	..	...	...
PEd118	Beginning Golf	1	7	84	12
...	...	...	..	...	...
...	...	...	..	...	...
Phl108	Philosophy	3	2	72	36
...	...	...	..	...	...
...	...	...	..	...	...
Soc105	Cultural Changes	3	3	75	25
...	...	...	..	...	...
...	...	...	..	...	...

Figure 9.5   Summary Report

porary, sorted copy or index). For a final example, a report is needed that summarizes the data by Course ID followed by Section Number sequence in order to find out how many students are enrolled in each section of each course. Once the data are temporarily sorted or indexed in this order, the software can aggregate data for a specific section of a specific course, printing a one-line summary of these data as the processing encounters a new section of the same course or a new course.

Summary report utilization is as diverse as people and businesses; therefore, one cannot definitively say that every conceivable summary report has been thought of. Summary reports are often the one-time reports discussed earlier in this chapter because a supervisor or manager wants to analyze a particular business phenomenon from a particular perspective.

Analytical reports can be either detail or summary reports but are most often summary. There are at least five variations of analytical reports as follows:

1. Horizontal
2. Vertical
3. Counterbalance
4. Variance
5. Exception

Both horizontal and vertical analytical reports present information in a form that allows the person reviewing it to analyze the information and establish similarities and differences in the information as presented. In Figure 9.6, a **horizontal** analytical report shows two years of balance sheet (an accounting term) information and the increase or decrease in dollars and percent from 1994 to 1995. Figure 9.7 shows a

1/27/96	Data Based Decisions, Inc. Comparative Balance Sheet — Horizontal Analysis For Fiscal Years 1994 and 1995 (values in millions)			
	1994	1995	Amount Difference	Percent Difference
Assets				
Cash	$ 0.6	$ 0.8	$ 0.2	33.0 %
Accounts Receivable	3.3	3.7	0.4	12.1
Office Equipment	5.2	5.5	0.3	5.8
Total Assets	9.1	10.0	0.9	9.9
Liabilities				
Accounts Payable	1.1	1.2	0.1	9.1
Long-Term Debt	3.2	2.8	(0.4)	(12.5)
Total Liabilities	4.3	4.0	(0.3)	7.0
Capital				
Common Stock	3.0	3.0	0.0	0.0
Retained Earnings	1.8	3.0	1.2	66.7
Total Capital	4.8	6.0	1.2	25.0
Total Liabilities and Capital	9.1	10.0	0.9	9.9

Figure 9.6   Horizontal Analytical Report

Data Based Decisions, Inc.
Comparative Income Statement - Vertical Analysis
For Fiscal Years 1994 and 1995 - values in millions

	--------1994--------		--------1995--------	
	Amount	Percent	Amount	Percent
Income:				
Hardware Sales	$ 1.6	21.6	$ 1.9	20.9
Software Sales	1.2	16.2	1.7	18.7
Supplies Sales	0.2	2.7	0.3	3.3
Consulting Services	4.4	59.5	5.2	57.1
Total Income	7.4	100.0	9.1	100.0
Expenses:				
Advertising	0.2	2.7	0.3	3.3
Office	0.3	4.1	0.4	4.4
Salaries	3.3	44.6	3.9	42.9
Hardware	0.9	12.2	1.1	12.1
Software	0.7	9.5	1.2	13.2
Supplies	0.1	1.4	0.2	2.2
Total Expenses	5.5	74.5	7.1	78.1
Net Income before taxes	1.9	25.5	2.0	21.9
Income Taxes	0.8	10.8	0.9	9.9
Net Income	1.1	14.7	1.1	12.0

Figure 9.7 Vertical Analytical Report

vertical analytical report of an income statement (an accounting term) for two years along with the percentage of net sales that each row represents. An accounting manager might use either of these two reports during a meeting of senior managers to show how the business did financially during two years.

The third type of analytical report is the **counterbalance** analytical report, which is most often used in projections or forecasts of the future. Figure 9.8 illustrates this with three projections—worst-case, middle of the road, and best-case scenarios. A group of managers could use such a report to help them decide about expanding into another market in the coming year or taking on another project.

The **variance** analytical report, as shown in Figure 9.9, is used to compare actual performance indicators with budget, forecast, standard or baseline, or quota indicators. If you have established a personal budget for yourself and you keep track of your actual expenses, then you can put the two sets of numbers together for comparison as a variance report. But be prepared for some disappointing results—personal budgets are often hard to adhere to!

The final analytical report is the **exception** report as shown in Figure 9.10. As its name implies, its purpose is to identify unusual or exceptional conditions. The definition of *exception* is user defined for each situation. It can be as simple as saying something like, "We need a report showing all weekly paychecks that are greater than $2,000" or "Give us a report showing all of the salespeople who have sold less than $10,000, or more than $25,000 this month" or "Provide us with a daily report during the course registration period showing courses that students tried to register for but could not because the course was full."

The exception report became popular with a management philosophy called

8/12/96	Data Based Decisions, Inc.		
North Island Networking Project Analysis			
Project Start Date: 10/1/95 - ($000 omitted)			

	Worst Case	Moderate Case	Best Case
Income:			
Hardware Sales	$ 800	$ 800	$ 850
Software Sales	300	300	350
Supplies Sales	050	050	050
Consulting Services	150	175	190
Total Income	1,300	1,325	1,440
Expenses:			
Project Overhead	200	200	200
Salaries	140	140	150
Hardware	750	700	650
Software	200	180	160
Supplies	40	40	40
Total Expenses	1,330	1,260	1,200
Net Income on the project	(30)	65	240

Figure 9.8   Counterbalance Analytical Report

11/9/95	Student Personal Monthly Budget			
Variance Analysis				
Month: October 1995				

	Budget	Actual	Variance	
Income	$ 700	$ 680	$ (20)	(2.9%)
Expenses				
Rent	150	150		
Telephone	20	24	(4)	(20%)
Car Payment	100	100		
Insurance	70	70		
Food/Drink	135	145	(10)	(7.4%)
Clothes	50	38	12	24%
School Supplies	25	8	17	68%
Savings	50	50		
Utilities	25	28	(3)	(12%)
Gifts	25	35	(10)	(40%)
Discretionary	50	55	(5)	(10%)
Total Expenses	700	703	(3)	(0.4%)
Balance Remaining	0	(23)	(23)	

Figure 9.9   Variance Analytical Report

management by exception. Typically, managers and other people only have so much time to devote to each activity that they are responsible for. Exception reports can save them time by providing them with a list of the potential problems or opportunities without all the other information that is within normal thresholds. Detail reports often consist of exception information, but it is buried deep within all of the acceptable transactions on the report, which makes it difficult to find. Video display of exception information is able to take advantage of the terminal characteristics such as sound,

```
┌─────────────────────────────────────────────────────────┐
│ 1/9/96        San Diego Weather Exception Report          │
│                   For Calendar Year 1995                  │
│            Lower Limit: 45 degrees; Upper Limit: 85 degrees│
├─────────────────────────────────────────────────────────┤
```

Date	Time of Day	Temperature
01/11/95	04:15am	42
01/12/95	04:12am	41
01/13/95	04:11am	41
02/08/95	03:50am	39
02/26/95	03:40am	40
03/14/95	02:57am	44
07/19/95	02:45pm	89
07/20/95	02:38pm	94
07/21/95	02:30pm	95
etc...	etc...	etc...

Figure 9.10   Exception Analytical Report

blinking, bolding, or presentation in a different color to set it apart from the normal and acceptable information. I have developed an information system which displays customer information and uses a bold red color to indicate information that is outside established tolerances.

Historical reports are historical records of activities of the business. Historical reports are not a unique type of report, but rather a report that was initially in the detail, summary, or analytical classification, and now is considered historical because the information on the report is old. You may ask, "How old is old?" Every situation is different. One report, printed today, may be historical tomorrow; another report printed today might not be considered historical by the people in the business for several months. Generally speaking, reports which are produced on a regular schedule, such as daily, weekly, or monthly, are usually considered historical as soon as the newest version of that report is available. This is kind of like your daily campus newspaper—today's is current; yesterday's is historical. My class rosters for this semester become historical when the semester ends.

OUTPUT: GRAPHS

Spreadsheet software was a major contributor to the acceptance of personal computers in business in the 1980s. Today's spreadsheet software's row, column, cell, and formula concepts are the same as the "father" of spreadsheet software, Visicalc, but the similarity virtually ends there. Today's spreadsheet software commonly includes multidimensional worksheets, aggregation of worksheets, graphs, what-if, and goal-seeking features.

Business analysts and managers constantly review and analyze numeric information. Several of the reports in the prior section of this chapter presented comparative numeric information in a traditional report format. Often it is more effective to present numeric information in graphic form because visualization of the numbers can enhance the analysis activity. Several types of graphs are available to present numeric information. The key is knowing which graph type to use for which type of analysis. Because graphs have a specific way of presenting information, it is possible to choose a less effective graph type for presenting the information. This can make the data analysis no easier than from a printed report. Usually, the user will assist with this important output decision.

Spreadsheet software is often presented in an introductory course and students are often given the opportunity to produce one or more graphs as lab assignments. The purpose here is to briefly describe the purpose of each of four graph types and provide an example of each. The four types are scatter diagrams, line diagrams, bar charts, and sectographs.

The **scatter diagram** is best used to reveal trends in data. Note in Figure 9.11 the list of X and Y values on the left side of the figure. Can you see the trend in the data? Probably not! It's not easy to see any trend just scanning rows or columns of data

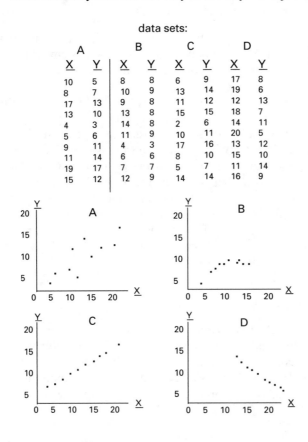

Figure 9.11   Scatter Diagrams

values, but graphing them in a scatter diagram, as on the right side of the figure, makes it much easier to see (1) if there is a trend, and (2) what type of trend it is. Business analysts often use this type of graph to forecast into the future based on a scatter diagram showing historical trend data.

The **line diagram** is best used to show data fluctuations over some time period as in Figure 9.12. This diagram, like the scatter diagram, also shows trends in the data but adds the time dimension into the graph so the trend can be seen with respect to time. Line diagrams are often used to compare two or more things over the same time period although they certainly can be used to show just a single thing. *Things* could be any two or more meaningful aspects of the business such as product sales, salesperson's sales activity, student registrations for courses, and so on. Note that the *x*-axis is usually labeled with the time periods and that the *y*-axis is labeled with some numeric value such as units, dollars, and so on.

The **bar chart** shows proportions or quantities as they relate to each other. The **horizontal bar chart** is best used to compare different items in the same time period, while the **vertical bar chart** is best used to compare the same item in different time periods, as illustrated in Figures 9.13 and 9.14. Notice that a weakness for these two graphs is that neither one directly shows the total of all the items in the same time period or total of the single item in the different time periods, respectively. A creative and clearer variation of either of these bar charts is to use a symbol or icon in place of the bars. For example, if we are talking about sales of automobiles, we would use several automobile icons to represent the length of the bar(s). The *USA Today* newspaper usually includes one or more of these creative bar charts each day it publishes the paper.

The final graph type is called a **sectograph** because its purpose is to take a total amount of something and split it up into its proportionate or sectional parts. Both a **pie chart** and a **layer graph** are included in this grouping and are illustrated in Figures 9.15 and 9.16. A pie chart segments a circle into two or more pieces of the pie. The

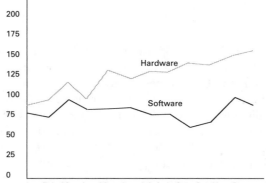

Figure 9.12   Line Diagram

California State University, Pasadena
College of Business Administration
Enrollment by Major - Fall 1995

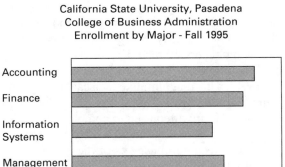

Figure 9.13   Horizontal Bar Chart

California State University, Pasadena
College of Business Administration
Total Enrollment by Semester

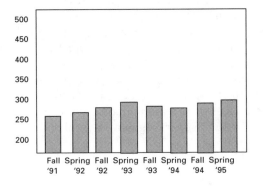

Figure 9.14   Vertical Bar Chart

size of each piece is proportionate to the percent of the whole (total) that each piece represents. The layer graph is similar to a line graph; however, two or more lines are stacked on top of each other to show the total. The space between the lines represents the proportion of the total that a line represents. As with the bar chart, neither of these graphs directly gives you the total of all the items being represented.

Many software packages provide three-dimensional (3D) capability for output graphics. Many of these 3D graphs are used to represent numeric data, but not always. 3D is perhaps more elegant and impressive looking than 2D graphs, but their readability is somewhat more difficult. Figure 9.17 is a 3D graph of the data presented in Figure 9.13. How many majors are there in Finance—92 or 95? Do you use the front of each bar on the chart for the reading or the rear of each bar? Looking at Figure 9.13's

California State University, Pasadena
College of Business Administration
Enrollment by Major - Fall 1995

Figure 9.15   Pie Chart

California State University, Pasadena
College of Business Administration
Total Enrollment by Semester

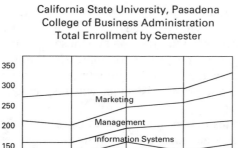

Figure 9.16   Layer Graph

bar for Finance, you can see that Finance had 95 majors, which is equivalent to the reading using the rear of the Finance bar in Figure 9.17. Care must be given when interpreting numbers on 3D graphs, otherwise distortion could take place.

OUTPUT: INTERNAL CONTROLS

The last topic to cover in this chapter briefly addresses the notion of internal controls for all types of outputs. If full attention were given to this topic, it could take several

California State University, Pasadena
College of Business Administration
Enrollment by Major - Fall 1995

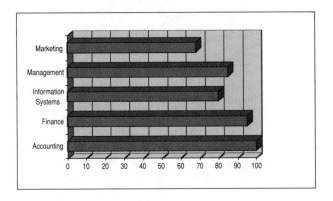

Figure 9.17   Three-Dimensional (3D) Horizontal Bar Chart

chapters in a book on auditing and controls. Certainly, there need to be limitations, constraints, or security placed on our information systems outputs. Not all information generated by an information system should be available to everyone in the world. There are general guidelines for addressing limitations, constraints, and security of output. However, no specific guidelines can be given here because every output of every information system is unique with respect to its limitations, constraints, or security precautions.

Some restrictions are mandated by law or policy. For example, in most universities, professors are not allowed to post student grades along with the student's name or identification number due to privacy laws. A corporation's personnel records and accounting information are confidential or secret as would be military personnel and equipment location information during wartime.

Once identified, each output whether display, hard copy, or other should be evaluated with respect to timing, volume, processing time, distribution, and access security. The **timing** of each output should be determined in consultation with the user. What is meant here is the frequency of the output—hourly, daily, weekly, monthly, and so on as a normal recurring frequency, or on demand, which means that users are free to request the output at a time suitable to their choosing as their needs dictate. Coupled with timing is the **volume** of the output. A single screen display is quite different than a 500-page report; however, the amount of computer **processing time** to produce either of these outputs could actually be about the same depending on the complexity of the output's content.

Distribution of output refers to the people who are the recipient users of the more permanent output such as paper, microfilm, and microfiche. Often several copies of the output must be produced in order to cover the distribution list of users authorized to receive the output. Depending on the confidentiality of the output's information, distribution lists of users are strictly monitored and adhered to. A person whose name is not on the distribution list for a specific confidential output should not have access to it.

The final control to discuss here is that of **access security** and this relates primarily to output display or sound information. In some environments, video display terminals are an invitation for someone to simply sit down and retrieve any output that is available in the system. Where confidentiality is important, authorization is enforced before anyone can request and receive a specific display or sound output. Of course, once the output display is on the screen, the user is expected to protect the information from being seen or distributed (via a print screen feature) to unauthorized persons. Password-protected screen-saver software is one simple way to prevent unauthorized viewing of information on a display while users are away from their desks.

Significant thought should be given to the security and controls for information systems, specifically its inputs, outputs, storage of data, and its processing.

THE FUTURE OF OUTPUT DESIGN

Output will still continue to be the dominant reason that information systems are created or purchased. Interaction with the user will continue to evolve the most as enhancements in technology allow improved graphical user interfaces, voice, and sound. Output design will continue to get help in the form of wizards or semi-intelligent agents, voice input, report and application generators, visual programming tools, and enhanced object-oriented objects, classes, components, and prototyping tools.

SUMMARY

This chapter has focused on information systems outputs, identifying their value to people and businesses. Some characteristics that contribute to information being of high quality and usable were also discussed. Knowing these characteristics will help you better address outputs with your users.

Some outputs are primarily to be used within a business, some outside the business, and some used by those outside the business and sent back to the business in the form of input. These types of output are internal, external, and turnaround, respectively. Two other characteristics of outputs deal with their frequency and were described as dynamic and static depending on consistent frequency or on-demand frequency, respectively.

Output must be "put on" some medium, such as paper, video display, microfilm, or microfiche. In addition, there are some general formats for output known as zoned, graphic, and narrative. Several report types were presented and discussed followed by a presentation and discussion of several graph types for the display of numeric data. The chapter concluded with a brief discussion of security and controls for output.

QUESTIONS

9.1 Discuss the characteristics that distinguish data from information.

9.2 What are some of the characteristics that make up quality and usable information?

9.3 What are internal outputs? What are external outputs? Briefly define each.

9.4 What is the difference between a static and a dynamic output?

9.5 Briefly define three types of output formats.

9.6 What are the different types of analytical output reports, and what are the advantages of each?

9.7 What are some of the advantages that business analysts and managers get from the use of spreadsheet software?

9.8 Discuss four types of graphical representations that are possible with the use of spreadsheet software.

9.9 What are some of the internal control aspects that must be considered with regard to output?

REFERENCES

KENDALL, K.E., and J.E. KENDALL, *Systems Analysis and Design* (3rd ed.). Englewood Cliffs, NJ: Prentice Hall, 1995.

WHITTEN, J.L., L.D. BENTLEY, and V.M. BARLOW, *Systems Analysis & Design Methods* (3rd ed.). Boston: Irwin, 1994.

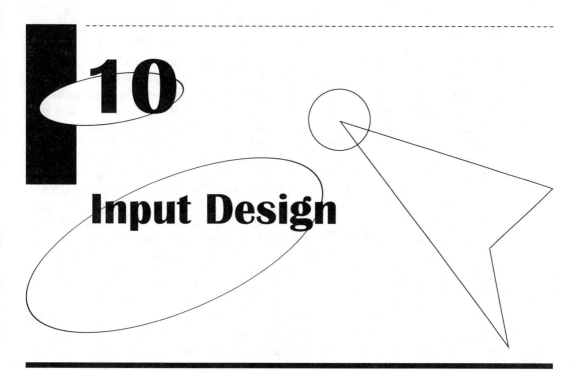

10

Input Design

CHAPTER OBJECTIVES (YOU SHOULD BE ABLE TO)

1. Describe some of the issues dealing with input data.
2. Describe different types of data validation and verification.
3. Define the different types of input data methods.
4. Identify the different types of input data devices.
5. Discuss several general guidelines for inputing data.
6. Define and discuss the different types of business codes.
7. Discuss control guidelines for batch and on-line input.
8. Discuss several input data validation techniques.
9. Describe the common windowing interfaces for display terminals.

INTRODUCTION

Have you ever heard these statements before? "Garbage In, Garbage Out!" (GIGO) and "You are what you eat!" They are a couple of old sayings that not only apply to humans but also to information systems. Sounds kind of strange, you say? Well, don't get me started about the daily "garbage" we humans take into our minds and the "garbage" that then comes out of our mouths and out through our actions! Just read the newspaper or watch the evening news and you know what I am talking about.

Information systems are only as good as the input data that are put into them. If my last name is entered into an information system as Norman, we have valid and error-free data to work with. If my last name gets entered as Norma, Nerman, Newman, or Normen, we have invalid and erroneous data to work with. Probably no big deal if we are developing a mailing list of 100,000 names and a few names get misspelled! How many times have you received some "junk" mail with your name spelled wrong or the wrong salutation such as Mr. when you are a Mrs. or Ms. or a Miss? Spelling your name incorrectly in a more important information system, such as a medical information system, social security information system, or university records information system, could cause serious problems for you sometime in your life. What if all my course grades and Ph.D. degree from the University of Arizona were associated with Ronald Newman instead of Ronald Norman? If I were to run for president of the United States and claim to have a Ph.D., this could cause a serious scandal and probably contribute to my downfall as a serious presidential contender! Joking aside, surely you get the idea.

Input (data) + processing = output (information)—the basics of computing once again. Even if the processing component is error free, it is absolutely impossible to get the correct information as output given the wrong or erroneous input. Oh, the output information will be correct based on the input entered and the error-free processing done on the input; however, the output information is supposed to be something else but isn't because the input data were in error. Consider the following example. You and a fellow student want to get together to study for an upcoming exam. You give him the written directions to your home or apartment. He follows your directions exactly but does not end up at your home or apartment. What went wrong? Let's look at the input, processing, and output for this example. The input data would be your written directions to your home or apartment; the processing would be your fellow student driving his automobile based on following your written directions; the output would be where he ends up after following your directions. If your friend followed your directions exactly and ended up in the wrong location, then your input (the directions) was in error. As human processors often do, your student friend might have noticed the error in your written directions. He could then take corrective action, which would allow him to end up at your home or apartment even though the directions were wrong.

Most information systems cannot make unplanned processing corrections or adjustments like human processors. However, information systems are becoming more sophisticated with some assistance or artificial intelligence built into them. For example, most of today's word processors have spelling and grammar checkers built into them. Even so, they still cannot detect every conceivable grammatical problem. The spell-checker feature is one that users believe really enhances the quality of their document.

The strategy used historically by most information systems to overcome the problem of "garbage in" (e.g., bad input data), has been to have elaborate error-checking or editing routines for each piece of data as it is being entered into the information system. This is like having your I.D. checked at the door of an over-21 establishment. Does this mean that absolutely everyone under 21 will be excluded from entrance? It should, but it doesn't always work. Perhaps you know some reasons why this is true or even know someone who has gotten in while being under 21 years of age.

THE MANY FACETS OF INPUT DATA

Data Validation and Verification

The error-checking or editing routines for input data are often referred to as **data validation and verification (V&V)** routines, and information systems devote a large amount of programming code to such routines. Users want their input data to be as "clean" or error free as possible. Sometimes the cost and human effort to have input data 100 percent error free are prohibitive. For example, in the mailing address example given previously, simple editing of the input names and addresses might be sufficient to get the job done, even though some errors will get into the mailing list database, such as a city or state with an incorrect zip code, an incorrect salutation, or a last name spelled incorrectly. On the other hand, a certain amount of human effort and money could be spent to improve the quality of the input data. To improve the incorrect zip code situation, a company could purchase a database of zip codes and associated cities and states for about $1,000 to guarantee that every mailing address has the appropriate city, state, and zip code. Ultimately, each company has to decide if these types of expenses and human time commitments to further validate the data are beneficial to it.

The systems analyst, working with the user, must decide exactly how much data validation is to be done on the various pieces of input data. This is not a decision left entirely to the systems analyst, although the user certainly expects basic editing to be done on all data. Alphabetic data are perhaps the most difficult type of data to detect errors in because we really can only spell check it if we have a clearly defined set of editing rules to follow. For example, all of the previous spellings of my last name could possibly be correct. No information system can detect this type of error and say something like "Excuse me, the name you just entered—Newsman—should be Norman" without a predetermined set of editing rules or an artificially intelligent editing component.

A few of the more common data validation and verification techniques are described here:

1. Self-checking digits or check-digit. Self-checking digits, also referred to as check-digits, are used most often on bank credit and debit cards, checking and savings accounts, inventory identification numbers, and customer and membership accounts. The concept makes use of an algorithm and some or all of the specific numbers in the card number, account number, inventory number, and so on. The result is a number that becomes a part of the original number. For example, an inventory identification number for a computer keyboard may be "425-102." The first five digits (425-10) may be arbitrarily assigned to the inventory item by a person or computer that generates inventory numbers. These five digits are input to a simple algorithm that determines the sixth digit—2 in this example—making the actual inventory identification number "425-102." A simple algorithm used to come up with the number 2 in this example is to sum the individual digits and divide the total by the number of positions in the inventory identification number before adding the calculated digit (5 in this example): $4 + 2 + 5 + 1 + 0 = 12 / 5 = 2.2$, which is rounded to 2. The digit 2 becomes the check-digit for this inventory identification number. Naturally, there are books that discuss self-checking digits at great length, since there are many different algorithms to do essentially the same thing that was just done with the prior example.

2. Combination check. A combination check makes use of two or more associated attributes to validate the data entered in one, two, or more of those attributes. A common example of this is city and state/province or state and zip code/postal code. Certain cities belong to a state or province and states or provinces have certain zip codes or postal codes. By cross-checking these two attributes, the information system can improve the accuracy of the input data.

3. Limit and range checks. Limit and range checks are used to validate numeric input data. Limits can be either low limits or high limits and the number entered must be greater than or equal to the low limit and less than or equal to the high limit. Picking a number between 1 and 10 is a good example of a limit check. Any number chosen must be greater than or equal to 1 and less than or equal to 10. Sometimes either the low or high limit is omitted, which gives more freedom to the input data. Picking a positive number sets a lower limit (zero) but no upper limit. A range check is the same as a limit check; however, some people find the word *range* to be more appropriate especially when the range does not start with zero or 1, such as picking a number between 3 and 9 or picking a number between 100 and 1000.

4. Completeness checks. Completeness checks can be established in many different ways. One of the most common ones is to have the information system establish a completeness check for each data entry screen in the system. Each data entry screen object should know what data are required to be entered in order to proceed with processing. For example, a data entry screen used to enter a person's name, address, city, state, and zip code information could use a completeness check to be sure that valid data had been entered for all of those attributes before proceeding with additional processing.

Input Data Methods

Information systems only have two methods for inputing data—batch and on-line, or sometimes called interactive. Both methods can be used in the same information system for different types of input data. In fact, with most business information systems, a combination of batch and on-line are utilized. What's the difference between batch and on-line? Traditional **batch** inputing of data is a three-part process of (1) collecting groups of related data, (2) entering the groups of related data onto some electronic medium such as magnetic tape or disk, and (3) processing the entire batch of related data as a group input into the information system. The first two activities—collection and entering activities—are done separately from the information system; it is only when the batch has been recorded on the medium that it is ready to be input into the information system.

There are some distinct advantages for the use of batch input, which are listed in Figure 10.1. The systems analyst and user should decide whether the information sys-

ADVANTAGES:

1. The first 2 activities, Collecting and Entering, can be done without impacting the performance of the computer system.

2. If there is a need for people to assist with the second activity, entering the groups of data, this can be done by specially trained data entry personnel making the data entry very efficient and accurate.

3. The final activity, Processing, can be done very quickly as the input is brought into the system in very rapid fashion since it is coming directly from an electronic media.

4. The final activity, Processing, can be done during times of the day or night or week-ends when there is less utilization of the computer resource, thereby leveling the impact of the information system more evenly over the time period.

DISADVANTAGES:

1. Data collection usually has to be a centralized activity.

2. Data entry usually needs to be done by specially trained personnel.

3. The processing activity is delayed, hence the possibility exists for data to be considered old or untimely when it finally gets processed.

4. Since processing is usually done during off-hours, input errors detected during processing would not get corrected until the next regularly scheduled processing of input data.

5. The off-hours computer operator may have to call the systems analyst or programmer if the program malfunctions.

Figure 10.1   Advantages and Disadvantages of Batch Input of Data

tem could benefit from these advantages, and if so, consider using batch input of data. The disadvantages of batch input are also listed in Figure 10.1, which need also to be considered during input design.

On-line inputing of data is generally considered the processing of input data as they occur in our environment. For example, an ATM machine accepts your keystrokes as on-line input to have the ATM machine perform a function for you; a university course registration system that allows you to use a telephone to add, change, and delete courses from your schedule is accepting your keystrokes as on-line input to the registration system. In both of these examples, you get immediate feedback, good or bad, from the ATM or registration system; no need to wait several hours or even overnight to see if your transaction request is accepted or rejected.

As with batch inputing of data, on-line input has its advantages and disadvantages, and some are listed in Figure 10.2. The systems analyst and user must also discuss these in light of the systems requirements.

There is also a hybrid data entry form that combines both on-line and batch. There is a significant amount of flexibility in how this is actually implemented in an information system. This is quite common in point-of-sale (POS) information

ADVANTAGES:

1. The data can be entered by its owners.

2. The data can be entered as close to their origination as possible.

3. Immediate feedback can usually be given regarding the correctness and acceptability of the data.

4. The input data can immediately update a database thus making it as current as possible.

DISADVANTAGES:

1. Equipment may be more costly to perform the input.

2. Users are not always well trained to input data.

3. User data entry procedural controls may be lacking.

4. Software must have additional controls to handle it.

5. Data is often only entered during business hours thus impacting the normal computer load.

6. The data entry activity could actually be slower than the equivalent batch processing for the same data.

Figure 10.2   Advantages and Disadvantages of On-line Input of Data

systems. For example, a checkout scanner at a supermarket is doing **on-line** entry of the product UPC codes you purchase in order to display and compute the cost of your trip to the supermarket. In addition, part of the information system associated with the scanner device might also be recording your purchases along with all other customers' purchases on a magnetic disk for use as a **batch** of input data into the supermarket's corporate inventory control system later that night to update quantities on hand and determine what products need to be ordered the following day. You interact with and see the on-line portion of this system but not the batch portion.

In the 1980s and 1990s, we have seen a major swing in information systems being developed with a significant on-line component in them. Today's technology is more capable and cost-effective, allowing systems analysts to do so more now than ever before. As programmers developing information systems in the late 1960s, we had on-line terminals to develop and test our programs, so the idea of on-line is not something new. What is new and exciting is the on-line display capabilities of our terminals, and the variety of different types of input devices available today compared to just a few years ago! Wow! And systems analysts have just scratched the surface of what will be possible in the future for on-line data entry. Be watching for virtual reality systems.

Input Devices

The number of different types of input devices has grown to double digits. Most of these should have been discussed in an earlier introductory course so their discussion will not be lengthy here. Each input device is listed in Figure 10.3 so you can discuss each one in more detail with your instructor and fellow students. As you discuss them, consider their use in any of the student information systems projects you may be working on for this or any other course.

General Guidelines for Inputing Data

There are only a few commonly accepted general guidelines for inputing data either in batch or on-line modes or both. The reason there are so few general guidelines is because we need significant flexibility for the inputing of data in order to meet the user's requirements for any type of information system. Four general guidelines for inputing data are discussed here.

The first general guideline is to **input only the data that are necessary.** This seems obvious, but in practice it is often not. There is really no need to enter data again once they are already in the information system. I don't know about you, but I

- Keyboard

- Magnetic Ink Character Recognition (MICR)

- Optical Character Recognition (OCR)

- Optical Mark Recognition (OMR)

- Digitizer

- Image Scanner & Facsimile (Fax) Machines

- Point-of-Sale Device (POS)

- Automatic Teller Machine (ATM)

- Mouse

- Track Ball

- Joystick

- Pens

- Scales

- Voice Recognition

- Touch Screen

Figure 10.3   Input Devices

get frustrated at the amount of **redundancy** in our world of information systems. Redundancy here refers to the amount of duplication of human effort or the duplication of information stored in different information systems. For example, every time I go to a different medical doctor, the office staff insists that I fill out their information forms. Wouldn't it be nice if this doctor's office staff could simply call my original doctor's office (or some other clearinghouse) and get that information electronically (of course with certain authorizations from me)? Then I would only need to enter new or changed information that would specifically apply to this doctor's office visit. It has only been in recent years that my university has implemented a student, faculty, and staff picture identification card with magnetic stripe. With this card, we can do almost any type of campus transaction—withdraw books from the library, make purchases at the bookstore or cafeteria (if our campus bank account has money), and even make photocopies of books and journals at the library. Prior to this information system, the faculty each had a faculty ID card, a photocopy card, a bookstore credit card, and a library card.

The second general guideline for inputing data is **do not enter data that can be derived or calculated by the information system.** This is similar to but different than the first guideline. The difference is that the information system is capable of creating or retrieving these types of data; therefore, let it. For example, let the computer's information system calculate arithmetic formula-driven items such as totals, subtotals, grand totals, amount of sales tax to pay, shipping charges, amortization amounts, loan principal amounts, interest charges, and so on.

In addition, try to take advantage of the information system's lookup capability to automatically provide other types of information such as (1) student's name, address, and other personal information after keying in a student identification number, (2) checking and savings account balances after scanning your ATM card's magnetic stripe, (3) name of product purchased and its cost after scanning the product's UPC barcode on the package, (4) the long state name after entering the two character abbreviation for it, such as CA for California, (5) the long name for a city airport after scanning the three character airport code on a barcoded luggage tag such as LAX for Los Angeles, and (6) filling in the correct city and state after keying in a zip code such as 92124 yielding San Diego, CA. By allowing the information system to retrieve or calculate information, you not only enhance the accuracy of the data, but also reduce the amount of time it takes to enter these data by the human user.

The third general guideline for inputing data suggests the **use of business codes** when and where appropriate. A **business code** is a group of one or more alphabetic, numeric, or special characters that identifies and describes something in the information system. Business codes, often just referred to as codes, have been a way of life for information systems for dozens of years. Telephone numbers, student identification numbers, social security numbers, state codes, UPC product codes, bank account numbers, credit and debit card numbers, driver's license numbers, vehicle identification numbers (VIN), and automobile license plate numbers are all examples. There are four distinct types of business codes—serial, sequential, block, and alphabetic, each described next. Following the discussion of these four codes is another discussion of a hybrid business code—called the group code—which is a combination of one or more of the four distinct types of business codes described before the group code.

The first type of business code is the **serial** code. Serial codes, as their name implies, consist of assigning a consecutive number to each "person, place, or thing" (referred to simply as "thing" from now on) in a group of things in whatever order the thing is encountered. In other words, serial codes are based on **arrival time** into the information system. For example, my daughter's high school student identification number is 134,435. This is a serial code example because there have been 134,434 students who have been admitted to the high school and assigned a student identification number prior to her being admitted and assigned a number. Over 30 years ago, the very first student to be admitted and assigned a student identification number at that high school was assigned number 1.

Serial codes do not necessarily start with the number 1 (or zero) and do not necessarily keep incrementing by one from their inception to the current time. However, the numbers should be sequential but may start at any meaningful place, say 10, 100, 1,000, and so on. In addition, the numbers could repeat over again based on some time limit policy. For example, an order entry information system started assigning invoice numbers to orders back in 1986. The information system started with invoice number 1. As of December 1994, the invoice serial numbers being assigned were in the 300,000 range. When the invoice serial number reaches 999,999 it will probably roll over to number 1 again rather than go on up to 1,000,000. In another information system situation, sales transactions are assigned a serial number starting with number 1 every month. Figure 10.4a illustrates a sample of incoming freshman students at your university being assigned a serial student identification number based on their arrival time into the university's information system.

The second type of business code is the **sequential** code, which is essentially the same in format as the serial code but differs in its application. Sequential codes are assigned to each thing in a group of things only after some meaningful organization or sort is performed on the group of things. Figure 10.4b illustrates the use of sequential codes on the same data. Notice arrival times are not considered. The data are sorted in ascending last name alphabetical order, then assigned a sequential student identification number.

To summarize these two codes before continuing, serial codes are based on arrival time into the information system and have no other meaningful association with the things the serial numbers are assigned to. Sequential codes are assigned after some meaningful sort is done on the things. This seriously limits the ability of the information system to accept new items after the sequential numbers have been assigned without resorting and reassigning sequential numbers based on the new sort sequence of the things. Sequential codes are best used in situations where the addition of new things after the assignment of the numbers is not allowed.

The third type of business code is the **block code.** Block codes are useful to create groupings of similar things. For example, a men's clothing store may decide to assign a number to each of the different clothing items in its inventory. For example, a pair of Levi's 502 blue jeans size 36x30 would be given a unique number, say 101. All

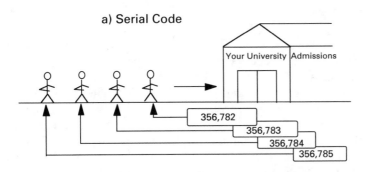

a) Serial Code

Your University Admissions

356,782
356,783
356,784
356,785

Student ID Assigned by Admissions based on student arrival time.
(assumes that there have been 356,781 previous students)

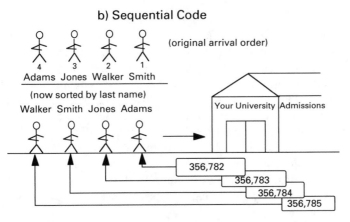

b) Sequential Code

(original arrival order)

4   3   2   1
Adams Jones Walker Smith

(now sorted by last name)
Walker   Smith   Jones   Adams

Your University Admissions

356,782
356,783
356,784
356,785

Student ID Assigned by Admissions based on Last Name
sort of all newly admitted students in a semester.
(assumes that there have been 356,781 previous students)

Figure 10.4   Serial and Sequential Business Codes

like inventory items of the same size would also have this same number—101. Another inventory item with the same name but in size 36x32 would also have a unique number, say 105. All pairs of the Levi's 502 blue jeans size 36x32 would have the same identifying number as assigned earlier—105. Certainly, using a serial code as discussed previously would do the job and might look like Figure 10.5. Notice the duplication of serial codes on equivalent inventory items.

Giving a little more thought to the information system, the clothing store manager felt it would be very useful for him to get sales and inventory information based on groupings or blocks of like clothing items such as slacks, shirts, shoes, ties, belts, and so on. Using a serial code for the clothing, as in Figure 10.5, would not allow him to easily get this grouped sales and inventory information. So the systems analyst and the manager working together created a block code scheme to help accomplish the manager's goal. Figure 10.6 shows a partial list of the block groupings they agreed to establish.

Each item within a specific clothing group would then be given a number within the block code range that best described the item from the manager's perspective. No-

Inventory Item	Inventory Number
Levi 502 Bluejean, 36x30	101
Brand 1 Silk Shirt, 16x32	102
Levi Dockers, 34x32	103
Brand 1 Dress Shirt, 15.5x32	104
Levi 502 Bluejean, 36x32	105
Brand 1 Silk Shirt, 16x34	106
Brand 2 Bluejean, 34x32	107
Brand 1 Dress Shirt, 16x30	108
Levi 502 Bluejean, 36x30	101 *
Brand 1 Silk Shirt, 16x32	102 *

* duplicate inventory serial number (same item)

Figure 10.5   Use of Serial Codes in a Clothing Store

Bluejeans	100 - 199
Slacks	200 - 299
Shirts	300 - 399
Ties	400 - 499
Shoes	500 - 599
Belts	600 - 699
Accessories	700 - 799
Underwear	800 - 899
Socks	900 - 999
Sport Coats	1000-1099

Levi 502, 34x30	100
Levi 502, 34x32	101
Levi 502, 34x34	102
Levi 502, 36x30	110
...	
Brand X, 32x30	130
Brand X, 32x32	131
Brand X, 34x30	137
...	

Figure 10.6   Use of Block Codes in a Clothing Store

tice that the numbers within the block code range can be used in any manner desired, even leave gaps, as illustrated, for future expansion of product lines. One disadvantage of block codes is that the block code ranges need to be large enough to accommodate expected growth. Otherwise you will run out of available numbers within the range to use at some future time. Another disadvantage deals with the situation when there is a need to change the block code number assigned to an item to put it into one of the other block code groupings. For example, the manager may have initially established a single block code grouping for all men's undergarments. Several months later he has determined that there should be two groups, one for undershirts and another one for underpants. This means renumbering at least one of these product groups, which would cause problems with historical sales and inventory data. The historical data are stored with the original codes. The final disadvantage of block codes is the situation in which an item does not conveniently fall into one specific block code grouping; in this case, you are forced to pick one specific block code group to assign it to.

The fourth type of business code is the **alphabetic** code. These codes are often exemplified by their tabular look. Perhaps the most common alphabetic code in the United States is the two character state abbreviation code table. Using CA for California, MI for Michigan, and so on is a convenient way of simplifying the entry and storage of state names. Many other examples of alphabetic codes exist, such as:

1. Unit of measure alphabetic code such as ga for gallon, qt for quart, pt for pint, and so on.
2. Airport codes, such as LAX for Los Angeles, SAN for San Diego, CHI for Chicago and so forth.
3. Academic degrees, such as BS for bachelor of science, MS for master of science, MBA for master of business administration, and Ph.D. for doctor of philosophy.
4. Periodic table of elements in chemistry, such as H_2O for water.
5. Street types, such as Blvd. for boulevard, Dr. for drive, St. for street, and Ct. for court.
6. Salutations, such as Mr., Mrs., Dr., and Rev.

The final business code is actually a hybrid of the other four, created by combining these codes to create a supercode known as the **group code.** Each position or positions of a group code is meaningful. For example, the VISA, Mastercard, and Discover credit cards use group codes for the credit card number. The first four digits in the number identify the card as being either VISA (4XXX), Mastercard (5XXX), or Discover (6XXX). Social security numbers utilize group codes also. The first three digits indicate the geographic region within the United States where you applied for the number. Other examples of group codes can be found in the Uniform Industry Codes (UIC) for businesses, vehicle identification numbers (VIN) for automobiles and trucks, and universal product codes (UPC) on products. Figure 10.7 gives an example of a group code in a paint store.

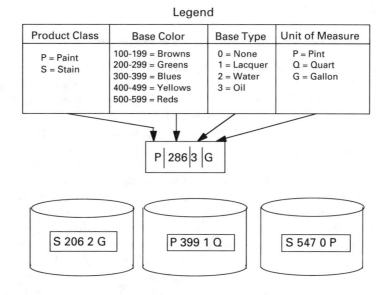

Legend

Product Class	Base Color	Base Type	Unit of Measure
P = Paint S = Stain	100-199 = Browns 200-299 = Greens 300-399 = Blues 400-499 = Yellows 500-599 = Reds	0 = None 1 = Lacquer 2 = Water 3 = Oil	P = Pint Q = Quart G = Gallon

P | 286 | 3 | G

S 206 2 G   P 399 1 Q   S 547 0 P

Paint Product Examples

Figure 10.7   Use of a Group Code in a Paint Store

Group codes are very useful, especially in inventory or product environments. One of their disadvantages is that they tend to be longer numbers (e.g., the VIN number on automobiles is 17 positions long, credit cards are often 15 positions long), which are difficult for people to work with. Another disadvantage is that they often require a *legend* to help you translate from the code to the item it describes.

To summarize business codes, they should conform to the following guidelines:

1. The codes selected should allow for expansion.
2. The codes selected should be unique.
3. The size of the code (number of positions) is important as the larger the number of positions in the code, the more difficult it is to use and enter into an information system by a person.
4. The codes selected should be convenient.

The final general guideline for inputing data suggests that alphabetic and numeric data entry performed via a keyboard and associated display terminal should be done **in a left-to-right, top-to-bottom motion** as displayed on the video display terminal because this is the accepted way that people are familiar with reading a page from a book or magazine. This guideline has been used for years in the design of text-based data entry screen displays. As we move through this decade, we are seeing more and more utilization of graphics on data entry display terminals. Even though the Xerox Star, 1981, the Apple Lisa, 1982, and the Apple Macintosh's introduction in 1984 revolutionized the human-computer interface, it has only been in this decade that graphical interfaces have reached worldwide utilization on personal computers.

Graphical User Interface (GUI) Design for Input

Designing input screens for display terminals used to be a relatively easy task; we only had 80 columns by 24 rows of text-based alphabetic and special characters and numbers for a total of 1,920 characters per screen. With the standard for today's display screens demanding resolutions of upwards of 1,024 x 1,024 pixels, we can create an endless number of characters, numbers, special characters using different font types and sizes, graphics, and video, all in brilliant color. Couple this with the commonly available graphic windowing software features of minimize, maximize, hide, pull-down menus, pop-up menus, nested menus, menu bars, menu pads, cascade windows, tile windows, icon images, touch screens, sound cues, push/radio buttons, spinners, and so on and we find that the robustness of our input screens is virtually limitless. The remaining figures in this chapter illustrate some of these commonly available windowing features.

However, with this robustness comes the need for more graphical layout design expertise and time needed to create such elegance. Therefore, several software vendors are selling or incorporating graphical user interface development kits as stand-alone software packages, bundled with programming languages, as part of prototyping software tools, or as part of their CASE software. What this means is that the commonly accepted graphical windowing guidelines for creating effective input screens are being automated so that all software running on a particular platform (e.g., hardware and operating system software) will have the same "look and feel" for the human user. This is very important for user acceptance and adaptability to new information systems. For example, a user of Novell's WordPerfect for Windows word processing software could probably sit down with Microsoft's Word or Lotus's AmiPro word processors and be reasonably proficient without even opening a user manual for these software products. This is a major victory for the user community even if the generally accepted "look and feel" of today's software isn't as perfect as it could be.

Graphical user interface (GUI) design has literally revolutionized the human-computer interface. From a practical point of view, it all started with the introduction of the Macintosh computer in 1984 (Super Bowl Sunday), and catapulted into prominence in the PC industry with the introduction of Microsoft's Windows 3.1 in the early 1990s. There are several other GUI-based operating systems on the market, but these two have about 75 percent of the GUI market. The **mouse** or its equivalent pointing device has become a valued part of every GUI-based PC. In fact, most users would have difficulty using a GUI interface without a mouse. GUI-based software also exploits a **dashboard** metaphor. The dashboard metaphor is borrowed from automobiles, airplanes, and other vehicles. The dashboard has gauges, dials, buttons, and so on that the operator of the vehicle observes and uses throughout the trip. Because of its familiarity as part of everyday life for most people, the dashboard metaphor has been adapted to the software GUI-interface.

The research community that created GUI-interfaces has created its own terminology to describe a GUI-interface, terms such as *scroll bars, button bars, spinners, pop-up menus, push buttons, check boxes,* and so on. Figure 10.8 lists many of the terms used in a GUI-based environment.

Terminology

Pop-up menus

Pull-down menus

Drop-down/List boxes   Modes of Operation

Radio Buttons
                             • Navigation
Check Boxes
                             • Data Entry
Spinners

Push Buttons

Menu Bars

Menu Pads

Maximize/Minimize

Button Bars

Tool Bars

Docking

Scroll Bars

Figure 10.8   Graphical User Interface (GUI) Terminology

A GUI-based PC environment has two basic modes of operation—navigation and data entry—as listed in Figure 10.8. **Navigation** is the way a user instructs the computer's software to do something, such as launching a software program like Excel, QuattroPro, WordPerfect, fax, file manager, or directing the computer to print or copy a file, and so on. Navigation is accomplished in two basic ways—functions selected from word menus or pushing iconic buttons that represent functions. For example, to print a page using WordPerfect software, the user can either select PRINT from the FILE menu or push the printer iconic button by pointing to it with the mouse. Some GUI-based software packages incorporate scroll bars and panning bars to help the user move forward and backward in the data, or left and right in the data, respectively. Examples of word menus, iconic buttons (icons), and scroll bars can be seen in Figures 10.9 and 10.10. Additional examples of pull-down, pop-up, and nested function menus can be seen in Figures 10.11, 10.12, and 10.13.

Data entry is the actual data that users input into the GUI-based software. For example, the text you are reading was input into WordPerfect via data entry. When I enter student names and test scores into Excel, a spreadsheet program, I am inputing data via data entry. Data entry can also be accomplished by pushing radio buttons for certain choices you want, pushing drop-down box buttons, pushing spinner arrows to

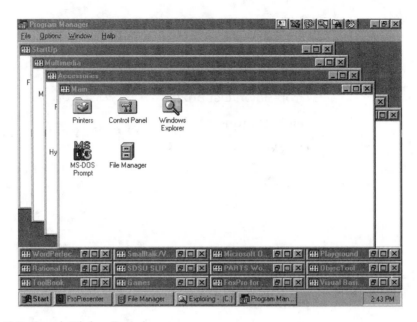

Figure 10.9   Cascade Windows Example

Figure 10.10   Tile Windows Example

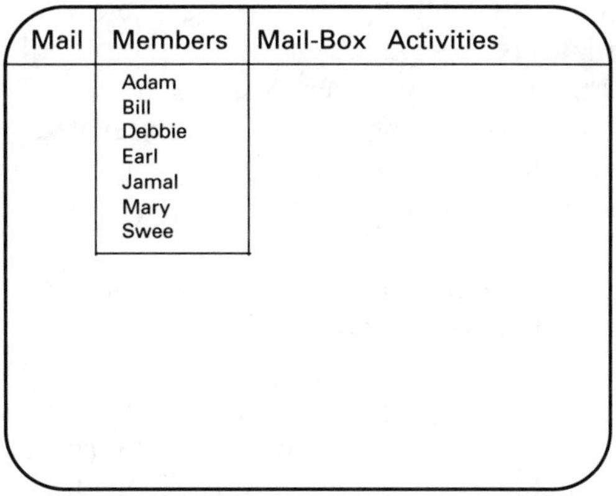

Mail	Members	Mail-Box	Activities
	Adam		
	Bill		
	Debbie		
	Earl		
	Jamal		
	Mary		
	Swee		

Figure 10.11   Pull-Down Menus

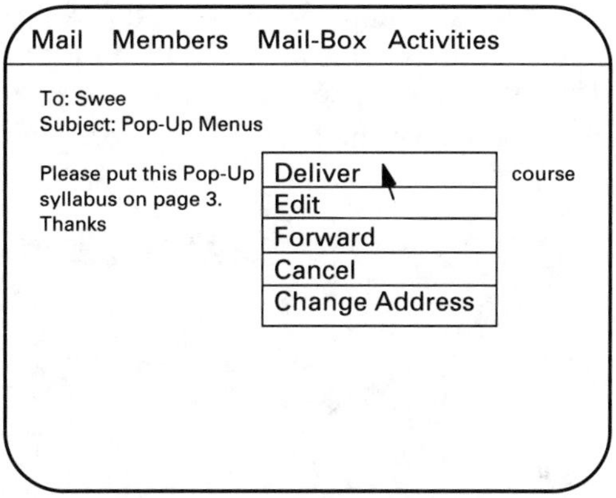

Mail	Members	Mail-Box	Activities

To: Swee
Subject: Pop-Up Menus

Please put this Pop-Up   Deliver   course
syllabus on page 3.   Edit
Thanks   Forward
                           Cancel
                           Change Address

Figure 10.12   Pop-Up Menus

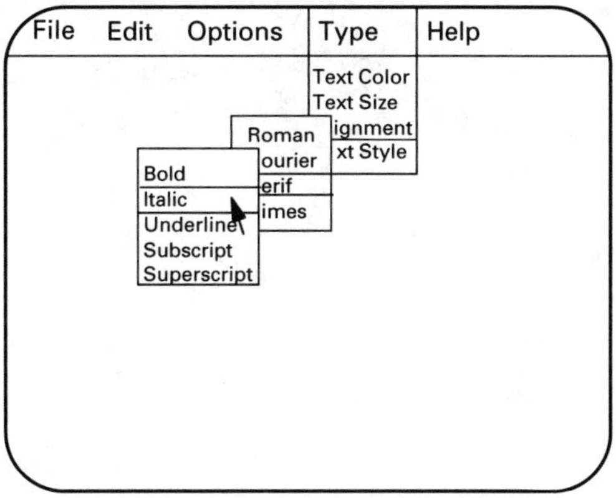

File	Edit	Options	Type	Help

Text Color
Text Size
Roman   ignment
ourier   xt Style
Bold   erif
Italic imes
Underline
Subscript
Superscript

Figure 10.13   Nested Menus

select a number you want, and making a selection from a list of check boxes. Data entry examples of these can be seen in Figure 10.14.

GUI-based display screen design is more difficult to do well than is traditional text-based display screens. Programmers devote a great deal more program code to handling the GUI navigation than the actual code to accommodate the business functions being performed. For the systems analyst who is designing a GUI-based display screen, three common tendencies are to (1) put too much on a GUI-based display, (2) use too many different colors at the same time, and (3) use smaller fonts and graphic symbols, which make readability difficult.

Icons can be intuitive to people and are used on a worldwide basis for many purposes. Some icons are universal and transcend language barriers. Can you recall the universal icons for food, lodging, and gasoline posted on freeway signs, rest room icons for men and women, and airport icons for transportation and luggage? Even traffic signs can be considered icons. You would probably recognize a stop or yield sign without the words written on the sign.

Today's personal computer GUI-based interfaces exploit the dashboard and icon metaphor. Push buttons with icon symbols on them are the norm. The hourglass, clock, and watch icons represent "waiting" time; the garbage can is symbolic for deleting things; and the printer is the icon that the user would push to start printing.

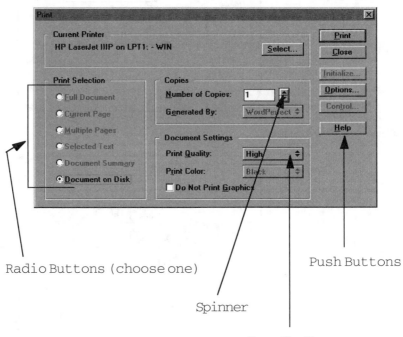

Figure 10.14   Data Entry Screen With Navigation Choices

Although GUI-based user interfaces have made a significant contribution to user acceptance of computers, they are not without shortcomings. However, the shortcomings are continually being addressed by systems analysts as newer versions of software attempt to improve on the shortcomings. One limitation of iconic push buttons is that of a user being able to remember what the iconic symbol represents. More recent software provides brief pop-up textual help next to the pointing device's arrow on the display screen as the user moves the arrow onto a push button. Another shortcoming is related to the size of the icon and any associated text. Some icons are so small that the user may have difficulty recognizing the icon symbol as well as any text associated with it. Some users have commented that they need to put their contacts or glasses on to clearly see the icons. One final shortcoming is related to each user's ability to cope with the large number of icons appearing on the display screen at any one time, for example, the word processor being used to type this page of text has close to 50 icons along the top and right borders of the display screen.

SUMMARY

This chapter has concentrated on the concept of input and some of its design considerations. The most common issues related to dealing with input data were discussed followed by coverage of the various methods for doing data input. The list of input devices is long, and a few of the specific and more common devices were discussed. When data are being input, there are some general guidelines for doing it efficiently and effectively and each of these was reviewed. Business codes have become a way of life for business information systems and each of the types was discussed and illustrated. Input controls are very important in order to avoid "garbage in." Several controls related to batch and on-line data entry were discussed. Several data validation and verification techniques were discussed, and the chapter concluded with a discussion of the windows-based graphical user interface and some of its design considerations.

QUESTIONS

10.1 What is meant by data validation in an information systems context?

10.2 Describe the types of data validation and verification.

10.3 What are the two methods of data input in information systems?

10.4 What are some of the advantages and disadvantages of group processing?

10.5 What are some of the advantages and disadvantages of on-line processing?

10.6 What are four general guidelines to keep in mind when entering data into a system?

10.7 Briefly discuss four guidelines that should be followed when using business codes as input data.

10.8 What are the different modes of operation in a graphical user interface PC environment?

10.9 What are some of the disadvantages associated with graphical user interfaces?

REFERENCES

KENDALL, K.E., and J.E. KENDALL, *Systems Analysis and Design* (3rd ed.). Englewood Cliffs, NJ: Prentice Hall, 1995.

NIELSEN, JAKOB, "Applying Discount Usability Engineering," *IEEE Software,* 12, no. 1 (January 1995), 98–100.

WHITTEN, J.L., L.D. BENTLEY, and V.M. BARLOW, *Systems Analysis & Design Methods* (3rd ed.). Boston: Irwin, 1994.

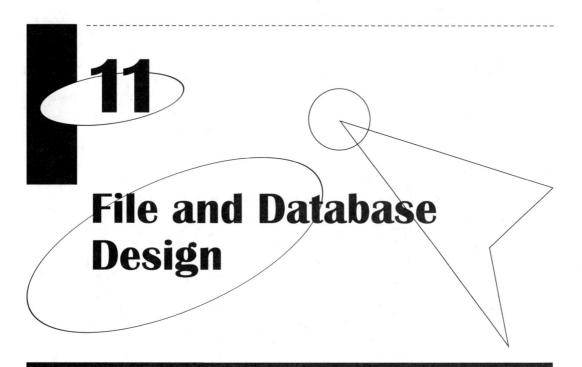

File and Database Design

CHAPTER OBJECTIVES (YOU SHOULD BE ABLE TO)

1. Define bit, byte, attribute, record, file, database, and folder.
2. Name and describe the two data structure objectives.
3. Name and describe the different types of attributes.
4. Name and describe the different types of files.
5. Name and describe file access and organization types.
6. Discuss the process and purpose of normalization.

Have you ever lost your wallet or purse? What a frustrating experience! After getting over the initial frustration of the lost money, you probably began to think about the items you will need to replace and the items that cannot be replaced. For many people, wallets and purses represent a more permanent storage location for personal items. People naturally store possessions that they have. Come to think of it, people store all kinds of things in their attics, basements, garages, closets, drawers, cupboards, cabinets, shelves, and desks.

Information systems, especially business information systems, also have the need to store some things on a more permanent basis. What is stored is referred to as data which have been entered into the system or have been created or organized through the processing done by the information system. Systems analysts have significant empirical evidence, accumulated over dozens of years, that suggests that **a business's data are the most stable part of its information system.** If we look at the history of student registration systems over dozens of years (or many other business information systems), we find that these systems basically deal with students, course schedules, instructors, and classroom availabilities. The data that are stored and used today in a student information system are probably very similar to the data stored and used in this type of information system dozens of years ago. The most significant change in student information systems (or many other business information systems) over the same time period has been the processing method used to do student registrations.

FILES AND DATABASES

As the discussion of files and databases begins, please understand that there are complete textbooks and perhaps semester courses dedicated to files, data structures, and databases. Therefore, this chapter is intended to be a basic primer or review of these topics only.

The data that exist in an information system are either input or calculated. Once input or calculated, the data exist in the computer's memory and will eventually be deleted or stored for more permanent use. To keep the input or calculated data, information systems store these data in **files** or **databases,** which are usually recorded electronically on magnetic disks. **Persistent data** is a term used in computer science or software engineering circles and is the equivalent of files and databases. Figure 11.1

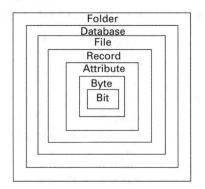

Bit (8 bits = 1 byte)

Byte (1 or more bytes = an attribute)

Attribute (1 or more attributes = a record)

Record (1 or more records = a file)

File (1 or more files = a database)

Database (1 or more databases = a folder)

Folder (1 or more folders = organized disk drive)

Figure 11.1   Hierarchical Components of Persistent Data

visually illustrates the conceptual template for files and databases starting with the innermost layer of the template being a bit, progressing through a byte, an attribute, record, file, folder, and ending with the database. As each of these layers are defined, remember that they represent a template or framework for the instances of actual data that exist in the information system.

A **bit** is a single binary digit—0 or 1. It takes several of these in combination, usually 8 as in 8-bit ASCII code which is common on most personal computers, to represent a character or byte. A **byte** is the term used to describe a single alphanumeric character, such as an A 1 8 w Y; even a space, comma, period, parenthesis and other characters. In addition, most ASCII character sets include a number of nonprinting characters, such as a smile face. An **attribute,** which is also referred to as a **field, data element,** or **object,** is made up of one or more bytes that together create a template for a meaningful piece of data. For example, combining 15 bytes and naming them LAST NAME creates a template for a meaningful piece of data to be associated with—specifically someone's last name. The information system will be able to assign actual instances of last names (e.g., Adams, Kumari, Newman, Norman, Smith, Yu, and so on) with the attribute template called LAST NAME as in Figure 11.2a. A **record,** also referred to as a **tuple** or even an **object,** combines one or more related attributes to form a record. For visualization purposes, it may be helpful to visualize attributes as columns and records as rows in a two-dimensional array as in Figure 11.2b.

a) Bits   0 1 1 1 0 0 0 1

b) Bytes   A, B, ... Z, 0,1...9, #, &, $, etc...

c) Attributes

Template

First Name	Middle Initial	Last Name	Social Security Number	State
Ronald	J	Norman	559-65-8213	CA

Values or Instances

d) Records (each row is a record)

First Name	Middle Initial	Last Name	Social Security Number	State
Ronald	J	Norman	559-65-8213	CA
Rashmi	B	Kumar	371-48-4562	MI
James	R	Logan	559-63-8472	OR
Susan	L	Johnson	243-74-5219	NY

Figure 11.2   Bits, Bytes, Attributes, and Records

A **file** consists of one or more similar type of records. For example, one file may consist of student records, another may contain course information, and a third file may contain faculty information as illustrated in Figure 11.3. A **database** consists of one or more related files as illustrated in Figure 11.4. Finally, a **folder,** also referred to as a **directory,** has become popular with personal computers and is used to organize one or more related files and databases as in Figure 11.5.

Data Structures

For our purposes, **data structures** are synonymous with records because they too are created by combining related attributes. The term *data structure* is often used in place of *records* when discussing the composition of a record (e.g., the attributes that make up the record) or the relationship between different record templates. For example, in the "old" days two systems analysts talking about student records and course records would refer to them as just that—student records and course records; today, two systems analysts would more than likely refer to these as a student data structure and a course data structure.

File #1 - Student Information

First Name	Middle Initial	Last Name	Social Security Number	State
Ronald	J	Norman	559-65-8213	CA
Rashmi	B	Kumar	371-48-4562	MI
James	R	Logan	559-63-8472	OR
Susan	L	Johnson	243-74-5219	NY

File #2 - Course Information

Course Number	Course Name	Units	Department
Act102	Accounting Principles	3	Accounting
Bio101	Intro to Biology	3	Biology
Chm109	Organic Chemistry	3	Chemistry
Eco104	Macro Economics	3	Economics
Eng100	Beginning English	3	English
MIS111	Intro. to Computers	3	M.I.S.
Mkt114	Principles of Marketing	3	Marketing
PEd118	Beginning Golf	1	Phys. Educ.
Phl108	Philosophy	3	Philosophy
Soc105	Cultural Changes	3	Sociology

File #3 - Department Information

Department	Department Head	Telephone	No. of Majors
Accounting	J. Morgan	594-2348	275
Biology	S. Tishman	594-4459	110
Chemistry	P. Dayson	594-7728	120
Economics	R. Kumar	594-0923	75
English	J. Amar	594-8276	60
M.I.S.	K. Kettleman	594-1010	175
Marketing	A. Winters	594-2034	140
Phys. Educ.	T. Tolner	594-2229	225
Philosophy	A. Hayley	594-9011	150
Sociology	B. O'Neal	594-3927	70

Figure 11.3   Files

Folder with Database #1   Folder with Database #2

Figure 11.4   Database

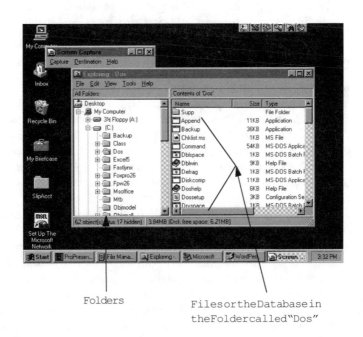

Folders   Files or the Database in
                           the Folder called "Dos"

Figure 11.5   Folders

When designing data structures or records, there should be two overriding objectives—simplicity and nonredundancy. **Simplicity** merely means that the data structures created should be as straightforward and user understandable as possible. Two reasons for this are that the simpler the data structure, the more easily it is maintained over the life of the information system, and the easier it is for users to retrieve their own output information using the data structures as input data to report generator software.

Nonredundancy is the second objective to achieve. This means no duplication of any attributes, records, or files within the business's databases. This is a very lofty goal. In well thought out information systems designs, nonredundancy has been achievable down to the attribute level. In business information systems especially, it is virtually impossible to completely eliminate all duplication of attributes and still end up with an efficient and effective information system because these two design objectives—simplicity and nonredundancy—are counteracting objectives. What this means is that simplicity and nonredundancy are like the opposite ends of a teeter-totter; that is, in order to achieve balance between the two, there is probably some compromise necessary. If the data structure is too simple, then there is probably a significant amount of redundancy, and conversely, if the data structure has no redundancy, it probably is more complex than it should be.

There is one final comment regarding data structures to be made before moving on. Using a structured methodology for information systems development, one would create one or both of an entity-relationship diagram and a data structure diagram to help visually show the relationships that exist among the data structures. In the object-oriented methodology, the same thing is accomplished through the use of generalization-specialization structures, whole-part structures, and instance connections.

Attribute Classifications

Attributes can be grouped into four general classifications—keys, descriptors, audits and control, and security. Looking at any data structure, record, or object (for example, the student record shown in Figure 11.6, which could have many attributes), a systems analyst can assign one of these four classifications to each attribute based on their definitions as described here. While the following discussion uses the word *record* due to its long history in information systems, please be aware that the words *tuple* and *object* could be substituted in keeping with a discussion on relational databases (tuple) and object-oriented software engineering (object).

A key attribute is an attribute that helps identify one instance of a record from all other instances in the file. Quite often in business information systems key attributes are characteristics such as student identification number, course identification number, social security number, part number, bank account number, ATM card number, universal product code, and so on as illustrated in Figure 11.7. Key attributes do not have to be numeric, but numbers have historically been very effective for allowing the information system to quickly retrieve a specific student, course, person, part, bank account, and so on. Some of the key attribute suggestions mentioned previously may actually contain alphabetic characters combined with numbers.

Attribute Type	Student Record Attributes
Key	Social Security Number
Descriptor	First Name
Descriptor	Middle Initial
Descriptor	Last Name
Descriptor	Street Address
Descriptor	City
Descriptor	State
Descriptor	Zipcode
Descriptor	Telephone
Descriptor	Grade Point Average
Descriptor	Grade Level
Audit & Control	Date Student Record Added
Audit & Control	Initials of Person Adding Record
Security	System Function Usage
Security	Person Authorizing Function Usage

Figure 11.6   Attribute Types—Key, Descriptor, Audit and Control, and Security

Key Attribute Name	Instance/Value Example
Student ID Number	68372
Social Security Number	559-68-0923
Vehicle ID Number	JA3XC52BONY002400
Course Number	MIS-111
VISA Card Number	4128 0022 2048 2552
Checking Account Number	128-0049
Video Store Account Number	Norm001

Figure 11.7   Key Attribute Examples

Keys are either primary, secondary, or foreign. A **primary key** has two purposes. First, it will uniquely identify one record instance from all other record instances in the file. In order to accomplish this, a primary key attribute must not contain any duplicate values across all records in the file. For example, student Tom Smith has a student identification number of 13,432. In order for the student identification number attribute to be the primary key for the student file, no other student can have this same student identification number—13,432. If there is no single attribute within the data structure (record) that has unique values such as the student identification number did in the example, then a primary key can only be created by the concatenation of two or more attributes. The combination of two or more attributes' values must be such that, together, these values would uniquely identify one record from all others in the file. For example, using a student Last Name attribute in a large school setting is probably not going to yield all unique values due to the abundance of names such as Smith, Yang, Chen, Yu, and Jones. Therefore, to create a primary key, one would probably need to concatenate a Last Name attribute with a First Name attribute and a Middle Name Initial attribute to arrive at a situation that might possibly be unique. As the number of student records increases, the probability of even these three attributes combined being unique diminishes. So the systems analyst and the user must give serious thought to which attributes are candidates for primary keys.

The second purpose of a primary key is to define the access path to each record within the file. Records usually need to be accessed in a logical and orderly way. That way is usually via a primary key. A discussion on data access can be found later in this chapter.

A **secondary key** provides an alternative access path to data records in a file. A secondary key attribute, like the primary key, is often used to access a specific record in a file too. Unlike the primary key, however, secondary key values do not have to be unique. Attributes designated as secondary keys are treated differently by the information system than those designated as primary. The information system software expects a primary key to yield a unique record from all others in the file, whereas the information system software must anticipate duplication of values when accessing records via the secondary key. Nonetheless, using secondary keys can be very helpful to reduce the potential number of records down to a manageable size before selecting the specific one you need. For example, a secondary key is declared for an attribute called Student Last Name. A faculty person wishing to see Tom Smith's course schedule this semester has access to an information system that can display this information provided that the instructor knows either the student's identification number (primary key) or the student's last name (secondary key). If the professor knows the student's identification number, she could simply type it as input to the information system and the information system would retrieve the student record that matched the student identification number input by the professor. Then the information system would, in a few seconds, display the student's course schedule. If the professor does not know the student's identification number, she enters the student's last name—Smith—and the information system displays a list of all the Smith last name records showing their first names and any other identifying information about each particular Smith. The professor simply highlights or points to the correct Smith shown on the display, and then the information system can display the course schedule for the highlighted or pointed to Smith. Because she did not know the student's identification number, it has taken the professor two steps to see the course schedule instead of one.

A **foreign key** is an attribute of the records in one file that is used to associate data records in another file with its records. In other words, a foreign key represents the access path that links more than one record together. An example is the student course schedule which consists of two files—student and course schedule file—as illustrated in Figure 11.8. Notice that the student identification number has been inserted as an attribute in the course schedule file, thereby allowing the information system to associate records in this file with a particular student.

The second type of attribute is the **descriptor.** The vast majority of all attributes are of this type. These attributes basically reveal more information about the particular

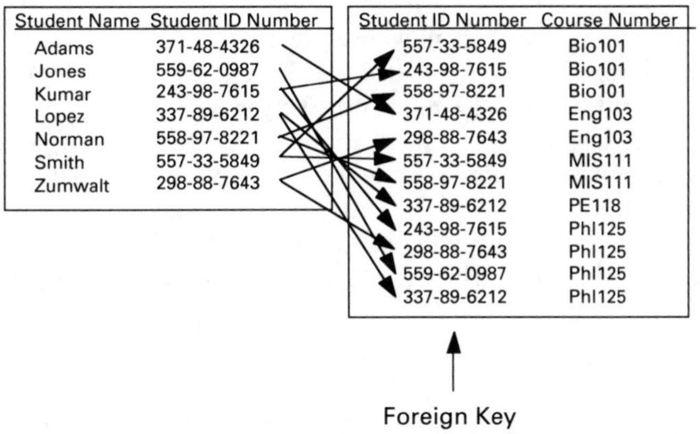

Student Name	Student ID Number		Student ID Number	Course Number
Adams	371-48-4326		557-33-5849	Bio101
Jones	559-62-0987		243-98-7615	Bio101
Kumar	243-98-7615		558-97-8221	Bio101
Lopez	337-89-6212		371-48-4326	Eng103
Norman	558-97-8221		298-88-7643	Eng103
Smith	557-33-5849		557-33-5849	MIS111
Zumwalt	298-88-7643		558-97-8221	MIS111
			337-89-6212	PE118
			243-98-7615	Phl125
			298-88-7643	Phl125
			559-62-0987	Phl125
			337-89-6212	Phl125

Foreign Key

* Note: both of these files would have additional attributes

Figure 11.8   Foreign Key Example

thing we are trying to describe. Some examples of descriptor type attributes would be Address, City, State, Zip Code, Telephone Number, Date of Birth, and Year Graduated from High School. Descriptor attributes can also be primary or secondary key attributes; however, when used as keys they are most often used as secondary keys due to the potential for duplication of attribute values throughout the file.

The third attribute classification is the **audit and control** attribute. As the name implies, this type of attribute is included as a part of the information system in order to either assist with the auditability of the information system's data or the control aspects of the information system, respectively. Often an internal auditor employee is part of the development team in business information systems development. The internal auditor's role in the development team is to ensure that the information system, when operational, will be more easily auditable by the auditing group within the business. Audit attributes are included in selected records and files to help accomplish the necessary auditing functions.

The control attributes are strategically placed within selected records and files of the information system by the systems analyst or programmer so that the system can meet its functional responsibilities. Historically, a control attribute has been referred to as a **flag** attribute. Flag attributes are usually not part of the original data structures from analysis, but are created during design to accommodate the processing necessary to perform user-required functions. For example, the sales manager at a clothing store wants a report each morning of each item sold the prior day. Recognizing that there are several ways to produce this report, the systems analyst decides to add an attribute to the inventory data structure and call it Date-Last-Sold. When an item is sold, the date is changed to be the current date. At night, the report is created by the part of the information system that scans the inventory file printing all items whose Date-Last-Sold attribute has today's date. This attribute is not part of the user's real functional requirements but is deemed necessary to produce one of the user's output requirements. Control attributes have been used in almost limitless ways in thousands of information systems over the years.

The last attribute classification is called a **security** attribute and represents attributes that are created during design to address all desired security issues within the information system. There are books written about the topic of information systems security; therefore, we will only say that the need exists to create attributes that can either authorize or restrict the use of certain aspects of the information system and its data by certain users.

File Types

Business information systems store permanent data in files to help it accomplish the functions it was created for. Relational database technology uses the word *table* as a synonym for file. Files can be classified as master, transaction, table, temporary, log, mirror, and archive, as listed in Figure 11.9. Each of the file types is discussed next.

Master File - contain the records for a group of similar
                things which collectively represent the
                foundational data component for the
                information system

Transaction File - contain the records for a group of
                     similar things which collectively
                     represent the business activity data
                     component for the information system

Table File - a type of Master File that is relatively
               unchanging over time and usually has a
               few data elements

Temporary File - created and used briefly or over an
                   extended period of time to help the
                   information system accomplish its
                   intended purposes

Log File - contains copies of Master and Transaction file
              records for audit, statistical, and recovery
              purposes

Mirror File - an exact copy of one of the other types of
               files used to minimize or eliminate information
               system downtime

Archive File - a historical copy of a master, transaction,
               table, or log file

Figure 11.9   List of the Different Types of Files

A **master file** contains the records for a group of similar things which collectively represent the foundational data component for the information system. An information system will have one or more master files. For example, a student course registration information system would have a student master, course master, and course schedule master file. A payroll information system would have an employee master file. A bank ATM information system would have a bank account master file. Finally, a seat reservation information system would have a seat availability master file, and so forth. A way of thinking about master files is that they are the core data component of the information system; without them the system simply cannot function. Figure 11.10a illustrates a master file.

A **transaction file** contains the records for a group of similar things which collectively represent the business activity data component for the information system. An information system will have one or more transaction files. For example, a student course registration information system would have a student course schedule transaction file which would contain all of the courses that the students have signed up for. A payroll information system would have a time card transaction file which would represent the hours worked for a time period for the employees. A bank ATM information system would have an ATM card usage transaction file which would represent the activity of the ATM cards used. Finally, a seat reservation information system would

a) Student Master File

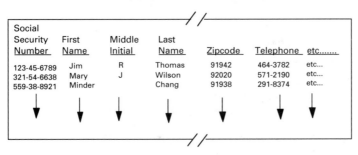

Social Security Number	First Name	Middle Initial	Last Name	Zipcode	Telephone	etc........
123-45-6789	Jim	R	Thomas	91942	464-3782	etc...
321-54-6638	Mary	J	Wilson	92020	571-2190	etc...
559-38-8921	Minder		Chang	91938	291-8374	etc...

b) Course Registration Transaction File

Course Serial #	Course #	Section #	Student #	Semester	Transaction Date/Time
10294	Eng100	5	559680843	Spr95	941115/1202
29832	MIS111	2	525987391	Spr95	941115/1202
42198	Act102	2	371234959	Spr95	941115/1202
17620	Soc118	1	559680843	Spr95	941115/1203
10294	Eng100	5	224942874	Spr95	941115/1203
28734	PhE119	3	104873298	Spr95	941115/1203
44398	Chm107	2	525987391	Spr95	941115/1204

Figure 11.10   Master and Transaction File Examples

have a seat reservation transaction file representing all the seats that have been reserved so far. A way of thinking about transaction files is that they represent the business activity that takes place on a moment-by-moment basis within the information
system. Figure 11.10b illustrates a transaction file.

A **table file** is a type of master file with the following characteristics. First, a
table file is virtually unchanging for long periods of time, and second, this type of file
has a tabular quality to it—several rows with just a few columns. For example, a U.S.
state code table file would contain rows of the two-character state codes in column 1
and the long name for the state as the second column. A U.S. state sales tax table
would have rows representing the dollar range as column 1, and the applicable sales
tax as column 2. Figure 11.11 illustrates these two table files.

Relational database technology refers to all files as *tables,* which can cause a bit
of confusion. For our purposes, table files are discussed here more for historical completeness and should not be confused with the relational database technology equiva-
lence of tables and files.

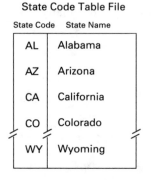

Figure 11.11   Table File Examples

A **temporary file** is just that—temporary. It can be created and used anytime an information system needs to do so. Temporary files can exist for a brief amount of time, such as a few seconds, or as long as several days. Why would an information system need to use temporary files? Well, there are all kinds of reasons for doing so. For example, data in the master or transaction files are often not in the necessary order (sort) that is needed for a report or display, so these files or parts of these files are re-arranged temporarily in order to produce the required output. Another example for using temporary files is when the information system's processing situation requires several steps before concluding with some output. The elapsed time to perform all of the steps may be very brief, such as a few seconds, or take several days. So temporary files are created, used, and deleted along the way in order to assist with the necessary processing. Temporary files can be used anytime the systems analyst believes they will help the overall required processing of data.

The next type of file, the **log file,** contains copies of master and transaction file records. A log file represents the chronological changes that are made to master files during a period of time. Often the log file contains both a before the change copy of the master file record and an after the master file change copy of the record along with the record of the transaction file change. The purposes of log files are to:

1. Allow the information system to be audited.
2. Allow the information system to be analyzed for a variety of statistical information about the processing done within the information system.
3. Provide a mechanism for recovery or reconstruction of the master and transaction files in the event of system failure.

A **mirror file** is a redundant copy of some other type of file. As its name implies, a mirror file contains the same data as its original file. A mirror file is used in information system situations where downtime cannot be tolerated. The mirror file is recorded on another media, such as a second disk drive, to help minimize the risk of downtime. In order to be effective in overcoming downtime, all files in the information system should have mirror copies on the second disk drive. Should a disk failure occur during the use of an information system containing mirror files, the information system will automatically switch to and use the other media containing the mirror files, thus no interruption in processing. A mirror file must be updated just like the original file in order for it to be an exact copy of the original. The time between updating the original file and updating the mirror file is the exposure period for the information system and should be kept as brief as possible. An airline reservation information system is the type of system that would be a good candidate for mirror files in order to minimize or eliminate downtime, since downtime could severely disrupt the airline's business.

The final file type, the **archive file,** is usually a historical version of master, transaction, table, and log files. Each information system has procedures that dictate what data and when the data in active files are to be moved to archive files and removed from the active files. For example, an accounts receivable information system may have a policy of archiving all completed purchase and payment transactions that have occurred more than 24 months ago. After moving these transaction records to an archive file, they are deleted from the active transaction file, thus reducing its physical number of records and size, and possibly allowing the accounts receivable information system to increase its processing speed, since "old" data have been removed from the active transaction file.

File Access and Organization

File access refers to the method of reading or writing records within a file, and **file organization** refers to the method of storing records within a file. There are basically only two methods for accessing records in a file, sequentially or directly, and only four ways of storing data in files—serial, sequential, relative or direct, and indexed. The needs of the user as documented in the requirements specification document should be the major determinant for a file's organization, and the file organization determines whether a file may be accessed sequentially, directly, or both.

Sequential access is access by physical record location or position within a file. It means starting at record 1 and sequentially reading or writing every record to the end of the file or some other stopping point within the file's records. Sequential access can be illustrated using a shopping mall situation. A shopper must enter the mall and then sequentially pass by all the stores between the mall's entrance and the store that the shopper wants to shop in. Sequential access is appropriate in today's information systems when there is a need to do something with the majority of the records in the file, such as updating all of them, doing a summary calculation using data in each record, and so forth.

Direct access is access by physical address. Direct access, also referred to as random, is most often used in on-line, interactive information systems because of the ability of the information system to go directly to certain records within files without having to sequentially scan from the first record in the file as with sequential access. An electronic shopping mall, such as those found on the Information Superhighway, can have its stores accessed directly without having to "pass" by all the other stores that come before the one you want to electronically shop in.

In order to use the direct access method, the information system must know some identifying piece of data such as an account number, student identification number, part number, product UPC code, student last name, and so on. Once that identifying piece of data is known to the information system, it should be able to go directly to the physical address in the file that contains these data.

Serial file organization is the storing of records in a file in chronological order. This is analogous to a first-in queue commonly associated with voice mail systems. For example, someone calling Microsoft's software technical support hotline will get placed on hold with a message stating, "your call will be answered in the order in which it was received." A log file is a good example of one that would store records in chronological order as shown in Figure 11.12. One final example of a serial file would be the recording of a bank's ATM transactions as they occur.

E-Mail InBox File

From	Date	Time	Subject
Dean	11/28/95	09:12	New Enroll
President	11/28/95	11:55	Discrim. Policy
JSmith	12/01/95	10:16	Grade in Class
MChen	12/01/95	15:43	Research Paper
Dean	12/01/95	16:28	Faculty Mtg.
KHaddad	12/02/95	07:48	Personnel Mtg.

Figure 11.12   Serial File Organization

Sequential file organization is the sequential storing of records in a file in some logical order. One or more of the record's data elements or attributes are identified as being the appropriate logical order, such as a person's name. This is similar to the previous discussion of key data elements. Once this logical order is determined, records are sorted or ordered in ascending or descending order and then stored in the file, as illustrated in Figure 11.13. When new records need to be added, they must be inserted between existing records that have lower and higher logical values than the new records. This insertion of new records is done by copying the entire file to a new file and inserting the new record at the appropriate record position during the copy procedure, as shown in Figure 11.14. Remember, the records in a sequential file must always be maintained in some logical order.

Relative or direct file organization is the direct storing of records in a file in a specific physical record location (address) based on a hashing algorithm that uses the record's key data element(s) as the input to the hashing algorithm, as illustrated in Figure 11.15. There are books written on the topic of relative file organization, and there are dozens of hashing algorithms from which to choose. Our purpose here is basic exposure to the technique using a very simplistic hashing algorithm.

When a record location is determined by inputing the key data element(s) value to the algorithm, an attempt is made to write and store the record at this address; however, if another nonempty record already exists at this location, then some procedure must be followed to move one of the two records to a new address while not losing the ability to find both of these records later during a read operation. A **collision** occurs when two or more records' key data element values yield the same record address after being input to the hashing algorithm and must be resolved in some way. A simple way of doing this is to create an audit and control type data element and store the address of the moved record in it, as illustrated in Figure 11.16.

When the need arises to find a record after it has been stored in the file, the same hashing algorithm, key data element values, and collision procedure must be used in order to find it again.

a) File ordered by Student ID Number

Student ID Number	Student Name
102-58-9762	Smith, Fred
204-78-7652	Baker, Jane
371-48-4133	Haddad, Kamal
450-22-9611	Chang, Minder
557-38-9120	Rice, Jerry
558-56-6749	Montana, Joe

b) File ordered by Student (Last) Name

Student ID Number	Student Name
204-78-7652	Baker, Jane
450-22-9611	Chang, Minder
371-48-4133	Haddad, Kamal
558-56-6749	Montana, Joe
557-38-9120	Rice, Jerry
102-58-9762	Smith, Fred

Figure 11.13   Sequential File Organization

The final file organization method, **indexed,** functions similarly to a book's index. The index for a book is sorted in alphabetical order of topics. Associated with each topic listed in the index is one or more page numbers where this topic is discussed. The book index actually increases the total pages in the book, but the utility of the index makes these additional pages worthwhile.

Similarly, an indexed file organization uses and maintains a sorted index in key order, which stores the record location of the actual data file record as part of the index. As with relative file organization, indexed file organization is also the subject of many books so we will simply illustrate the concept here for basic familiarity.

The index file organization usually requires two files, the index file and the actual data record file. However, it is possible to combine the two into one physical file. The index file is maintained in key order using one of several strategies, while the actual data file is probably created and maintained in serial data file organization; that is, new records are simply added to the end of the file while the index file is updated and maintained in key order. Figure 11.17 illustrates a conceptual view of index file organization.

Student Master File ordered by Student ID Number

Student ID Number	Student Name
102-58-9762	Smith, Fred
204-78-7652	Baker, Jane
371-48-4133	Haddad, Kamal
450-22-9611	Chang, Minder
557-38-9120	Rice, Jerry
558-56-6749	Montana, Joe

Insert new students:

298-73-0912	Jackson, Janet
557-93-8247	Carey, Mariah

NEW Student Master File ordered by Student ID Number

Student ID Number	Student Name
102-58-9762	Smith, Fred
204-78-7652	Baker, Jane
298-73-0912	Jackson, Janet
371-48-4133	Haddad, Kamal
450-22-9611	Chang, Minder
557-38-9120	Rice, Jerry
557-93-8247	Carey, Mariah
558-56-6749	Montana, Joe

Figure 11.14   Insertion of New Record in Sequential File

Normalization

Normalization is the process of simplifying complex data structures so that the resulting data structures will be more easily maintained and more flexible to meet present and future needs of the user. Normalization focuses its attention on the data elements or attributes of the data structure. The term *normalization* became popular with relational databases in the 1970s, but the concept has been practiced in business information systems even before this time. Normalization research has identified seven consecutive levels of normalization—first, second, third, fourth, Boyce-Codd, fifth, and domain-key. Each of these is called a normal form, such as first normal form, Boyce-Codd normal form, and so on. Even though there are seven normal form levels, data structures in most business information systems are only normalized to the third normal form for optimization reasons. Therefore, we will only discuss the first three as well.

Each normal form has associated with it a rule for simplifying the data structure. Worst case for a data structure is to not be in any normal form at all, called **unnor-**

Student Master File using Direct Organization

Pre-allocate
the maximum
number of
student
records the
file can hold
(12 in this
example)

Student # Name Address Phone GPA..etc..
1
2
3
4
5
6
7
8
9
10
11
12

Divide each
Student ID #
by 12 and
add 1 to the
remainder to
get the "home"
address record
for each student

Smith, Fred   102-58-9762   / 12 =10 + 1 = 11

Baker, Jane   204-78-7652   / 12 =8 + 1 = 9

Haddad, Kamal   371-48-4133   / 12 =1 + 1 = 2

Chang, Minder   450-22-9611   / 12 =3 + 1 = 4

Rice, Jerry   557-38-9120   / 12 =4 + 1 = 5

Montana, Joe   558-56-6749   / 12 =1 + 1 = 2 (duplicate)

Locate each
student record
in its "home"
address within
the Student
File.
All is well
until 2 or more
student records
have the same
"home" address.
See next figure
for solution to
this problem.

Student # Name Address Phone GPA..etc..
1
2 371-48-4133 Haddad, Kamal
3
4 450-22-9611 Chang, Minder
5 557-38-9120 Rice, Jerry
6
7
8
9 204-78-7652 Baker, Jane
10
11 102-58-9762 Smith, Fred
12

Figure 11.15   Direct File Organization
Example

malized. In other words, the data structure does not conform to the rule associated with first normal form even if the data structure by chance happens to conform with one or more higher-level normal form rules. A data structure is said to be in "x" normal form if and only if the data structure conforms to the rule associated with "x" normal form and all lower-level normal form rules. Figure 11.18 illustrates the normalization process for the first three normal forms.

Each conceptual file or data structure identified during analysis and design should pass the normalization test for ensuring high-quality information system database design. We would like to proceed with an example of normalization by starting with a user-required input document, ABC Incorporated's Sales Order Form, Figure 11.19, that is in a hard-copy paper form here but could also be designed to be a video display screen input form as well. This Sales Order Form represents a complex data structure because there are data elements on the form that:

1. Relate to the sales order, such as order number, order date, order total, tax, shipping, and grand total.

Montana, Joe   558-56-6749   / 12 =1 + 1 = 2 (duplicate)

"Home" address (2) for Joe Montana is already in use.

So, find next available (vacant) record and put his record in that address (3).

Then, go back to "home" address (2) and put the record number for Joe in the Pointer attribute.

	Student # Name Address Phone GPA..etc..	Pointer
1		
2	371-48-4133 Haddad, Kamal	3
3	558-56-6749 Montana, Joe	
4	450-22-9611 Chang, Minder	
5	557-38-9120 Rice, Jerry	
6		
7		
8		
9	204-78-7652 Baker, Jane	
10		
11	102-58-9762 Smith, Fred	
12		

When trying to find "Joe":

Algorithm will yield 2 as the 'home' address.

You compare the student ID number in 2 and it is not Joe's.

You then follow the Pointer address to 3 and the student ID number in record 3 is the same as Joe's....so you found him.

Figure 11.16   Direct File Organization Collision Resolution

2. Relate to the customer, such as customer number, name, address, city, state, and zip code.

3. Relate to the products being ordered, such as product number, name, color, unit price, quantity, and total price.

Using an object-oriented methodology during analysis, a SalesOrder class could have been created to represent this complex data structure, as shown in Figure 11.20, which has been expanded to show the attribute section. If not already done, now is the time to use normalization on this class to simplify it. In doing so, we will no doubt create some new or repetitious class symbols. The ones created through the normalization process could be repetitious because the same or similar class may have already been defined as part of the overall information system model.

Please direct your attention to a few of the attributes in Figure 11.20. First, note the attribute called OrderNumber. This attribute has been designated as the primary key for the class. Remember, primary keys must have values that are not duplicated in any other object within the class. Sales order numbers are usually consecutively numbered, so duplication should not occur. In other words, there will only be one instance of an object called SalesOrder that has order number 134678 or any other number for that matter. Second, note the group of attributes indented under the words *For each*

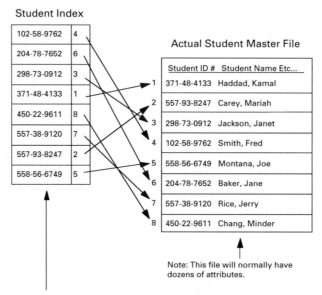

Student Index

102-58-9762	4
204-78-7652	6
298-73-0912	3
371-48-4133	1
450-22-9611	8
557-38-9120	7
557-93-8247	2
558-56-6749	5

Actual Student Master File

	Student ID # Student Name Etc...
1	371-48-4133 Haddad, Kamal
2	557-93-8247 Carey, Mariah
3	298-73-0912 Jackson, Janet
4	102-58-9762 Smith, Fred
5	558-56-6749 Montana, Joe
6	204-78-7652 Baker, Jane
7	557-38-9120 Rice, Jerry
8	450-22-9611 Chang, Minder

Note: This file will normally have
dozens of attributes.

Search Student Index File to find Student ID Number.

Get Pointer Value and access that record in Student Master File to
find the actual student record.

Figure 11.17   Conceptual View of Index File Organization

product ordered (up to 7). This group of up to seven different products being ordered is considered a repeating group and will be discussed in more detail later. Looking back to Figure 11.19, you will note that a sales order has room for up to seven products. Third, in Figure 11.20 notice the attributes that have the word *derived* to the right of them in parentheses. All five of these attributes can be calculated on demand; therefore, they should not really be a part of the attribute lists in the final normalized version of this class. It is a good idea, however, to include them during analysis and design for completeness purposes along with the comment about being derived.

To put the Sales Order Form complex data structure into **first normal form,** we must remove all repeating groups from the structure, creating a new structure for each distinct repeating group that may exist. In object-oriented notation, this means creating a new class for each repeating group. Figure 11.21 shows the results of applying rule 1 to the original Sales Order Form. There was one repeating group, products ordered, which is extracted to become its own ProductsOrdered class with associated attributes. Now both of these classes are in first normal form and appear as in Figure 11.21.

Looking at the ProductsOrdered class's attribute list in Figure 11.21, you will notice that in order to create simpler, normalized, data structures, the OrderNumber attribute has been duplicated in this class, adding a small amount of redundancy to the model. Remember the earlier discussion about balancing simplicity and redundancy? It is necessary to include the OrderNumber attribute as a foreign key in the ProductsOrdered class because it is now physically separate from the SalesOrder class, yet there

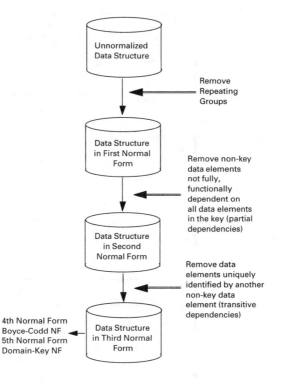

Figure 11.18   The Normalization Process

is a definite need to be able to relate products ordered to the sales order that they were ordered on. The OrderNumber attribute is that link between the two classes. This is no different than you taking your soiled clothes to the cleaners or your broken VCR in to get repaired. The store attendant will ordinarily give you a tear-off claim check for the item(s) which has an identifying claim check number on it. The portion of the claim check that stays with the clothes or VCR has a matching claim check number on it. Hence, the claim check number is what allows you to be matched up with your clean clothes or repaired VCR when you return several days later.

The rule for **second normal form** says to remove all nonkey data elements that are not fully, functionally dependent on all data elements that make up the key. In other words, this rule deals with partial dependencies that may exist in the attribute list of each class being normalized. This rule only applies to data structures already in first normal form, and more specifically only to those data structures that have two or more data elements or attributes that make up the primary key for the data structure.

It's at this point that many normalization novices start to get confused, so let's fill in an actual Sales Order Form with some sample data as in Figure 11.22. Now we will transfer these sample data into records (objects) within the first normal form data structures from Figure 11.21, giving Figure 11.23. Notice that there is only one record (object) in the SalesOrder class, and that there are five records (objects) in the Prod-

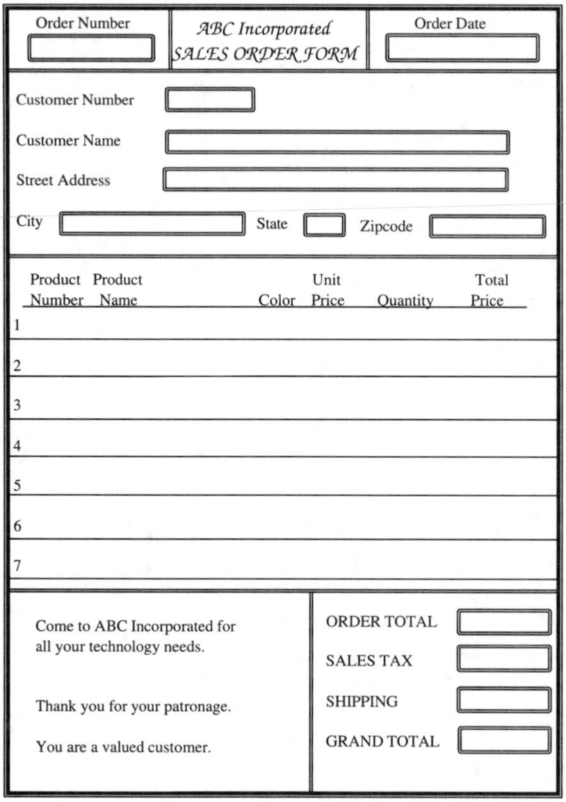

Figure 11.19   Sales Order Form (Unnormalized)

uctsOrdered class each having the OrderNumber associated with it in order to be able to match each one with the data in the SalesOrder class (just like we did with the claim check for your laundry or VCR repair earlier). Remember it is absolutely essential that the order number be duplicated for each product ordered, otherwise the information system will not be able to relate products ordered with the specific sales order that they were ordered on.

Again, if you are having difficulty visualizing the need for this, picture two buckets, one for SalesOrder records (objects) and one for ProductsOrdered records (objects). Take some 3x5 cards and on the first one write all the information from the attributes for SalesOrder, then take the next five cards and write the products ordered information for each of the five products on a separate card. Put the single SalesOrder 3x5 card in the first bucket and the remaining five ProductsOrdered cards in the other bucket. Now imagine the first bucket containing hundreds of these 3x5 cards, one for each sales order taken during a week's time at ABC Incorporated, and imagine the second bucket containing thousands of 3x5 cards, one for each product ordered on all of the hundreds of orders in the other bucket. How in the world can you associate the

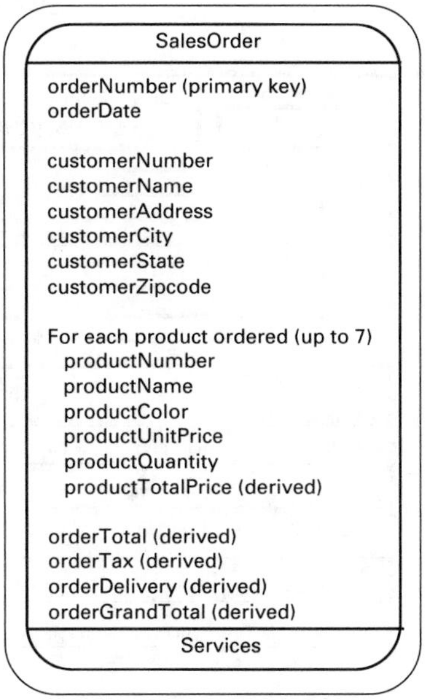

SalesOrder
orderNumber (primary key)
orderDate
customerNumber
customerName
customerAddress
customerCity
customerState
customerZipcode
For each product ordered (up to 7)
productNumber
productName
productColor
productUnitPrice
productQuantity
productTotalPrice (derived)
orderTotal (derived)
orderTax (derived)
orderDelivery (derived)
orderGrandTotal (derived)
Services

Figure 11.20 Sales Order Class

SalesOrder
orderNumber (primary key)
orderDate
customerNumber
customerName
customerAddress
customerCity
customerState
customerZipcode
orderTotal (derived)
orderTax (derived)
orderDelivery (derived)
orderGrandTotal (derived)
Services
1,7
1
ProductsOrdered
orderNumber (primary key)
productNumber (primary key)
productName
productColor
productUnitPrice
productQuantity
productTotalPrice (derived)
Services

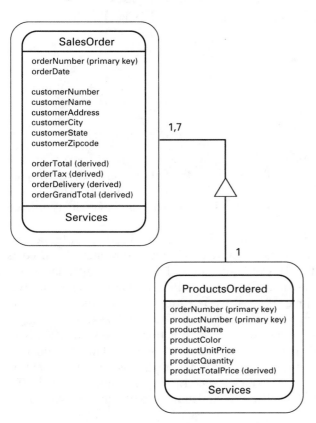

Figure 11.21 SalesOrder and ProductsOrdered Classes in First Normal Form

Order Number	ABC Incorporated	Order Date
34820	SALES ORDER FORM	12/02/95

Customer Number [534]

Customer Name [Norman Business Systems, Inc.]

Street Address [7150 University Blvd., Suite 218]

City [San Diego] State [CA] Zipcode [92108]

	Product Number	Product Name	Color	Unit Price	Quantity	Total Price
1	IC-PENT	Intel Pentium CPU	Bn	$675	1	$675
2	PS-220	220 V. Power Supply	Sl	$150	1	$150
3	KB-102	102-key Keyboard	Tn	$ 75	1	$ 75
4	MO-675	Mouse - Serial	Tn	$ 65	2	$130
5	HD-550	550 MB Hard Disk	Sl	$325	1	$325
6						
7						

Come to ABC Incorporated for all your technology needs.

Thank you for your patronage.

You are a valued customer.

ORDER TOTAL	$1,355
SALES TAX	$ 95
SHIPPING	$ 25
GRAND TOTAL	$1,475

Figure 11.22   Completed (Filled-in) Sales Order

thousands of 3x5 cards representing products in the ProductsOrdered bucket with the hundreds of 3x5 cards representing sales orders in the SalesOrder bucket? Only by looking at the OrderNumber that is on every 3x5 card in both buckets! The same is true for the information system since the systems analyst split up the products ordered from the rest of the sales order.

Now back to second normal form. Remember, we only need to apply the second normal form rule to data structures that have two or more attributes identified as the primary key. The SalesOrder class in Figure 11.21 has only one attribute as the primary key so it is now in second normal form. The ProductsOrdered class has two attributes designated as the primary key so we need to apply the rule to this data structure. We do so by examining all of the nonkey attributes and asking the following question for each one, "Is ProductName fully, functionally dependent on both OrderNumber and ProductNumber or is it dependent only on OrderNumber or ProductNumber but not both?" It turns out that ProductName is only dependent on ProductNumber. In other words, if you know a specific ProductNumber, such as IC-PENT, then you can look up this specific ProductNumber to obtain the name of the

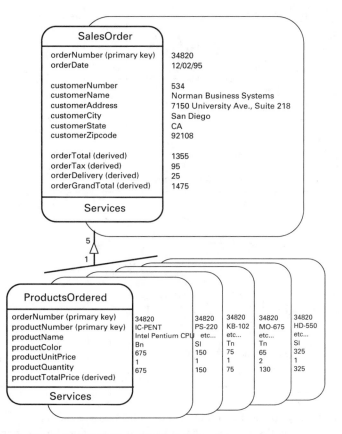

Figure 11.23   Sample Objects for SalesOrder and ProductsOrdered

product, Intel Pentium Microprocessor Chip. The product's name is the same regardless of what the OrderNumber is. In fact, the OrderNumber could be anything, but anytime someone orders product IC-PENT he or she gets the same product—the Intel Pentium Microprocessor Chip. Look at Figure 11.24, which is an expanded version of Figure 11.21 incorporating a few more sales orders and products. Note the same product numbers on several different sales orders.

The same question should be asked for each of the remaining nonkey attributes. When doing so, you find that ProductColor is the only other attribute that is also not fully, functionally dependent on the full primary key. So a third class is created calling it Product. The Product class has attributes ProductNumber, ProductName, and ProductColor for now. As the second normal form rule question was asked for the attribute ProductUnitPrice, it was discovered that this attribute really belongs with ProductNumber in the Product class but it also is fully, functionally dependent upon the full key because the user had specified that product unit prices can change every few weeks, and the user only wishes to retain the most current unit price for the product. Since this situation is true, then the information system needs to retain the Product-

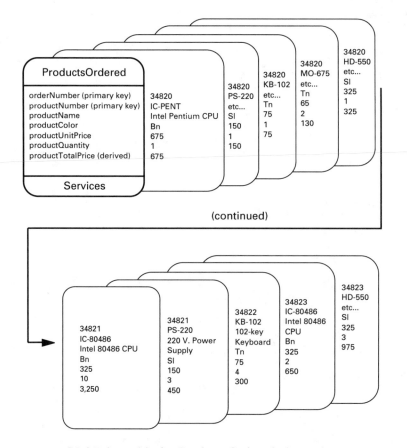

ProductsOrdered						
orderNumber (primary key)	34820	34820	34820	34820	34820	34820
productNumber (primary key)	IC-PENT	PS-220	KB-102	MO-675	HD-550	etc...
productName	Intel Pentium CPU	etc...	etc...	etc...	etc...	SI
productColor	Bn	SI	Tn	Tn	SI	325
productUnitPrice	675	150	75	65	325	1
productQuantity	1	1	1	2	1	325
productTotalPrice (derived)	675	150	75	130	325	

Services

(continued)

34821	34821	34822	34823	34823
IC-80486	PS-220	KB-102	IC-80486	HD-550
Intel 80486 CPU	220 V. Power	102-key	Intel 80486	etc...
Bn	Supply	Keyboard	CPU	SI
325	SI	Tn	Bn	325
10	150	75	325	3
3,250	3	4	2	975
	450	300	650	

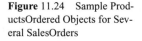

Figure 11.24   Sample ProductsOrdered Objects for Several SalesOrders

UnitPrice with the ProductsOrdered class, since our user needs a "snapshot" of the unit price for the product at the time of the order. The information system ends up with ProductUnitPrice being duplicated as an attribute in both the ProductsOrdered and Product class. This is okay since the value for any product, say the IC-PENT product, will change as we move through time, but when this product was ordered on a sales order, the information system has a permanent record of the unit price that ABC Incorporated sold this product for at that time.

To summarize for second normal form, after applying the rule for second normal form, the information system ends up with three data structures (classes) as illustrated in Figure 11.25, each having some sample data that could look like those in Figure 11.26.

The final normal form rule discussed here is for **third normal form,** and it states that data elements (attributes) that are uniquely identified by another nonkey data element (attribute) need to be removed. In other words, this rule is dealing with the notion of transitive dependencies. The third normal form rule is applied to all data structures (classes) normalized through second normal form. In the Sales Order Form example this includes SalesOrder, ProductsOrdered, and Product. Once again, the systems analyst considers every nonkey attribute and asks the following question, "Is this attribute dependent on the primary key or is it dependent on some other nonkey attribute?" If the

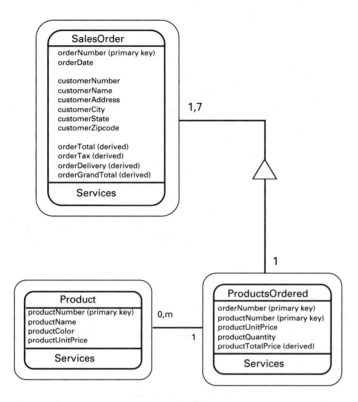

Figure 11.25 Sales Order Data Structure in Second Normal Form

systems analyst explores this question with the user, it will be found that both the ProductsOrdered and Product classes are in third normal form just as they are because all nonkey attributes are dependent on the full primary key. In examining SalesOrder, however, the systems analyst would find that CustomerName, CustomerAddress, CustomerCity, CustomerState, and CustomerZipCode are uniquely identified by another nonkey attribute—namely, CustomerNumber. This is very similar to what was discovered when considering partial dependencies with the rule for second normal form. Once again, if the information system knows a customer number, it can look up its name, address, and so on, so these attributes should be removed from this class and put in their own, as shown in Figure 11.27. Customer number must be retained (and duplicated) as a foreign key in the SalesOrder class for the same reason that OrderNumber must be retained as a foreign key in the ProductsOrdered class as discussed earlier.

After applying the third normal form rule, we end up with classes that are optimized with respect to simplicity and nonredundancy. In the Sales Order Form example, the normalized information system ends up with two master files—Customer and Product—and two transaction files—SalesOrder and ProductsOrdered. Figure 11.28 illustrates the interaction between these four classes using some sample data.

Figure 11.29 represents a template for the application of the rules for the first three levels of normal form, which may help you visualize the application of the normalization rules.

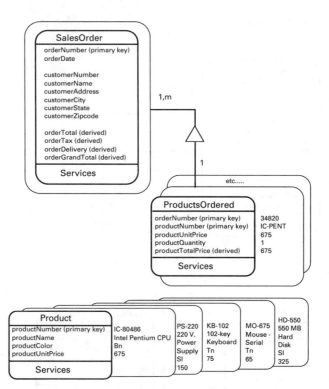

Figure 11.26   Sample Objects for Second Normal Form Sales Order

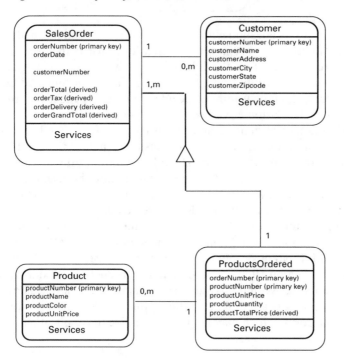

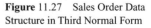

Figure 11.27   Sales Order Data Structure in Third Normal Form

Order Number	Order Date	Customer Number	OrderTotal (derived)	OrderTax (derived)	OrderDelivery (derived)	OrderGrand Total (derived)
34820	12/02/95	534	1355	95	25	1475
34821	12/02/95	871	7200	504	15	7719
34822	12/02/95	290	300	21	17	338

OrderNumber	ProductNumber	ProductUnitPrice	ProductQuantity	ProductTotalPrice (derived)
34820	IC-PENT	675	1	675
34820	PS-220	150	1	150
34820	KB-102	75	1	75
34820	MO-675	65	2	130
34820	HD-550	325	1	325
34821	IC-80486	325	10	6750
34821	PS-220	150	3	450
34822	KB-102	75	4	300

ProductNumber	ProductName	ProductColor	ProductUnitPrice
IC-PENT	Intel Pentium CPU	Bn	675
IC-80486	Intel 80486/DX4 CPU	Sl	325
HD-550	550 MB Hard Disk	Sl	325
HD-1GB	1-GB Hard Disk	Sl	550
KB-102	102-key Keyboard	Tn	75
MN-209	NEC .29 Monitor	Tn	375
MO-675	Mouse - Serial	Tn	65
PS-220	220 V. Power Supply	Sl	150

Customer Number	Customer Name	Customer Address	Customer City	Customer St	Customer Zipcode
107	Chips 'N Bits	824 E. Main Street	Pasadena	CA	92875
290	Computers 4 U	925 W. Broadway Avenue	Tucson	AZ	85721
534	Norman Business Systems	7150 University Ave., Suite 218	San Diego	CA	92108
871	Computers Unlimited	2978 So. Grand Avenue	Lansing	MI	48286

Figure 11.28 Sample Data or Sales Order in Third Normal Form

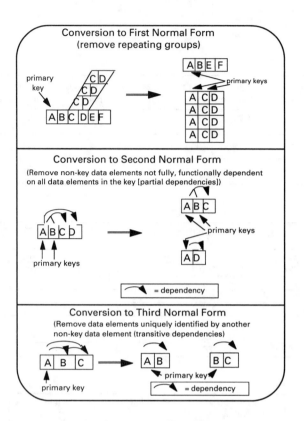

Figure 11.29   Normal Form Rules Template

- Database technology had gained a better understanding of transactions, recovery, memory management, indexing schemes that makes designing the architecture for any database system less tedious.

- OODB supported applications that no database system was well suited for at that time.

- Object-oriented languages were rapidly gaining acceptance, and OODB proved to be able to support the persistence data needs better than record-based database models.

- The majority of conceptual language-design work from object-oriented programming languages carries over easily to OODB.

- Information systems are becoming more and more complicated and sophisticated.

Figure 11.30   Reasons for the Emergence of OODB

Object-Oriented Database

An object-oriented database (OODB) is an integration of object-oriented concepts combined with traditional database capabilities. In most information systems, database functionality is needed to assure persistence and concurrent sharing of data in applications. With the assistance of an object-oriented database, an object-oriented information system can have the state of its objects persist and be updated between various program invocations. In addition, numerous users can also concurrently share the same data as in traditional database management systems. Therefore, the definition of an object-oriented database can be summarized as follows:

Object-Oriented Database = Object Orientation + Database Functionality
whereby
Object Orientation = Complex Objects + Object Identity + Encapsulation + Types of Classes + Inheritance + Overriding Combined with Late Binding + Extensibility + Computational Completeness
Database Functionality = Persistence + Concurrency + Transactions + Recovery + Querying + Versioning + Integrity + Security + Performance

An object-oriented database is an architecture that represents its major items as class objects and their interrelationships and uses an object model as its foundation. An OODB should be a representation of a network of class-objects, which in themselves are encapsulations of state, data, and processing as a single unit. Each record/tuple should represent an instance. The tuple becomes a template for describing its structure. All instances that share the same set of data items and methods are grouped within a class. Classes may be interconnected to one another as a network.

Evolution of Object-Oriented Database

Object-oriented concepts came initially from object-oriented programming (OOP), which was developed in the 1960s with the Simula language as an alternative to traditional programming methods. Little came of this work until Xerox's Smalltalk was introduced around 1980. Before OOPs, data and procedures were isolated from each other causing data to be passive and the procedures to be active. In stark contrast, in an OOP environment the programmer asks objects to perform operations on themselves.

The initial concept for OODB seemed to arrive on the coattails of Smalltalk and PS-Algol in the early 1980s. There were several reasons for the rapid emergence of OODBs as summarized in Figure 11.30 and discussed here. First, database technology had gained a better understanding of transactions, recovery, memory management, and indexing schemes that makes designing the architecture for any database system less tedious. Second, the concept of OODB supported applications that no database system was well suited for at that time. Third, object-oriented languages were rapidly gaining acceptance, and OODB proved to be able to support the persistence data need better than record-based database models, such as hierarchical, network, and relational. Fourth, the majority of conceptual language-design work from object-oriented programming languages (OOPLs) carries over easily to OODB. Fifth, information systems are becoming more and more complicated and sophisticated.

As information system complexity increases, it makes future modifications and the ability to anticipate ripple effects of small changes more difficult. The need for a new generation of database technology can be attributed to the need to reduce the cost of developing and maintaining the newer applications. Some database researchers and vendors believe that this situation directs the need for a fundamental change in database technology. They believe a paradigm shift is required rather than an incremental extension of the capabilities of existing database management systems.

There are two reasons why the object-oriented paradigm appears to be a sound basis for a new generation of database technology. First, the object-oriented paradigm can be the basis for a data model that embodies the relational data model. The object-oriented data model is designed to handle arbitrary, user-defined data types. Thus, the new feature will overcome the relational model's perceived limitation of only being able to handle limited types of data, such as integers, floating-point numbers, and strings.

Furthermore, an object-oriented data model allows not only the representation of data, as the relational data model does, but also the encapsulation of data and programs. Through the notions of encapsulation and inheritance, the object-oriented data model is designed to reduce the difficulty of developing and evolving complex information systems or their design models.

Nonetheless, before object-oriented technology can deliver technology advances to the database market, new database systems that incorporate an object-oriented data model must also support an SQL-compatible database language. SQL-compatibility is a key to the adoption of new database systems that incorporate an object-oriented data model. The importance of having SQL-compatibility is to help minimize the training time necessary for current relational users and will facilitate the migration of relational applications to the new systems.

The object model's ability to support complexity attempts to relax the perceived restrictions of the relational model but at the same time tries to retain the strong theoretical foundation of the relational model. In addition to inheritance and object identity, OODB also incorporates abstract data typing, another fundamental concept of object orientation. As shown in Figure 11.31, OODB's concepts were derived from traditional database systems, semantic data models, and object-oriented programming.

Characteristics of an Object-Oriented Data Model

Object-oriented data model standards are emerging and the Object Management Group (OMG) consortium of businesses is playing an important role in the OODB standards arena. There appears to be a minimum set of characteristics that a data model must possess before it can be considered an object data model as shown in Figure 11.32. These include:

1. Support the representation of complex objects.
2. Extensibility; allow the definition of new data types as well as operations that act on them.

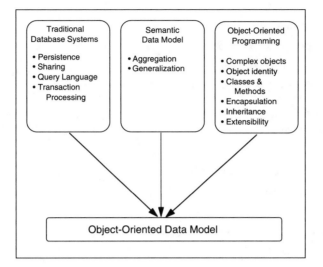

Figure 11.31   Object-Oriented Data Model

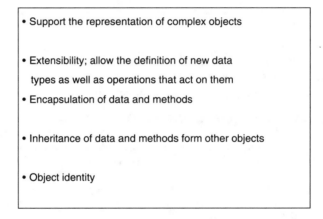

Figure 11.32   Common Characteristics of an Object Data Model

3. Encapsulation of data and methods.

4. Inheritance of data and methods from other objects.

5. Object identity.

The first comprehensive attempt to define a list of required object-oriented database management systems (OODBMS) features was published as "The Object-Oriented Database System Manifesto" in 1989 as part of the proceedings for the First International Conference in Deductive and Object-Oriented Databases in Kyoto, Japan. The "manifesto" illustrates thirteen mandatory features and other optional

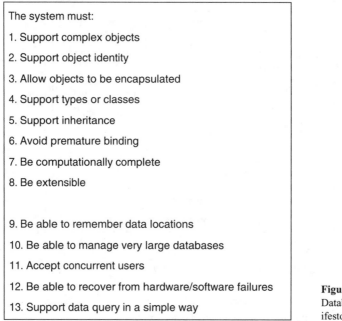

The system must:

1. Support complex objects

2. Support object identity

3. Allow objects to be encapsulated

4. Support types or classes

5. Support inheritance

6. Avoid premature binding

7. Be computationally complete

8. Be extensible

9. Be able to remember data locations

10. Be able to manage very large databases

11. Accept concurrent users

12. Be able to recover from hardware/software failures

13. Support data query in a simple way

Figure 11.33   The Object-Oriented Database Management System Manifesto Rules

characteristics of OODBMS. The thirteen rules are listed in Figure 11.33. The first eight are characteristics of an object-oriented system and the remaining five are characteristic of a traditional database management system.

Strengths of an Object-Oriented Database

There are several strengths that are associated with object-oriented databases. Figure 11.34 lists them and each is discussed here.

1. Data modeling. The object-oriented data model provides capabilities to represent object identity, abstract data, and object relationships. It gives users flexibility and extensibility in handling complex data. It also supports object-oriented analysis and design models very well.

2. Heterogenous data. The relational data model puts heterogenous tuples in separate table relations. An object-oriented database groups similar objects as a class. However, the data structure of a class is not flat. It can contain multivalued attributes that break the first normal form rule of the relational data model. Since an object-oriented database stores classes in the form of a hierarchy, a subclass can extract its characteristics from its superclass.

3. Variable-length and long strings. In an object-oriented database, variable-length strings can be defined as a new data type (class). A string can be accessed through its object identifier or its contents. A special string search can be defined as a method of its class. Very long strings can be indexed using schemes such as B-tree.

Strengths

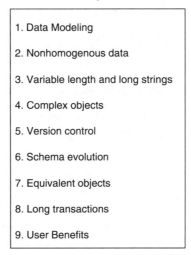

1. Data Modeling

2. Nonhomogenous data

3. Variable length and long strings

4. Complex objects

5. Version control

6. Schema evolution

7. Equivalent objects

8. Long transactions

9. User Benefits

Weaknesses

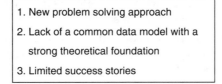

1. New problem solving approach

2. Lack of a common data model with a

   strong theoretical foundation

3. Limited success stories

Figure 11.34   Strengths and Weaknesses of an OODB

Moreover, to save memory space, the system allows retrieval of parts of a long string rather than retrieval of the whole object.

4. Complex objects. A complex object can be retrieved as a single unit. Therefore, users do not have to collect tuples from different relations to reconstruct the complete object.

5. Version control. Object-oriented database controls versions by using the previous-version and next-version attribute.

6. Schema evolution. Conventional DBMSs do not support efficient mechanisms to support schema evolution mainly because most business applications do not require frequent schema changes. On the other hand, the object-oriented database makes schema evolution easier by allowing user-defined operations. The OODB schema evolution includes changes to the definition inside a class, thus changing the structure of the class lattice.

7. Equivalent objects. OODB uses generic objects to represent the semantics of equivalent objects explicitly. A generic object contains attributes that identify different representations of the same object. Constraints can be added to the definition of the generic object so that modifications in one representation are reflected in others.

8. Long transactions. OODB supports the notion of long transactions which could span several seconds, minutes, or hours in order to accommodate concurrent access and recovery.

9. User benefits. Object-oriented programming discipline promises significant benefits to developers as well as end users.

Weaknesses of an Object-Oriented Database

OODB is not without its shortcomings either. Figure 11.34 lists a few of them and each is discussed in more detail here.

1. New problem-solving approach. The hardest thing for a systems analyst and a programmer to learn when developing an object-oriented system is to identify objects. There are a lot of techniques but it requires months of practice, experience, and tutoring before one is able to readily recognize objects in a system.

2. Lack of a common data model with a strong theoretical foundation. OODB technology is at a very young stage. It still lacks a strong theoretical foundation when compared to the relational databases where theoretical research has given a clear specification of a relational database system in terms of the data model and query language. There is no consensus on a single object-oriented data model at this time. However, industry standards are emerging.

3. Limited success stories. Currently, a systems analyst has to not only develop the information system but has to also learn a significant amount about object technology and its problem-solving strategy. This makes systems development more difficult during this transition period.

The object-oriented database is being challenged by an extended relational database model which is capable of supporting much of what the OODB is capable of supporting. Only time will tell which, if either, is the overwhelming winner in the marketplace. Since the relational database is the dominant incumbent in most information systems development organizations, it should have the inside track in the OODB community.

SUMMARY

This chapter has focused on persistent data, data that must be stored on a more permanent basis in files and databases. The chapter started out by defining and discussing the hierarchy of components that are the building blocks for persistent data—bits, bytes, attributes, records, files, databases, and folders. Next the chapter described the objectives for creating optimal data structures—simplicity and nonredundancy. This discussion was followed by a discussion of the different types of attributes—key, descriptor, audit and control, and security.

Seven different types of files—master, transaction, table, temporary, log, mirror, and archive—were defined, discussed, and illustrated with an example. After this, file access types—sequential and direct—were defined and discussed, followed by the different file organization strategies—serial, sequential, direct, and indexed. A discussion of the first three levels of normalization for relational databases was presented. Finally, an overview of an object-oriented database was presented covering its history, characteristics, strengths, and weaknesses.

QUESTIONS

11.1 Describe what is meant by the term *persistent data.*

11.2 Discuss what distinguishes a bit, a byte, an attribute, and a record. Be specific.

11.3 Why is it so important that a data structure be simple?

11.4 What is a key attribute, and what is its purpose in file and database design?

11.5 What is a master file? Discuss its importance within a database environment.

11.6 What purpose does a temporary file have in a database environment?

11.7 Discuss the purposes of a log file.

11.8 What distinguishes sequential from direct access?

11.9 What is the role of normalization in file and database design?

11.10 Briefly discuss the first three normal form levels.

11.11 Briefly discuss the history of OODB.

11.12 Discuss the characteristics of an OODB.

11.13 List the thirteen rules found in the OODB "manifesto."

11.14 Discuss several strengths and weaknesses of OODB.

REFERENCES

CATTELL, R., *Object Data Management: Object Oriented and Extended Relational Database Systems.* Reading, MA: Addison-Wesley, 1991.

DAVIS, WILLIAM S., *Business Systems Analysis and Design.* Belmont, CA: Wadsworth Publishing Company, 1994.

EDELSTEIN, H., "Relational vs. Object Oriented," *DBMS,* 14, no. 12 (November 1991), 68–78.

KENDALL, K.E., and KENDALL, J.E., *Systems Analysis and Design* (2nd ed.). Englewood Cliffs, NJ: Prentice Hall, 1992.

KHOSHAFIAN, S., *Object-Oriented Databases.* New York: Wiley, 1993.

LOOMIS, M.E.S., "OODBMS: The Basics," *J. Object-Oriented Programming,* 3, no. 1 (1990), 79–81.

LOOMIS, M.E.S., "OODBMS vs. Relational," *J. Object-Oriented Programming,* 3, no. 2 (1990), 79–82.

LOOMIS, M.E.S., A.V. SHAH, and I.E. RUMBAUGH, "An object modeling technique for conceptual design," *Procs. ECOOP '87.* New York: Springer, 1987, pp. 192–202.

ROB, P., and C. CORONEL, *Database Systems* (2nd ed.). New York: Boyd & Fraser, 1995.

WHITTEN, J.L., L.D. BENTLEY, and V.M. BARLOW, *Systems Analysis & Design Methods* (3rd ed.). Boston: Irwin, 1994.

12

Software Construction and Testing

CHAPTER OBJECTIVES (YOU SHOULD BE ABLE TO)

1. Describe the general software design principles.
2. Describe the general software construction framework.
3. Describe the object-oriented software construction framework.
4. Name and describe the software construction strategies.
5. Name and describe the levels of cohesion and coupling.
6. Describe the general principles for software testing.
7. Name and describe the software testing strategies.
8. Name and describe the generic software testing methodology.
9. Briefly describe the purpose of application and code generators.

INTRODUCTION

In his 1979 book, **Structured Design,** Ed Yourdon tells the following tale about an information systems development manager:

> The boss dashes in the door and shouts to the assembled staff: "Quick, quick! We've just been given the assignment to develop an on-line order entry system by next month! Charlie, you run upstairs and try to find out what they want the system to do—and in the meantime, the rest of you people start coding or we'll never get finished on time!"

Today, most men and women in the information systems development (software engineering) profession get a good laugh out of this brief tale, but the sad part about it is that this type of scenario was painfully true of most information systems development practices in the 1970s. The boss really was under a significant amount of pressure to deliver information systems as quickly as possible, and they had the misguided belief that programmers were not productive unless they were coding.

If that same tale were to be told about today's software engineering practices, in the mid- to late 1990s, it might be jokingly told like this:

> Ms. Smith, the software engineering manager, dashes in the door and shouts to the assembled team of culturally diverse and gender balanced staff members: "Quick, quick! We've just been given the assignment to develop a GUI-based, client-server order entry system by next week using our state-of-the-art integrated CASE tools! Charlie, you run upstairs and try to find out what our customer wants the system to do—and in the meantime, the rest of you wonderful staff people start coding or we'll never get finished on time!"

The players have changed, the technology has changed, the development tools have changed, but the development mindset has not! Almost 20 years later, software engineers and their managers still have a "shoot from the hip" development mentality worldwide! In all fairness and sincerity, this scenario, although still prevalent, is not as common as it was in the 1970s so there has been some amount of progress. It has literally taken a generation to overcome the dominant mindset of the 1970s when it comes to creating software. It would be one thing if the problem existed only at the managerial level, but the sad fact is that many software engineers prefer to develop software "from the hip" and view their work as a creative effort rather than an engineering discipline.

The quality and variety of information systems being built today were, at best, only dreams 20 years ago. In the 1970s, for example, most businesses were either using all IBM, Digital, Unisys (Burroughs and Univac back then), or some other vendor-dominated hardware and associated software. Today that scenario is quite different. Most large businesses have a heterogeneous hardware and software environment. Their user workstations are a combination of dumb terminals, IBM PC or clones, Macintosh, or Sun or other RISC-based workstations. The operating systems utilized in these heterogeneous environments include, but are not limited to, Unix (several variations), Windows (several variations), XWindows, Motif, OS/2, and so on. Virtually all of these heterogeneous environments have the need to exchange information between different types of computers and workstations in the form of letters, voice mail, e-mail, video, fax, spreadsheets, and so on.

With the foregoing heterogeneous environments in mind, this chapter has been written from a generic perspective, one that might be useful in any software development environment. Not knowing the exact environment that you might be working in while reading this book or the one you will get hired into after graduation, I am only providing some general principles for sound software construction and testing. Much of what is presented here applies to the smallest projects as well as the largest ones you may be involved in someday. "How large is large?" you might ask. Well, a presentation was given recently in which a software engineer from IBM shared software lines of code (LOC) information for his IBM "Space Transportation System" project with the audience. This project started prior to the first space shuttle flight in 1981 and continues today even though IBM is no longer responsible for it. As of the presentation, there have been less than two dozen software engineers working on this project, with incredibly low employee turnover rates. Obviously, the software for the space shuttle must be defect free. (This does not mean that there are no bugs in the software; it means that there are no known bugs.) One way of improving the odds of this is to have low turnover of software engineers on the project. The LOC numbers, stated in thousands (K), looked something like this:

1. Flight Software System in the Space Shuttle   500 KLOC
2. Software Production Facility Simulator   1,700 KLOC
3. Mission and Payload Control Centers   4,000 KLOC
4. Launch Processing System   2,500 KLOC
5. Space Lab (prototype version)   800 KLOC

TOTAL Lines of Code: 9,500 KLOC

There are almost 10 million lines of code in this system, which is still growing as the Space Lab portion advances. The IBM software engineer said that every space shuttle flight since the first one in 1981 has had some portion of its software changed, and every time there is a change, the entire software suite for the space shuttle has to be re-certified (tested). This is the real stuff, folks!

GENERAL SOFTWARE DESIGN PRINCIPLES

The information systems created today usually include a significant amount of software. Remember, information systems are generally made up of software, hardware, data, people, and procedures. This chapter is focusing primarily on the software component of an information system, so the word *software* will be used in the discussion of software design principles. Keep in mind that you could substitute the words *information system* in place of the word *software*. However, the emphasis is on the software component, hence the use of that word.

Software must conform to quality and cost-effectiveness standards. If you give me an infinite amount of time and money to develop an information system, I can probably guarantee you a high-quality information system, but is it one that is cost-effective and timely? Probably not! Having infinite development time and money is an example of an information systems development utopia, but as we all know, utopia only exists at Disneyland (or was that Autopia?). Regardless of time and money, the following principles should be adhered to for the software development portion of the information system.

1. The software should work correctly. Just before I started to write the first draft of this chapter, I upgraded my word processing software. Fortunately for me, I was saving my work after every paragraph. After finishing the preceding IBM space shuttle paragraph, I saved the chapter again. Then I made a few minor edits to the same paragraph and clicked the "disk" (save) icon once again. This time I got a "General Protection Failure" and had to restart not only the word processing software but also Windows. As I continue writing this chapter, each time that I click the "disk" icon and get the hourglass icon on the screen, I will hold my breath! This error never happened during the year that I used the prior version of the word processing software! The "software should work correctly" principle has been driven home in real life for me today! (Update: As I write my final draft of this chapter, the word processing software has become very stable for me.)

2. The software must conform to the requirements specification document created during the analysis portion of the project and any addendum added later during design. This principle is often overlooked in small to medium-sized information systems projects. The main reason is that as development proceeds into design and construction, new documentation and models have been created to assist the software engineers (programmers) to perform their detailed tasks more effectively. In so doing, the original specification document tends to become somewhat obsolete, since changes that occur are rarely reflected back into that document. Older methodologies contribute to this problem as they call for new documents and models to be created later on during design. The object-oriented methodology attempts to reduce or eliminate this problem by utilizing the same documentation and models from start to finish, enhancing and expanding them with more details as necessary.

3. The software must be reliable over time while processing a potentially limitless combination and types of data, as well as user keystrokes and pointing device (mouse) clicks. This principle is one of the most difficult to completely address. For example, this problem is exposed all too often with new releases of shrink-wrap software. Within weeks of newer versions of software being released to the general public, reports appear in the trade journals describing the "bugs" that were discovered in the software by some users. You may ask, "Why didn't the software development company that created the software find those bugs during their testing?" A good question. Part of the answer relates to this principle. There is absolutely no way that testing can possibly test the infinite combinations and types of data input, user keystrokes, and mouse clicks. Doing so, would reminisce back to the unlimited time and unlimited money scenario discussed earlier, which never exists! As I write the first draft of this chapter, once again I experienced the reality of another software design principle. However, the principle was experienced with my personal computer hardware, not software. Here's what happened. I purchased a 386-to-486 upgrade microprocessor chip for my computer—guaranteed to work by the chip manufacturer. After taking most of my computer apart to get to the CPU chip on the mother board, I replaced the 386 CPU chip with the new 486 CPU chip. When I tried to boot the computer, it didn't work! A 486 CPU chip manufacturer's telephone support person told me to send it back and he would send me another one. Same problem with the new chip. It didn't work! Finally, I had to return the upgrade chip for a refund and be satisfied to trod along at 386 speed as I write the draft today. How disappointing! I guess the manufacturer never tested the chip with my brand of computer. (Update: As of the latest draft of this chapter I now have a newer and much faster computer—yea!)

4. The software must be maintainable and evolutionary over time. The IBM Space Transportation System, discussed earlier, is a good example of this principle. There has been some empirical evidence suggesting that object-oriented information systems are easier to maintain than traditional information systems. The current dilemma is that relatively few business information systems have been developed using an object-oriented approach.

As information systems are being developed to meet the current needs of the user, software engineers should give some thought to the use of evolutionary tactics when building the systems. For example, let's say that you are creating a telemarketing name and address mailing list information system for a friend of yours who owns her own company and sells a line of personal exercise machines. Your friend tells you that she would like to have the information system keep track of the names, addresses, and telephone numbers of people who live in your city. Even though your friend says something like this, "I only want the system to keep track of the people that live in our city," you might consider the inclusion of a telephone area-code (even country code) as an evolutionary tactic. In addition, you might want to design the address to be a generic address that would handle foreign addresses that may not use the U.S. zip code or other mailing standards. Both of these ideas are examples of evolutionary tactics. You are designing the system to meet your friend's current requirements—keeping track of people in your own city—as well as building in some evolutionary elements that would accommodate the addition of people who live beyond the boundaries of your city, even beyond the boundaries of your country.

5. The software must be easy to use (user friendly). This is a relative statement depending on the complexity of the information system. Nonetheless, most users are not as eager and adept with computers as you may be, so you should take that into account when developing software. The shrink-wrap personal computer software as well as the custom developed workstation software of the mid-1990s is far more friendly and easier to use than that of the late 1980s, but there is still plenty of room for improvement. New and hopefully better user interface paradigms continue to be researched and implemented.

6. The software should be easy to test and implement. Again, as with the preceding principle, easy is relative. Test plans should be prepared very early in an information systems development project. In so doing, care will be taken to develop information systems that can be more easily tested. However, even the best test plans will not cover every conceivable processing path within the information system, every conceivable combination of data input to the information system, and every conceivable user keystroke press and mouse click.

Implementation of an information system addresses a number of issues, among them being the one of simply installing the new system on the hardware platforms intended for it to run on. When I installed my upgraded version of the word processing software, I simply inserted a diskette, typed a simple command, and clicked my mouse to commence the installation. The installation software did all of the rest for me—easy! There are additional implementation issues to address, and these are covered in a subsequent chapter.

7. The software should use computer resources efficiently. Although today's personal computers are fast and inexpensive, software engineers should still be conscientious stewards of computer resources when creating information systems. There are a number of issues that accompany this principle. One example I recall from years ago is a software program that one of my users had written to do a sort (he claimed to be a computer hacker!). The sort took over 24 hours to run on his company's mini-computer—wow! He would start it when he came into work on Monday morning, and pray that the computer would operate correctly for the next 24+ hours until his sort finished around 10 a.m. the following day. I looked at his software and discovered he was using one of the least efficient algorithms for sorting—a bubble sort. I provided him with another algorithm for a very efficient sort. He coded this algorithm and then ran the software. This time the sort only took a few hours! He could actually have his results before he went to lunch on Monday!

There is a tendency today to minimize the need to utilize computer resources efficiently. In the old days when CPU cycles were expensive and programmers were inexpensive, efficient utilization of computer resources was given high priority. With today's CPU cycles being inexpensive and software engineers being expensive, utilization of computer resources is given low priority. This principle should still have a place among software design principles.

SOFTWARE CONSTRUCTION FRAMEWORK

As with homes, apartments, and buildings, software is built from the ground up. What is meant by this is that software is developed one programming language statement after another, as in Figure 12.1. Combining several related programming statements together yields a module. A module may be a routine that simply prints a heading on a report, displays a tool bar on a menu screen, calculates total pay, controls a heater's on/off switch, or performs some other identifiable function. Even though the construction crew starts building a home from the ground up, they always have the construction drawings and blueprints that show all the details of the entire construction project prior to the actual construction grading, trenching, footings, and so on. Having a plan and the blueprints prior to beginning to build a building is a solid construction industry principle.

Software construction should have the same principle. Actual software construction should be preceded by the preparation of software blueprints in the form of the requirements specification document and any other supporting documents which contain the model of the system. What is being said here is that even though Figure 12.1 shows that software construction starts with a programming statement, please bear in mind that the modules, programs, subsystems, and system have already been planned for and their functionality has been included in the preparatory documentation prior to writing that very first line of programming code.

Software modules, as illustrated in Figure 12.1, can be stored in software libraries for reuse purposes in potentially any current or future software projects. The notions of **software libraries** and **reuse** continue to be hot topics. Many of their advocates believe that these concepts can significantly contribute to the reduction of the software indus-

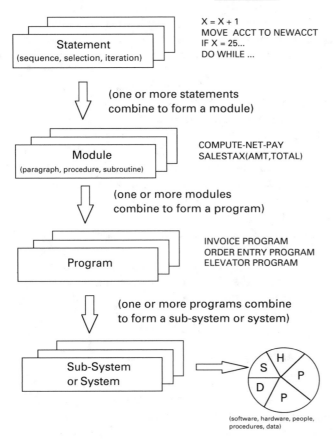

EXAMPLES

Statement
(sequence, selection, iteration)

X = X + 1
MOVE ACCT TO NEWACCT
IF X = 25...
DO WHILE ...

(one or more statements
combine to form a module)

Module
(paragraph, procedure, subroutine)

COMPUTE-NET-PAY
SALESTAX(AMT,TOTAL)

(one or more modules
combine to form a program)

Program

INVOICE PROGRAM
ORDER ENTRY PROGRAM
ELEVATOR PROGRAM

(one or more programs combine
to form a sub-system or system)

Sub-System
or System

(software, hardware, people,
procedures, data)

Figure 12.1   Hierarchical View of
Software Construction

try's large software development backlog. Advocates believe that software (module) reuse will reduce the time it takes for a software engineer to create the equivalent software module to just library retrieval time to locate the appropriate reusable software module. Reuse would also eliminate the need to test an already tested (and certified) module. The software engineer would still need to do integration testing to verify that the reusable module works correctly in the intended information system.

One form of reuse that has been around for dozens of years is the use of programming statements such as COPYLIB in COBOL and INCLUDE in Fortran. These and other similar statements in other languages reuse software modules that have been placed in a software library.

Many software engineers practice partial reuse by taking an existing module, usually one they have personally written, making a copy of it and modifying the copy to fit the current need that they have. Of course, they now have two, albeit similar, but different modules to maintain over time. Also the resulting changed module will have to be retested so some of the reuse benefit is lost.

Groups of software modules are assembled together to make up a program, as shown in Figure 12.1. Programs are usually associated with one or more specific user functions, such as print paychecks, register for classes, withdraw money from an ATM, control a street traffic light, and monitor a hospital patient's vital signs. The information system user can usually relate to these functions, and software engineers tend to group them into a program unit.

Combining one or more programs yields the software component for a subsystem or an entire system depending on the system's complexity, as illustrated at the bottom of Figure 12.1. For example, there are usually several programs that make up a bank's ATM machine subsystem. Combining the ATM subsystem's programs with others that the bank tellers, assistants, and managers use within the bank's branch make up the bank's information system. A program to add classes to your personal class schedule, another to delete a class from your class schedule, and a third to verbally tell you your current class schedule over the telephone make up the student registration subsystem, which is part of the larger admissions and records information system at your university.

Finally, as illustrated in Figure 12.1, the software subsystems and systems get integrated with the other four components of an automated information system—hardware, procedures, people, and data—to complete the system.

OBJECT-ORIENTED SOFTWARE CONSTRUCTION FRAMEWORK

Object-oriented software construction principles are quite similar to the ones discussed in the preceding section. Construction would proceed as in Figure 12.2 with the understanding that we already have well-defined classes from our analysis and design efforts. Statements are created in an object-oriented programming language, such as Smalltalk, C++, Eiffel, and so on, and groups of statements form a service (method). Services are associated with a specific class, and each service represents a necessary operational method that will allow the class to accomplish its responsibilities within the system. Services are similar to the modules shown in Figure 12.1. Classes are combined to form the subsystem or system software which then integrates with the other four components that make up an information system.

SOFTWARE CONSTRUCTION STRATEGIES

Over time students develop their own exam-taking strategy. Similarly, students of software engineering need to understand the different choices or strategies available for constructing software. The assumption here is that some or all of the software for an information system under development will be created by the software engineers. There are several software construction strategies from which to choose. Among them are top-down, bottom-up, middle-out, and any hybrid combination of these.

Top-down software construction follows the functional decomposition approach to problem solving. Using this strategy the software engineer starts with the design overview or high-level view of the system and then creates the program code for it. The higher-level program modules are usually control or management oriented, such as menus and calling modules. Next the software engineer works his or her way down into the details of the system writing the program code, layer by layer, until the

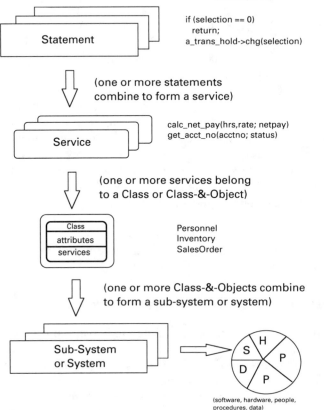

Figure 12.2   Hierarchical View of Object-Oriented Software Construction

bottommost modules are programmed. The lowest-level modules are usually the operational or worker-oriented ones, such as those which calculate results, update databases, print lines on a report, and display text or graphics on a video screen.

Bottom-up software construction strategy starts with the software engineer writing the program code for the lowest-level details of the system and proceeds to higher levels by a progressive aggregation of details until they collectively fit the requirements for the system. For example, a software engineer may begin by coding a low-level module that opens a file, or calculates sales tax, both of which were considered to be bottommost modules in the system. Next the software engineer would construct the code for the module or modules that calls these bottommost modules. He would continue this development strategy working in an upward or outward manner until reaching the topmost, control-oriented modules.

Middle-out software construction starts somewhere in the "middle" of the system that seems convenient, and proceeds both upward to higher levels and downward to lower levels as appropriate. The **hybrid** combination of top-down, bottom-up, and middle-out software construction is equivalent to a "home-grown" methodology and may be suitable for certain environments or projects.

The moral of the software construction strategy is to recognize that there is more than one acceptable way to construct software. Most software engineers tend to favor one strategy over the others. However, more experienced software engineers should be able to work with any of them if the project necessitates doing so.

COHESION AND COUPLING

Have you ever wondered just how "good" the software is that you write? Many professional programmers have the misguided belief that if their software functions correctly, then it must be "good." Certainly, having your software perform correctly is essential, but remember the other principles of software covered earlier in the chapter. Each of these should have some effect on software "goodness."

Coupling and cohesion have an effect on both software maintainability over its life and its efficiency in utilizing computer resources. Many businesses utilize the **chief programmer concept.** The chief programmer is a person or group of persons who actually reviews the code of junior programmers with the intent of offering improvement suggestions, among them being cohesion and coupling measures.

Cohesion and coupling are associated with modules and services. **Cohesion** is a measure of the strength of the interrelatedness of statements within a module (service). **Coupling** is a measure of the strength of the connection between modules (services). A programmer should try to maintain a high degree of cohesion and a low degree of coupling. Doing so is similar to the balancing act that was discussed in the database chapter of this book. The discussion in that chapter addressed the balance software engineers strive to achieve between simplicity and redundancy for data structures. It was also pointed out in that chapter that simplicity and redundancy often are counterbalancing measures in that too simple data structures lead to too much redundancy and too complex data structures lead to no redundancy at all. Cohesion and coupling, too, represent a balancing act.

Although they each measure different aspects of constructed software, cohesion and coupling tend to work well together. A programmer who is trying to accomplish the ultimate in cohesion usually ends up accomplishing the ultimate in coupling as well and vice versa.

The ultimate in cohesion is for the programmer to maximize it according to the chart in Figure 12.3, which is intended to show relative ranking of each of the cohesion levels with each other. Remember, cohesion measures the relatedness of statements within a single module. As each cohesion level is discussed, moving from worst to best, keep in mind that a software module exhibiting more than one of these levels of cohesion is (1) probably too large of a module for sound software construction, and (2) essentially exhibiting the lowest or worst level of cohesion identified in the module.

Coincidental cohesion, the worst type of cohesion, exists when a module performs multiple, virtually unrelated, actions. This would be analogous to writing a computer program from start to finish using just one "main" module which has all of the programming statements in it that are necessary to accomplish the job of the program. I have to admit that some of my early attempts at programming in college demonstrated coincidental cohesion. How about yours?

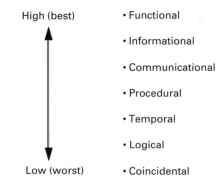

> COHESION: The measure of strength of the
>
> interrelatedness of statements within a module.

High (best)   • Functional

                     • Informational

                     • Communicational

                     • Procedural

                     • Temporal

                     • Logical

Low (worst)   • Coincidental

Figure 12.3   Cohesion Chart

Logical cohesion exists in a module when it performs a series of related actions, one or more of which are requested by the calling module. This is like going into an ice-cream parlor and selecting your favorite ice cream from dozens of flavors on display. Each customer (calling module) selects his or her own choice from the available flavors (series of related actions). In information systems software, logical cohesion exists in a module that calculates net pay for hourly, salaried, and piecework employees. The calling module tells this module what type of employee to calculate the net pay for, and then the module uses only one of the three calculation actions available.

Temporal cohesion exists in a module when it performs a series of actions related by time. Software is often written with one module selected to do "open housekeeping" at the beginning and another module to do "close housekeeping" at the end of the program. Housekeeping at the beginning of a program may involve initializing variables and/or opening files. Housekeeping at the end of a program may involve closing files. These types of modules exhibit temporal cohesion.

Procedural cohesion exists in a module when it performs a series of actions related by the sequence of steps being performed. For example, a module that retrieves a student's record via his or her identification number from the student file and then updates the number of students enrolled in a course file exhibits procedural cohesion because these two actions are related by the sequence of steps necessary for keeping track of how many students are signed up for a course. Procedural cohesion is better than temporal because the actions being performed are related to one another in order to get the job done, not merely because they are done at the same time.

Communicational cohesion exists in a module when it performs a series of actions related by the sequence of steps to be followed and, in addition, all of the actions are performed on the same data. For example, a module that retrieves a student's record via his or her identification number from the student file and then changes the address and telephone number exhibits communicational cohesion. This example is similar to the preceding procedural cohesion example. However, the difference is that both actions are performed on the same data (student file) in this example, whereas the other example performed actions on two different files (student and enrolled course). Once again, this is better cohesion than the prior ones because the actions being performed are related and affect the same data.

Informational cohesion exists in a module when it performs several actions, each having (1) its own entry/exit point, (2) independent code for each action, and (3) actions that are performed on the same data structure, as illustrated in Figure 12.4. Notice how each section of code has its own entry, exit, and code, making each section of the module independent but related to the other sections of the module.

The final and best level of cohesion is functional. A module that performs exactly one action or achieves a single goal has **functional cohesion.** Some examples of modules like this would be "get student ID number," "enter your ATM card," "print heading on top line of report," and "compute sales tax." Because these modules perform one task or have one goal, they can more readily be reused in a variety of situations, but even these modules must be written with care to allow reuse to really work. Functionally cohesive modules are usually more easily maintained than all other types of modules, again due to the concentration of code related to just one task or goal.

As stated earlier, **coupling** is a measure of the strength of the connection between modules (services). The ultimate in coupling is to minimize it according to the chart in Figure 12.5, which is intended to show relative ranking of all of the levels with each other. The **best coupling is none** at all. This indicates that all modules do a single task and do not depend on any other module. For software of any consequence,

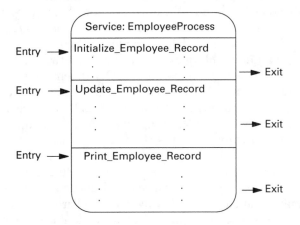

Figure 12.4   Informational Cohesion

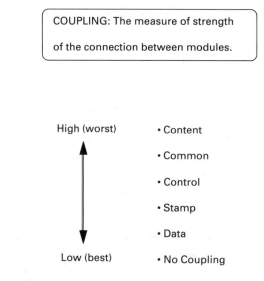

Figure 12.5   Coupling Chart

this level is very unusual to achieve. No coupling would be similar to you going through a day in your life without depending on anything or anyone. As each of the coupling levels is discussed, moving from worst to best, keep in mind that a module exhibiting a worse (higher) level of coupling is more difficult to maintain and reuse over its life and the life of the information system it is a part of.

The worst coupling is known as **content coupling,** and it exists when one module directly references the contents of another module. There are several variations of content coupling and these content-coupled modules are very difficult to maintain. For example, the branching from one module into the middle of another module, instead of its beginning, is very difficult to maintain. Content coupling exists less frequently today because few languages allow it, and programmers have learned of its limited maintainability and avoid it. For example, COBOL used to have an ALTER verb as part of its syntax, but it is no longer available.

Common coupling exists when at least two modules have access to the same global data. Use of Fortran's COMMON statement, COBOL's GLOBAL statement, and Visual FoxPro's PUBLIC statement are examples of this. This practice has proven to be very unmanageable, since having the ability to change global variables in local modules is very difficult to debug and maintain over time.

Control coupling exists if one module passes an element of control to some other module which, in turn, explicitly controls the logic of the second module. For example, passing a variable from a calling module to a called module whose value determines the logic path to be taken by the called module as it executes exhibits control coupling. A called module having code similar to "IF X=1 DO A, IF X=2 DO B, IF X=3 DO C" where the value of X is passed to this module by the calling module also exhibits control coupling.

Stamp coupling exists if one module is allowed to pass an entire data structure or record to another module. This is potentially difficult to maintain over time because one might assume that all data elements in the data structure are being utilized in some way by the second module when, in fact, only one or a few of the data elements are used.

Finally, if coupling has to occur, and it does in most software modules, then **data coupling** is preferred. Data coupling is more reasonable to expect and achieve in software construction than is no coupling at all. Data coupling simply views other modules as *black boxes* that either receive input data variables from other modules or provide output data variables to other modules or both. For example, a module that multiplies two numbers together receives two data variables as input from a calling module, does the multiplication on these two inputs, and passes back one answer variable to the calling module. This module exhibits data coupling.

Object-Oriented Cohesion and Coupling

Cohesion and coupling concepts are just as important in object-oriented information systems development as they are in the structured or other information systems development methodologies. However, better levels of cohesion and coupling are usually achieved as a by-product of using object-oriented concepts for analysis, design, and programming. Nonetheless, the potential exists for object-oriented software to have poor cohesion and coupling, just as it does in other software development methodologies. One of the many benefits of object-oriented software is its class paradigm which encapsulates both attributes and services. By default, attribute and service encapsulation lends itself to higher-quality cohesion and coupling. In addition, one object-oriented tenet is that the only way that new objects get created, existing objects get deleted, and existing class's attributes get modified is through the services encapsulated within the class. This tenet also contributes automatically to improved cohesion and coupling.

As software engineers and users identify services during analysis, they are expected to perform one task or have one goal. In so doing, the model is already exhibiting a reasonable level of cohesion. This is not guaranteed, but it is a serious possibility as a by-product of utilizing object-oriented methods and programming concepts. Further refinement of the services can take place during design, keeping cohesion and coupling in mind to again improve the level of both cohesion and coupling. While adhering to object-oriented programming concepts, it is virtually impossible to have coupling degrade beyond control coupling, and cohesion should not drop below the procedural level.

SOFTWARE TESTING

Armed with a knowledge of the previously discussed software development principles, software engineers must also give considerable attention to the software testing activity within software engineering. Software testing should conform to at least three principles. **The first principle is that software testing begins with plans for the tests and**

ends with the user accepting the tested software. Software test plans should be developed at the same time the requirements specification document is being developed and can be refined and updated during design as well. The plans for the user acceptance test should conform to the user requirements as documented in the requirements specification document. By doing so, the development loop is completed, and the software is certified to be in compliance with the requirements specification document. User acceptance tests include more than just the software; they include the other four information systems components—hardware, procedures, people, and data.

The second software testing principle is that testing's intent is to cause and discover errors. This is a very different mindset than one that thinks testing is to prove the program works. The first mindset is a healthy destructive one. The second one is really an unhealthy mindset because there is an assumed emphasis on the capability of the programmer rather than on the quality of the software product being produced. Suffice it to say that there is a psychological difference between these two mindsets. Many programmers exhibit this double mindset when testing software. For example, when programmers test another programmer's software, an accepted activity in many software development organizations, they take on the first mindset—to cause and discover errors. But when testing their own software, they approach it with the second mindset—to prove the software works. I've been there and done that too! After all, there is pride of ownership in our own software, and many programmers develop a mindset of "my software is bug free." At the same time, programmers are often skeptical and cautious when it comes to someone else's software, yet very confident of their own software. Testing should be done with as little pride of ownership as possible. In fact, it is much better and less embarrassing for programmers to thoroughly test their own code with the destructive mindset prior to letting anyone else test their code.

The third software testing principle is that rules of reasonableness should prevail. Testing could be never ending if software developers try to test every conceivable combination of input data, keystrokes, mouse clicks, and so on. Instead, testing should follow commonly accepted testing techniques designed to provide a certain threshold of statistical confidence in the tests. For example, rather than testing every record in a large database for a certain sequential processing situation, one commonly accepted testing technique is to test the first record, a few records in the middle of the database, and the last record. Many programmers have small test databases (a few dozen records, for example) that they use rather than testing with the huge actual database that may be several hundred thousand records. Research has shown that it is common for software to have processing flaws when dealing with the initial record or ending record in a sequential processing situation so this technique tests for these situations. There are many other testing techniques like this one, therefore, warranting an entire book and course on the subject.

Software Testing Strategies

Software testing strategies are the same as the strategies discussed in an earlier section on software construction. Namely, top-down, bottom-up, middle-out, and any hybrid combination of these. In addition, white box, black box, alpha and beta testing are all additional strategies that are utilized to complement top-down, bottom-up, middle-out, and the hybrid testing strategies.

Top-down testing follows the functional decomposition approach to problem solving. It starts at an overview or high-level view of the system to be tested, then works its way down into the details of the system, layer by layer until reaching the bottommost programmed modules. The higher-level program modules are usually control or management oriented, such as menus and calling modules, while the lowest-level modules are usually the operational or worker-oriented ones, such as ones that calculate results, update databases, print lines on a report, and display text or graphics on a video screen.

Bottom-up testing starts with the details of the system and proceeds to higher levels by a progressive aggregation of details until they collectively fit the requirements for the system. For example, testing may begin with a module that opens a file, or calculates sales tax, both of which are bottommost modules in the system. An expanded example of this is discussed and illustrated in the next section.

Middle-out testing starts somewhere in the "middle" of the system that seems convenient, and tests both upward to higher levels and downward to lower levels as appropriate. **Hybrid testing** is a combination of top-down, bottom-up, and middle-out is equivalent to a "home-grown" testing methodology and may be suitable for certain environments or projects.

Testing strategies usually become a function of the software construction strategy, and the testing type (discussed later). Regardless of which testing strategy is chosen, software testers view portions of the system, subsystem, programs, and modules as either a *white box* or a *black box* as they test. **Software testers** are people involved in the software development process. Programmers, software engineers, and users who test would be considered software testers in the discussion here.

The portion of software that software testers view as a **white box** is the portion of the system that we are testing from an inside perspective. Basically, the software tester doing the testing has the ability to actually look at the code statements within the white box portion being tested and make changes to them should they not work correctly. Most often, programmers do white box testing on code that they write and subsequently test themselves.

The portion of software that software testers view as a **black box** is the portion of the system that they are testing from an outside perspective. For this black box portion of the system, the software tester does not have the ability to look at the code statements and change them. Most often, software testers do black box testing on code that someone else wrote. If programmers are doing black box testing, it is often because they need to integrate that part of the system with the part of the system they are creating and testing. Software engineers and users usually do black box testing on software that someone else wrote; therefore, they do their testing from a black box perspective providing feedback to the software author for any problems discovered. For example, a software engineer or user may be testing a sales tax calculation module that someone else wrote. The software engineer or user knows what data have to be passed to the sales tax calculation module, such as dollar amount and sales tax percent. The software engineer or user should be able to determine what the sales tax calculation module should give them back as the correct sales tax dollar amount. If the module does not work correctly, the software engineer or user cannot make changes to the sales tax calculation module, but must report it to the project leader or programmer who wrote it for review and possible correction. Sometimes the testers make mistakes themselves. For example, in the sales tax calculation module test, the software engineer or user may inadvertently pass the wrong input to a black box module, so the module is correct but their test input is incorrect.

Personal computer software manufacturers such as IBM, Computer Associates, Microsoft, Hewlett-Packard, Novell, and hundreds of others spend millions of dollars testing new releases of their commercial software products before the product is released to the market for sale. That's staggering! Some of these companies have unbelievable testing facilities equipped with dozens and dozens of clone computers. Smaller software manufacturers simply cannot afford such an extensive testing budget or testing facility. They tend to "rent" some testing facility, such as a university or consortium lab and, in addition, rely quite heavily on a select group of customers to participate in the testing to reduce their expense for testing. For example, Microsoft "sold" hundreds of thousands of copies of a beta version of its Windows '95 operating system early in 1995.

When testing is done by in-house testers, such as programmers, software engineers, and internal users, this level of testing is referred to as **alpha** testing. In other words, a first cut at the acceptance tests is done in house (or in a rented test facility) by in-house staff. Once satisfied with the results of the alpha tests, which doesn't necessarily mean that the software is bug free, the software manufacturer moves to the next level of testing called **beta.** Beta testing usually involves a select group of customers to perform tests on the software as they would use it in their normal environment. In exchange, the customers get an advance copy of the soon to be released software, which usually has new features in it (and potential bugs!). For example, a bank might be willing to beta test a new release of Lotus 1-2-3 or Microsoft Excel spreadsheets. Other times, beta test customers have plans to use this new or revised software as part of their own product or to facilitate the creation of their own software product and are, therefore, very interested in testing an advance or beta copy of the software. For example, consider a company that develops and sells an income tax software package and uses C++ as the programming language and Windows as the operating system. This type of customer would then get an additional window of time to work with the soon to be released software product, and perhaps be able to put their own software product (income tax package) into the market for sale ahead of their competition, which has to wait for the official release of the new C++ or Windows software to use them on or with their products.

So alpha testing is usually done by software development's own in-house staff, and beta testing is usually done by customers. This is not always the case but is the generally understood meaning for these two testing types. Once again, as with many of the software development concepts, the software industry has many variations on how these two terms are utilized.

Software that is created by internal software development staff for use within its own class and not for general public sale may also use the alpha and beta terms for tests at various levels among the ones discussed as the generic testing methodology discussed in the next section. Most large companies have internal software development groups that fit this situation.

A Generic Software Testing Methodology

The following generic testing methodology, illustrated in Figure 12.6, can be utilized regardless of whether the resulting software is to be used in house or is to be available for public sale at computer stores or through mail order. Note the feedback or fallback arrows in the diagram. Whenever testing discovers errors at any level of the methodology, the programming staff will need to make coding changes followed by a trip back through the layers to ensure defect-free code in all test levels. As testing takes place, keep in mind that code changes made by programmers in one module or section of a program may fix the identified bug but may cause an error in some other seemingly unrelated section of the program when it is tested again. Never assume that a program change in one section or module of code could not possibly affect another section of the software!

Each of the testing levels will be discussed in detail below. Even though the discussion will progress from small tests to large tests and include general characteristics of each of these tests, testing flexibility and variations on these generic tests are common in the software industry.

Unit/module testing is usually done by the programmer who authored the software, and it is usually done while the programmer is in the process of creating the software module or unit. Some programmer workbench environments are much more

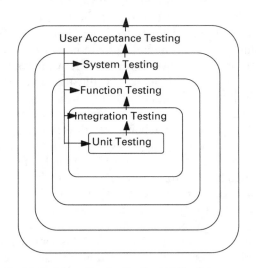

Figure 12.6   Software Testing Methodology Showing Feedback/Fallback Loop

conducive to this low-level testing than others. For example, many programmers find that interpretive programming languages tend to promote this level of testing more than do compiled languages simply because of the time, effort, and convenience involved to do the module test. An interpretive language, such as Visual Basic, Smalltalk, Visual FoxPro, and many database management packages allow a programmer to enter one or more programming statements and test them as soon as they are typed into the programming workbench environment. An interpretive programming workbench environment is waiting to automatically "execute" or test one or more statements "on command" of the programmer who typed them in or gave some simple command such as "do." Test results and feedback under this condition are almost instantaneous, which can lead to increased programmer productivity. Languages that require compilation and linkage prior to execution, such as COBOL and many other programming languages, take considerably more time compared to interpretive environments to compile and link the statement(s) created by the programmer in order to conduct the test. In some situations, a significant amount of the program must be developed before the compiler will even give the programmer an error-free compilation, which means that the programmer cannot test prior to an error-free compilation.

Finally, unit/module testing is usually done with test data that are created by the programmers themselves. For example, if the module being tested expects a keyboard entry of a person's last name, the programmer, when testing, may enter the following as potential "last names"—Xxxxx, Wtyeos, Smith, Jones, Yen, and so on. Basically, any printing character may be entered by the programmer to simulate an actual person's last name.

Integration testing is the combining of several units or modules for the purpose of testing them as a whole. As software is being constructed and simultaneously being tested for integration with other software modules by the programmer, a technique known as stub testing is often used. **Stub testing** is a test performed on individual modules as they are being integrated in a top-down, bottom-up, or middle-out manner. Using the stub testing technique, a programmer can simulate the presence of a calling (higher-level) or called (lower-level) module while testing the logic of a specific module above or below the stub module, as illustrated in Figure 12.7. The stub module has no logic in it; it merely simulates calling or being called by the module the programmer is testing at the moment. The programmer does this by simply hard coding a calling module to call the module being tested, as in Figure 12.8, or by simply hard coding a called module and having it display some simple message, such as "hello; in module x now," as also illustrated in Figure 12.8. Finally, integration testing frequently uses test data created by the programmers, similar to those used in unit/module testing.

Function testing is the combining of one or more integration tested groups of modules that collectively perform a user identified function, as documented in the requirements specification document. For example, making a withdrawal of cash from an ATM or adding a class to your list of classes via a telephone student registration system would be considered user functions and so treated for testing purposes.

At this level of testing, the programmer may be given a suite of test data from

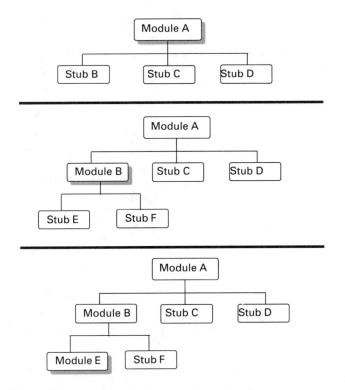

Figure 12.7 Stub Testing (Shaded module is the current one being tested)

the data administration group or the person/group responsible for creating the test data. The programmer should, in turn, test the functions using the prescribed test data.

At this point, you may be wondering why we continue to test primarily with test data rather than the actual, real data. The closer the testing gets to the outer layer of tests, shown in Figure 12.6, the more realistic the data should be. If software testers were to test using real data or a copy of the real data, their testing time may be significantly expanded due to the volume of real data in the information system under development. For example, an airline's frequent flyer bonus information system may have a database with millions of people's names in it who have accumulated frequent flyer miles. There is often no need to have such a database be involved in the early testing efforts when a few hundred names would suffice. Just the sheer size of the real database may preclude testing, since many environments use personal computers to do software development and testing and simply do not have the disk capacity to hold such a large-volume database.

System testing combines all of the user-defined functions together to form the subsystem or system. Just because independent functions work correctly during function testing doesn't guarantee that they will work correctly as a subsystem or system. All kinds of problems can surface when they are joined together, especially if the

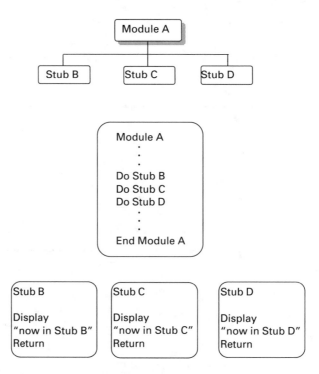

Figure 12.8   Calling Stub Modules During Construction and Testing

functions were developed by different programmers or teams of programmers. System tests are usually performed by programmers, software engineers, and project leaders and are often labeled as the alpha tests we discussed earlier.

At this level of testing, the software testers are given a suite of test data from the data administration group or the person/group responsible for creating the test data. In addition to this, a system test using real data may also be conducted. This can be done by making a copy of real databases and using them in the testing.

Acceptance testing is similar to system testing. However, the user is the one who usually does the testing. For example, a colleague of mine is employed by a government contracting company that has installed information systems in government-owned hospitals around the world. Prior to any new release of the software to the hospitals, a team of government testers comes to the software development company and thoroughly tests the new releases. Upon acceptance, the software development company then releases the software for implementation at the hospitals. In another scenario which was discussed earlier in this chapter, commercial software companies, such as Microsoft, send out beta copies of new software releases to "preferred" customers who in turn test and represent the acceptance testing prior to the public release of the software. By the time people like you and I purchase the software at a retail store it has had countless hours of testing, significantly raising the level of confidence on the part of the software development company in its software performing correctly when you and I use it. Nonetheless, within weeks of new software releases, you can read about potential problems or bugs in the software industry trade press newspapers. There is no 100 percent guarantee that software will be bug free! At best, software will have "no known bugs."

Consumers like you and I expect the products we buy in retail stores to work correctly—TVs, stereos, VCRs, CD players, washing machines, can openers, and so on. This same expectation carries over to the purchase of computers and software, which really forces a software development company to do as thorough testing as possible. Just as consumer hard goods like the ones cited earlier have consumer reports and rankings along with *Good Housekeeping* type "seals of approval," shrink-wrap software has or should have the same kind of recommendation.

Acceptance tests are most often conducted with well-defined test data, as well as the actual, real data. Real data are used to demonstrate the actual response time for the system in the real environment and demonstrate whether the software is still bug free.

This concludes the discussion of a generic software testing methodology. Figure 12.9 represents a framework for the methodology and how it interrelates with the other activities and deliverables of information systems development. The overriding goal in information systems development is to meet user requirements within an acceptable time frame and at a minimum cost. Hopefully, such a framework gives guidance and vision to the realization of that goal.

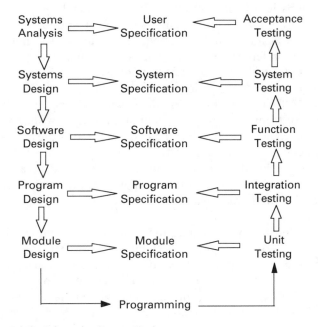

Figure 12.9   Framework for Information Systems Testing

APPLICATION AND CODE GENERATORS

Application and **code generators** are software engineering workbench tools that, as their names imply, generate complete or partial information system applications and program language code for portions of an information system application, respectively. These tools are considered high-level tools as they allow the software engineer the ability to work with computer-assisted models of an information system, database definition, video screen layout, and report painter tools that automatically generate the program code for the information system model, database, screen, or report being designed. Generators make a significant contribution to some software engineering environments because they generate/create the program code rather than having a programmer write it by hand.

Testing must still be done with generated software to ensure the operability of the software. Software testers may be able to reduce the amount of testing done but, most likely, they will want to thoroughly test because the generator will "blindly" create code based on the model or design layout created by the software engineer—no matter whether the model or design is right or wrong!

SUMMARY

As the discussion of software testing comes to an end, a final comment is in order. The diversity of information systems that can be created directly translates into the same diversity of testing methodologies that exists for these systems. This chapter has covered the basics of software testing with the knowledge that a given testing methodology in any information systems development project is unique and depends on a number of factors including but not limited to budget, time, personnel, skills, system type, user preference, hardware and software platforms, telecommunications, and so on. This chapter's discussion of testing has been designed to expose you to a generic testing methodology as discussed previously.

The first topic covered in the chapter was general software design principles. Several of them were presented and discussed. As with other principles in life, these should become a part of the makeup of future software engineers. The next discussion topic dealt with a general software construction framework. Software construction, like physical building construction, must be planned, organized, and carried out in the most effective and efficient manner possible. Software reuse was also discussed as a part of this framework. An object-oriented software construction framework was also presented and discussed.

Following the foregoing topics, the chapter proceeded with a discussion and illustrations of commonly accepted software construction strategies—top-down, bottom-up, and middle-out. Software cohesion and coupling principles were presented and illustrated in order to demonstrate sound programming principles.

Software testing was presented and noted that the most common testing strategies are similar to construction strategies—top-down, bottom-up, and middle-out. A generic software testing strategy was discussed and illustrated to provide the reader with an example of how one goes about testing software. Finally, the chapter closed with a brief discussion of application and code generators.

QUESTIONS

12.1 Briefly discuss some of the standards to which "good" software/information systems must conform. What is the main reason for this need to conform?

12.2 What is the first step necessary in software construction? Why?

12.3 What are the uses and implications of software libraries and software reuse?

12.4 List and briefly describe three different approaches to software construction.

12.5 What is meant by the *chief programmer* concept?

12.6 What are cohesion and coupling and what is their purpose in software construction?

12.7 What are the different types of cohesion? Which are more desirable? Which are less desirable?

12.8 What are the different types of coupling? Which are more desirable? Which are less desirable?

12.9 What are the three main principles to which software testing must conform? Why are they important?

12.10 What distinguishes the two terms *white box* and *black box*?

12.11 What are the differences between alpha- and beta-level testing?

12.12 Discuss the importance of the feedback/fallback loop in the generic software testing methodology.

12.13 What distinguishes acceptance testing from system, function, integration, and unit testing?

REFERENCES

DAVIS, WILLIAM S., *Business Systems Analysis and Design.* Belmont, CA: Wadsworth Publishing Company, 1994.

EL EMAM, KHALED (ed.), "IBM Federal Systems (Loral) - Space Shuttle Program," *Software Process Newsletter,* Committee on Software Practices, Technical Council on Software Engineering, IEEE Computer Society Press, no. 1 (September 1994). Note: The story in this chapter actually came from a presentation at the 1993 Software Maintenance Conference, San Diego, CA.

MYERS, G.J., *Composite/Structured Design.* New York: Van Nostrand Reinhold, 1978.

PRESSMAN, ROGER S., *Software Engineering: A Practitioner's Approach* (2nd ed.). New York: McGraw-Hill, 1987.

SCHACH, S.R., *Software Engineering* (2nd ed.). Boston: Irwin, 1993.

STEVENS, W.P., G.J. MYERS, and L.L. CONSTANTINE, "Structured Design," *IBM Systems Journal,* 13, no. 2 (1974).

Implementation

CHAPTER OBJECTIVES (YOU SHOULD BE ABLE TO)

1. Name and discuss the three phases of implementation.
2. Describe the conversion process.
3. Name and describe the conversion strategies.
4. Describe information systems failure perception and reality.
5. Discuss the stages of organizational change and how they affect information systems implementations.
6. Discuss action research strategy for identifying and dealing with user resistance to change.
7. Discuss the force field analysis tool for gathering user data.
8. Discuss the critical success factors for information systems implementation.

INTRODUCTION

Have you ever seen a ship christened? Have you ever seen the grand opening of a new shopping mall? How about the summer television commercial teasers for the new fall television shows? How about the introduction of a new line of clothing or perfumes? You are probably wondering what these events have in common with information systems engineering, right? Well, very little frankly! But the one thing that they do have in common is that each of these events represents the beginning of an implementation of something new. The launch of a new ship, opening of a new shopping mall, beginning of new television shows, and the introduction of new clothes and perfumes.

Each of these four events and thousands more just like them have three generic phases to their "kick-off" or new beginning. The phases—install, activate, and institutionalize—as shown in Figure 13.1 are identical to the implementation phases for a new or changed information system as well. The differences between the implementation of these events and information systems implementation are in the implementation details.

Before discussing each of the phases and information systems implementation activities, there are three important thoughts to keep in mind:

1. Implementation marks the end of development for now, and begins the user's realization of the new or changed information system.
2. Implementation is a critical time for users because the user must switch from the old to the new system, and the user must take responsibility for the successful utilization of the new system.
3. The information systems development team must anticipate, even expect, problems during implementation. Hopefully, the problems can be turned into opportunities.

Now let's look at each phase in more detail drawing examples from the four events cited earlier, showing the parallel between them and information systems implementation.

INSTALL: THE FIRST PHASE OF IMPLEMENTATION

Phase 1, **install,** is putting the new _____ in place. You can fill in the blank with ship, shopping mall, television show, clothes, perfume, or information system. Each of the ships, shopping malls, . . . , information systems had to be created, and no two of these were created with the exact same technique or process, or used the same types of materials throughout, or had the same type of personnel create it. In other words, each had to be created, but there is significant variation in the details of how each is created.

Just as putting a new ship in place means to have it floating in the ocean, and putting a new shopping mall in place means to build it, furnish it, and stock it with an

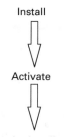

Install

Activate

Institutionalize

Figure 13.1   Three Phases of Implementation

inventory of goods and services, an information system being put in place means to physically install all of the five components—hardware, software, people, procedures, data—of an information system. We don't really install people, but we do train or prepare them as part of the install phase.

A very significant part of the install phase has to do with conversion. **Conversion** is the switching from one thing to another. Perhaps the new ship is replacing a retiring ship, so that many of the physical items on the old ship must be brought over to the new ship. Perhaps a store in the new shopping mall is being relocated from a few blocks away. Therefore, some or all of its furnishings and inventory must be moved to the new location. Telephone long-distance companies, such as AT&T and MCI, are constantly trying to get you and I to "convert" to their long-distance service. They usually even waive the conversion cost as an incentive for us to do so.

Most often a new information system is replacing an obsolete manual or automated information system. All or some of the data from the old system must be converted to become part of the new information system. Figure 13.2 illustrates how conversion affects each of the components of an information system.

Plans for an information system conversion must begin long before actually getting to the install phase of implementation. In fact, for the conversion to take place during the install phase, some of the activities to be performed as part of the conversion must begin during design and continue through programming and testing. The conversion plan may call for one or more conversion programs to be designed, written,

Hardware
The range of potential conversion extremes is to switch to a new computer from a new manufacturer all the way to no hardware upgrades at all (e.g. use the exact same hardware you are currently using). Somewhere in the mid-point we might need to upgrade the computer to one that is faster or has more disk capacity.

Software
One extreme end-point for conversion might be a completely different set of software e.g. going from DOS to Windows or Unix; changing your database software from dBASE to Paradox, changing from ABC Accounting software to XYZ Accounting software, etc.); the other extreme end-point might be a new version of the software you are already using which has a new feature in it. Somewhere in the mid-point for conversion might be a new version of software having many new features, revisions to some of the existing features, and removal of some unused features.

People
One extreme end-point here could be the creation of a whole new job due to the new information system, or a new way to do the existing job, or the reclassification of a job, or the elimination of a job. The other extreme end-point for conversion could be some small amount of training to familiarize the user with the new features of the system.

Procedures
One extreme end-point for conversion would be a completely new set of procedures for users to learn (e.g. such as the use of a mouse when the user is only familiar with a keyboard). On the other extreme end-point, the existing procedures have been changed ever so slightly to conform to the new or revised information system.

Data
One extreme end-point for conversion would be the reorganization of data from the old system so that it conforms to the needs of the new system. In addition, the old data may have to be purified or conditioned in an endless variety of ways. Also, there may need to be a lengthy data entry activity to get the initial data into the new system. An example of this could be a medical office putting all of its thousands of hand-written, manual patient charts onto CD ROM technology for visual display on a computer. The other extreme could be that the data are already in the proper format and condition for the new or revised system.

Figure 13.2   Conversion Effects in Information System Components

and tested that will be used during the conversion to purify and reorganize data for the new system. Sometimes this is a major effort on the part of the information systems development team and the time and effort to do it must be included in the budget for the project, otherwise severe time delays and cost overruns could occur. Over the years, there have been many projects that were on schedule but had overlooked the conversion activity during project planning, thereby causing the project to actually overrun its projected time and cost budgets. Conversions, when needed, simply cannot be overlooked or skipped; they must happen! And they take time and human resources.

There are basically two conversion strategies; however, both have variations. The two strategies are cutover and parallel. A **cutover** conversion strategy, also known as an **abrupt** conversion, is one in which the user uses the old system up to a specific point in time after which the user starts using the new system. A **parallel** conversion strategy is one that allows the user to continue to use the old system and the new system simultaneously for a period of time, eventually discarding the old system. Both cutover and parallel conversion strategies can be done for everyone using the system or for select users according to some predetermined order. Figure 13.3 illustrates examples of this for your consideration. Although other texts mention pilot and phased conversions, I believe these two basic strategies with variations cover these as well.

Notice in the figure that the groups of users can be based on any grouping the user and development team agree would be best for the user. Some suggestions for groupings could be all freshmen, all seniors, all students in a particular dorm, all students with 75 or more completed units, all students who have their home residence west of the Mississippi, and so on. Notice also that the primary difference between a

Last semester all students did their course registrations by mail or by waiting in line. This semester:

Conversion Strategy	All users	Group of users
Cutover	All students do their course registrations by using a telephone registration system.	Students whose last name starts with N-Z do their registration the same as last semester; students with last names starting with A-M use the new telephone registration system.
Parallel	All students will need to mail in their course registration (or stand in line) as well as call in and register using the new telephone registration system.	All students will need to mail in their course registration (or stand in line); students living in the STORM HALL dorm will also need to register by using the new telephone registration system.

Figure 13.3
Conversion Strategies

cutover conversion and a parallel conversion is simply that a parallel conversion implies that one or more users will be using both the old and the new information system for a period of time. In Figure 13.3 the users in Storm Hall needed to use both the old and the new information system this semester.

What do software engineers base their conversion strategy decision on? Well, there are three primary factors:

1. The design of the information system. A few examples of this follow. Sometimes the design of the new system so closely emulates the old system that operating and using both systems may make very good sense and be quite reasonable; other times the new information system is so different from the old one that operating both simultaneously would be difficult and possibly unreasonable. Sometimes the input and output of the new system are so different from the old system that it would be very difficult to have the user enter input into two systems and review very different output from two systems. A final example is that a new system sometimes incorporates new organizational policies and procedures not allowed or supported in the old system; therefore, the user must only use the new system.

2. User needs and preferences. Some examples for this follow. The user may simply prefer a particular type of conversion strategy and is willing to pay for it. In one example, user management figures that their employees would have to work significant amounts of overtime to operate two information systems together, so they opt for the cutover strategy. In another example, user management prefers to have their employees use the old and the new system simultaneously in order to increase the comfort level of the employees who are being asked to use the new system.

3. Risk factors. The risk profile of the business, information system, and user are considered here. For example, a bank may require a parallel conversion of its foreign currency exchange system due to its conservative nature. A sales organization may decide the new sales system can sustain a cutover conversion; if it fails they could fall back to the old sales system. Finally, a university decides it is too risky to let all students use the new telephone course registration system the first semester it is in operation so opts for a particular conversion strategy.

Both of the conversion strategies discussed previously, regardless of variation, have potential advantages and disadvantages. Figure 13.4 presents the most frequently cited ones. There is no guarantee that an advantage or disadvantage listed in the figure will actually occur, but, on average, these are the most frequently cited reasons for or against such conversions.

ACTIVATE: THE SECOND PHASE OF IMPLEMENTATION

The second phase of implementation, the **activate** phase, deals with getting the user to use the new or changed information system. This is often a very exciting and anticipated time for the user. Depending on the selected conversion strategy, the user could be one person, a group of people, or the entire business's people. The activate phase

Conversion Strategy	Advantages	Disadvantages
Cutover	1. No duplication of effort for users 2. No transition costs with the old system 3. Learning advantage for groups converted later, if cutover is done by groups	1. High risk 2. Cannot compare results 3. Sense of user insecurity
Parallel	1. Low risk 2. Sense of user security 3. Ability to compare results with old system	1. Duplication of effort for users 2. Transition costs 3. Additional processing strain on the computer

Figure 13.4 Conversion Strategy Advantages & Disadvantages

and the final phase, institutionalization, tend to blur or run together whenever a business chooses a full cutover conversion for all the users. However, for larger and more mission critical information systems, the user is usually some smaller unit of people who are the "guinea pigs" for the new system.

Over the years of doing implementations, software engineer experience has been that a group of initial users of a new or changed information system is usually, although not always, eager, excited, and looking forward to using the new or changed system. Perhaps that is why they were selected to be the initial users, or perhaps the development team did such a good job of preparing these initial users for the system that they simply cannot wait to have it. In either case, this is the ideal kind of users to have waiting to use the new system because their enthusiasm will go a long way to ensuring success for the system. Users can make or break a new system depending on their readiness or resistance level, much of which comes right down to the user's attitude toward using the new system. More will be said about this in the institutionalization section of this chapter.

Once the information system has been installed and the conversion performed, the implementation moves naturally into the activate phase as the predetermined users begin exercising and using the new system. During this critical phase the development team may continue training the users on an as-needed basis. The team should also anticipate and expect problems and be prepared to correct them. The team should also monitor the user's usage of the information system, taking note of what the users like and dislike, patterns of user usage, shortcut opportunities to improve the system's effectiveness, and so on. In short, the development team needs to pay serious attention to the information system and its use by the users during this phase and be prepared to maintain, fix, and enhance it as the need and opportunity to do so become available. On the flip side, the development team may be restricted from making any changes that are not corrective in nature during this phase due to the project's leadership or contractual arrangements with the user. Working under these conditions can be frustrating for both software engineers and the user.

INSTITUTIONALIZATION: THE FINAL PHASE OF IMPLEMENTATION

Mark Twain once said, "Nobody likes change except a baby." In order to move into the final phase of implementation, institutionalization, there must be a high degree of success with the new information system during the activate phase as defined by the users. Lacking success, momentum and progress will stall short of institutionalizing the new information system. Momentum is mostly measured by the number or percentage of users that are using the new or changed information system. Getting people to use the new or changed system is not always easy, as noted by Mark Twain.

Institutionalization means that the new or changed information system becomes the new status quo within the business. Status quo does not guarantee 100 percent use by the users, but it does imply that the new information system is now the standard for the business and that a certain percent of user critical mass has adopted and is using the new information system. For example, my information systems department within the university has decided that a particular word processing software package, such as Word, WordPerfect, or AmiPro, would be the department standard. It did not take long for the majority of faculty to adopt it; however, there were a few hold-outs. Another example involved a much more complicated announcement by management in a software development organization that CASE technology would be the standard in the business for developing information systems. Several years later CASE was still not the status quo because the information system developers were still using their old methods and tools to develop the systems. In this example, the software engineers were to be the users of CASE technology and got a good "taste" of how it feels to be a user.

Figure 13.5 illustrates what can happen to an information system that meets with significant resistance from the users. Resistance can manifest itself in many different ways, but all have the underlying theme of sabotaging the system. Sounds like something right out of a Hollywood movie or a best-selling novel, doesn't it? Well, it is a fact that some users will either directly or indirectly resist a new information system. Why? There are all kinds of reasons that this happens, some being:

1. Fear of job loss, failure, change, uncertainty.
2. Prefer old system over the new one.
3. Politics.
4. Vested interest in the current status quo rather than changing.
5. Pride.
6. Loss or reduction of control (e.g., "big brother" phobia).

Figure 13.5   Implementation Meets with Resistance

7. Management does not allow sufficient time to learn the new system.

8. User deadlines for current projects that use the new software do not allow an increased time factor for the user to become comfortable with the new system and its operation.

Think about your own resistance to certain changes that have occurred in your life. It is a natural reaction and one that must never be underestimated when dealing with implementation of new or changed information systems and technology that involve people.

Most often users resist in a passive or indirect way because this type of resistance is much more difficult to detect. Very few users would directly resist because there could be serious consequences for doing so, such as loss of job, for example. I have often cited a real example of user sabotage that made the headlines in the San Diego newspaper several years ago. The new information system in this article had been under development for over ten years, and the users kept asking for additional features, which caused serious time delays in the implementation and increased the cost by a factor of ten over the original $30 million dollar price tag. What looked on the surface to be complete involvement by the users turned out to be a covert plan to continually delay the new system. Users were allowed to submit their requests for new features and enhancements as the system was being designed. They did so with the hope that the changes would delay the system; thus, they would not have to use it. It was discovered that many of the users thought that the new information system was an attempt to reduce or eliminate the control that they had in their environments. After spending several hundred thousand dollars, the system was eventually canceled. Someone on the project team was quoted in the article as saying something like, "There's a lesson in this somewhere!"

Very few books talk about implementation failure, as if it doesn't exist in the real world. Sadly, countless information systems development projects have either failed or have gotten so far off track that they would have failed had not significant resources been applied to "save" the project. Information systems implementation failures are most often blamed on the technology portion of the newly installed system, as illustrated in Figure 13.6a. The perception and mindset are that the technology is the only thing that has changed in the environment; therefore, it is the source of the failure. Less blame is apportioned to the business's infrastructure and even less to human factors that surround the implementation of a new system. Several research studies have revealed that this perception is opposite to reality, as illustrated in Figure 13.6b. The users, when given the opportunity to talk about the failure, attribute the major failure cause to neglect of human factors, a lesser amount of blame to the business's infrastructure, and even less to the technology itself.

What this suggests is that information systems development teams are usually quite thorough and good at implementing the technology (hardware, software, and data) components of the system but are weaker at identifying, recommending changes in or to, and accomplishing changes in the business's infrastructure and human

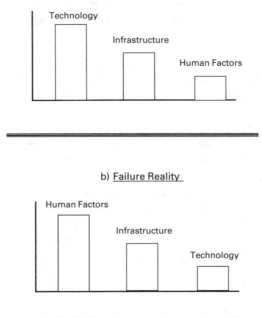

a) Failure Perception

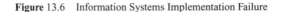

b) Failure Reality

(Note: Bars in both graphs are not to scale; representative only)

Figure 13.6   Information Systems Implementation Failure

factors. These two topics are addressed in the next section, which deals with organizational change issues and how such issues affect information systems implementation.

ORGANIZATIONAL (PLANNED) CHANGE FOR INFORMATION SYSTEMS

Organizational or planned change is a subject that warrants a full semester or more of study in itself. There are books devoted to the topic, so the coverage of it here will be brief and focused on applying it to information systems.

Organizational or planned change is a systematic effort to improve the functioning of some human system. An information system of any consequence could qualify as a systematic effort to improve the functioning of some human system. Oh, most managers and users don't think of it necessarily in those terms because they usually think of a new information system in terms of objectives, such as increasing revenues, decreasing costs, or improving customer service. The reality for accomplishing one or more of these three objectives usually involves an improvement in the functioning of some human system; hence, the information system and the information systems development process should both be considered an organizational change and treated as such. When an information system is viewed this way, attention is not only given to the technology but also to the business's infrastructure and human factors.

Organization infrastructure refers to any policies, procedures, culture, or structure that exist or should exist at the time of system implementation. Culture refers to the "way we do things around here." For example, IBM had a dark business suit and white shirt culture for dozens of years; it also had a promotion from within policy for its top executives until 1993 when it hired Louis Gerstner, an outsider, to be its chief executive officer (CEO). Contrast this with the shirt sleeves and flex-work hours culture of Apple Computer or Microsoft. **Structure** refers to the official, published organization chart as well as the unofficial, unpublished power structure within a business. New information systems often impact the business's infrastructure in one or more ways, and the business's leaders need to know this in advance and deal with it prior to the actual system implementation in order to improve the system's chances for success.

Human factors refer to preparing the business's people for the new information system prior to, during, and after its full implementation. Human factors include, but are not limited to, training users to be able to effectively use the pending information system, job retraining, identifying and addressing user resistance to the pending change, encouraging user movement toward the change, and reinforcement of positives for the change. As presented in the prior section, human factors have been identified as being very important for improving the probability of an information system's success. Human factors alone cannot guarantee success, but if ignored, they can cause it to fail. The definition of failure is diverse; however, the most obvious case is simply that users do not use the new system.

In a typical information systems development project within a business such as an accounting, manufacturing, CAD/CAM, order entry, or other system, the users and the information systems staff both have parallel activities to complete in order to converge at the same time with the information system's implementation. This is generically shown in Figure 13.7. If the user is ready for the information system ahead of the information system actually being ready, frustration and disillusionment might manifest themselves among the users. In addition, if the users have received some or all of their training for the new system, there may be a loss of the newly acquired skills due to nonuse. Conversely, if the information systems development team has the information system ready ahead of the users, users may not be able to fully utilize the new system due to their lack of readiness. The key is to try and coordinate convergence of both sides.

The Stages of Organizational Change

Three stages have been associated with organizational change—unfreezing, moving, and refreezing—as illustrated in Figure 13.8. Researchers have found a parallel or correlation existing between each of these three stages and the three phases of implementation. Unfreezing can be associated with installation; moving can be associated with activation; and refreezing can be associated with institutionalization.

The stages apply both to the business as well as to the individuals within the business. What is meant by this is that it is possible with diagnostic instruments to see

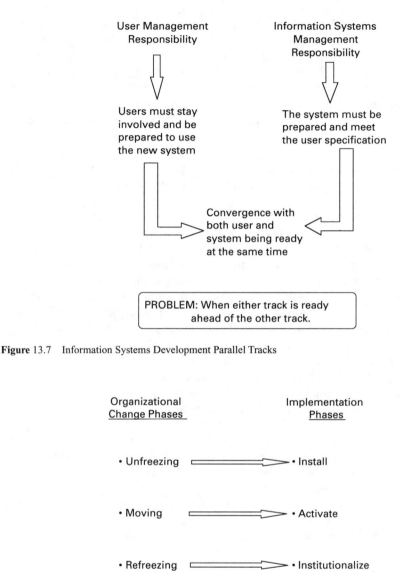

Figure 13.7   Information Systems Development Parallel Tracks

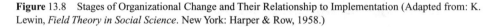

Figure 13.8   Stages of Organizational Change and Their Relationship to Implementation (Adapted from: K. Lewin, *Field Theory in Social Science*. New York: Harper & Row, 1958.)

the movement of individuals as well as the business itself (or groups of individuals within the business) as they move through the stages of organizational change. In almost all situations, individual movement by people within the business is measured through diagnostic instruments. When used within the framework of **action research,** the results obtained through use of the diagnostic instruments guarantee anonymity for the individuals. This is extremely important in order to protect them from potential political ramifications for their honesty. Violate user trust as you assess their movement through unfreezing, moving, and refreezing, and you will never again get the opportunity to measure their movement accurately! Oh, they may respond in the future, but don't count on it being true! Confidentiality and anonymity are essential for action research to work effectively.

At this point a walk-through the three stages of organizational change in a very simplistic fashion may be helpful so that you can get a feeling for each of the stages. Keep in mind that in any business there are dozens of different organizational change activities and projects going on at any moment in time due to the dynamic nature of businesses. Even though our example is simplistic, real-world organizational change is continuous and never ending. In fact most large businesses have one or more organizational change agents in their human resources department or hire outside consultants to assist with organizational changes across the business.

Many adults have a habit of overeating that tends to get out of control during the year-end holidays. Combine that with reduced exercise due to colder weather, parties, shopping, and time off from work, and it leads to extra pounds around the waistline and other unsightly areas of the body. After all the football bowl games on New Year's Day, people's "forces" for remaining status quo (overweight) are equal to or greater than their "forces" to change (lose weight). This is the beginning state, isolating it from all else that is going on in a person's life at this time. For simplicity, let's say that this overweight person is me. Within a few days, I begin to think seriously about reducing my food intake and increasing my exercise in order to lose the extra weight I gained during the holidays. While I think about doing this, perhaps even read an article about dieting or talk to friends about it, I am considered to be in the **unfreezing** stage. Until the forces for change (to lose weight) equal or exceed the forces for remaining status quo (overweight), I will not begin the process of losing weight. Oh, I may think I am doing something simply by reading about diets or talking to someone about it, but until I actually begin doing something, I am not moving.

Once I commit to reduce my food intake, change what I eat, or increase my exercise, I am in the **moving** stage of organizational change. This is a transitional period and may last for quite a long time. There are two important things to keep in mind during the move. First, there is usually a high degree of ambiguity during this time for the person doing the moving because he or she is dealing with the problem and trying to adjust to something new or different. Second, there is a high degree of energy being expended by the person because he or she may have to devote active mental energy, physical endurance, and will power to stay in the moving stage. Occasionally those going through the moving stage will slip (blow it) and have to get back on track again. Sometimes we actually fall back into our old habits and regress back to stage 1 (unfreezing) or even further, that of doing nothing or even increasing our food consumption. There is no guarantee that an individual or a business will "pop out" the far end of moving into the refreezing stage.

If I am successful in the moving stage and lose the desired weight, then I move into the **refreezing** stage where the changes I have made during moving now become the new status quo. If I merely start overeating again or limit my exercise, the cycle will probably start all over again. Right? Right!

Now let's relate the three stages of organizational change to an actual information systems development department within one division of a worldwide corporation. Simplification of this real situation is in order due to space considerations here. The department has about 60 information systems professionals—software engineers and programmers—referred to as staff. Management wanted to upgrade the development environment for each of the 60 staff members from a dumb terminal connected to a mainframe computer for systems development to a client-server environment using CASE technology for each of them who would have their own PC workstation.

Well, none of the staff had PC experience (at least none at their place of work) since there were no PCs available in their department. The unfreezing stage of organizational change can be seen in management's decision to modernize, followed by the actual installation of all the hardware and software (over $250,000). Next the business moved into the activate phase of implementation and the moving stage in organizational change. Management provided training to the staff but it was overwhelming—PCs, DOS, networks, word processors, e-mail, new language compilers for PCs, and so forth. Even more overwhelming was the use of the CASE software along with learning a new way to do their jobs, since CASE necessitated a new methodology for them.

To get to this point in the implementation process, several months had passed. Management thought the staff was now ready to use the new development environment. However, they were wrong! There was a problem. The problem was that only one very small project of about six staff was allowed to use the new environment up to this point. In addition, they were given a project deadline that had no allowance for the newness of all the client-server hardware and software. Out of frustration with the new technology and the deadline, the staff reverted back to their old ways. The old "dumb terminals" were still available for them to use. The new development environment had "hit a brick wall." The software development organization could not move on to the refreezing stage of organizational change and the institutionalization phase of implementation. Had everything gone as management had planned, the staff would all be using the new technology and development environment, and it would become the new status quo for information systems development within the department. As it was, no one was using it and management was very frustrated and wondered why.

The next section presents a strategy for organizational change and the resistance that often accompanies it, such as described in the preceding scenario. The strategy is usable either before problems arise or after they have arisen, as happened in the situation just described.

ACTION RESEARCH AND FORCE FIELD ANALYSIS

Action research is one method for identifying the data associated with the degree of conflict or resistance in a business's environment and then dealing in a real-time manner with it. As Figure 13.6 showed, software engineers' and software engineering man-

agers' perceptions of human factors and organizational infrastructure issues far out-
weigh the technological issues when implementing an information system. It is pre-
cisely these issues that most often are the major causes of implementation failure.

Using the action research method, an organizational change agent administers a
survey or conducts one or more small group meetings at planned points in time with
selected and voluntary staff. The results of this process are provided to the participants
and to management in an anonymous manner, allowing the participants to validate the
data gathered as being representative of their concerns, and allowing management to
see documented issues (with proposed solutions) with which they may make decisions
and take corrective action. Instead of denying or ignoring conflict, the technique is designed to seek out and address conflict so that it can be resolved.

An effective instrument used with either surveys or small groups is a tool called
force field analysis, illustrated in Figure 13.9. In general, this tool is used to identify
an individual's perceived positive and negative aspects about some current or pending
organizational change, such as the example in the preceding section. The data gathered are summarized so as to maintain individual anonymity and presented as de-

a) Blank Force Field Analysis Form

Restraining Forces (Negative)	Driving Forces (Positive)

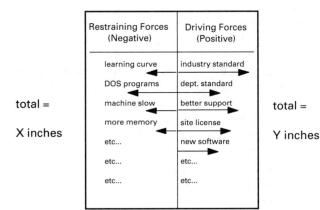

b) Completed Force Field Analysis Form

total =

X inches

Restraining Forces (Negative)	Driving Forces (Positive)
learning curve	industry standard
DOS programs	dept. standard
machine slow	better support
more memory	site license
etc...	new software
etc...	etc...
etc...	etc...

total =

Y inches

Figure 13.9   Force Field Analysis Tool

scribed in the preceding paragraph. The organizational change agent cannot simply assume a generic list of benefits and drawbacks because each group may have a unique set of these that they feel are important to them.

Using force field analysis (FFA) as a survey tool, each person is given a blank FFA sheet and asked to identify barriers (negatives) to the pending change and benefits (positives) of the pending change and list them in the appropriate column of the tool. The barriers and benefits can be anything real or perceived by the individual filling in the tool. The barriers and benefits can, but do not have to, have a direct effect on the person personally or professionally. As each person creates his or her lists, a line is drawn below each item indicating the strength or importance of the item to the person filling out the survey. Looking at Figure 13.9b, notice that the lines drawn under each item move from the center line outward toward the left or right edge of the FFA form. The center line dividing driving and restraining forces symbolically represents the status quo, or point zero (0). Lines moving to the left away from the status quo represent negative numbers, and lines moving to the right away from the status quo represent positive numbers. The change agent will measure the length of each line yielding a list of negative and positive numbers, which are then algebraically added together to arrive at one final number. This number will either be negative, zero, or positive. A negative or zero number indicates that the individual is probably not ready to "move" into this change situation. A positive number indicates that the individual is probably willing to "move" ahead with the pending change.

Putting "strength or importance" lines under each item is equivalent to rank ordering both the positive and negative lists and also yields a negative or positive number, which indicates the relative strength or importance of each item within each list to the individual. Once completed, the FFA surveys can be collected and aggregated for group rank-ordered results for both the barriers and the benefits of the pending change. The rank-ordered lists can then be presented back to the group in aggregate form for verification and clarification. In addition the lists can be used as well by the group for addressing potential solutions to the top 3–5 barrier items. Finally, the rank-ordered lists with potential solutions can be presented to management for their consideration and, hopefully, their action.

Using FFA as part of a group meeting, the organizational change agent facilitates the elicitation of barriers and benefits from the group. This can be done in one of two ways: (1) verbally with the facilitator writing the responses on large sheets off flip-chart paper (positives on one sheet; negatives on another) until the group members begin to show signs of saturation, or (2) through the use of electronic meeting software, which allows each person to anonymously submit barriers and benefits. With either technique, the individuals are asked to cast votes in a variety of ways once the lists are complete. A simple voting strategy is to give each person votes to cast using a simple $n/3$ formula for each of the two lists. For example, if there are 25 barriers identified, each person could vote for eight items ($25/3=8$) that he or she feels the strongest about. If there are 20 benefits identified, each person could cast seven votes ($20/3=7$). The items in each list are then rank ordered from most votes to least. The rank-ordered lists can then be presented to the group for addressing potential solutions to the top 3–5 barrier items. Finally, the rank-ordered lists with potential solutions can be presented to management for their consideration and, hopefully, their action.

Simplified versions of force field analysis are done by most of us every time we have a significant decision to make in our personal lives. Basically, we mentally think of the positives and negatives of the pending decision, even write them down, then decide based on the two lists and their impact on us.

The FFA tool can be used quite effectively with homogeneous groups. Their results can be aggregated for anonymous presentation of the positives and negatives of the group as a whole. I have even worked with the groups to identify possible solutions to the top negative factors that surface using this diagnostic tool. For example, one group identified "insufficient training time" as a negative for introducing the client-server and CASE environment discussed earlier. When asked how management could address this issue in order to reduce it as a resistance barrier, the group suggested that they would be willing to come into the office on the weekends on their own time for additional training if management would allow them access to their work environment on the weekends (currently, the office is closed for security reasons). Management was impressed that the employees were willing to invest some of their own time to address this negative factor. The office was made available for weekend access.

IMPLEMENTATION CRITICAL SUCCESS FACTORS

From an organizational change perspective, successful information systems implementation depends on the following critical success factors, which have been identified in a number of empirical research studies. Implied of course is the fact that the technology portion of the information system be consistent with the user requirements.

1. User commitment. The user must be the champion or advocate for the new information system. If no user is willing to take a stand and strongly support the new system and have organizational influence, problems could arise. Likewise, if no user is willing to take ownership and responsibility for the new system, it is probably destined for failure.

2. Organizational trust. Management and staff within the business, both information systems types and user types, must have a high degree of trust in one another; otherwise the development process and/or the implemented information system would more than likely fail.

3. Open communication between information systems management and staff and user management and staff. Communication problems and breakdowns can cause significant harm to an information systems development project and the information system's implementation.

4. Financial commitment from user management. Lack of or limited funding for the new system can cause significant problems during the development process and during the system's implementation. For example, taking shortcuts to save time or money by reducing a part of the testing or training most often returns to haunt the information system at a later time.

5. Common view of the information system's development and implementation strategy between management and staff. All participants must buy into whatever strategy is chosen for the development and implementation of the information system. If the strategy is confusing, misunderstood, or ignored by all or parts of management and staff, the development and/or the implementation of the system can be seriously jeopardized.

When these critical success factors are supported and encouraged during information systems engineering projects, it significantly enhances the probability of success of the project.

SUMMARY

This chapter explored the intricacies of information systems implementation. Three phases of implementation—install, activate, and institutionalize—were described along with a discussion of the conversion process that accompanies virtually all implementation projects. The conversion activity can become quite involved, which is why there are two primary conversion strategies—cutover and parallel.

Over the years technology has taken a lot of "heat" over implementation failures. Unfortunately, the heat is most often misfocused. The fact is that human and business infrastructure issues contribute more to implementation failure than do technology issues.

An important organizational change concept introduced by Lewin in the 1950s can have a significant effect on information system implementation. Lewin's three phases of organizational change—unfreezing, moving, and refreezing—are strongly correlated with the three phases of implementation mentioned earlier. Studies have shown that information systems implementations have a higher probability of success when organizational change issues are considered and addressed sometime during the information systems development and implementation.

Action research is an effective strategy to use to identify and deal with organizational change issues. The force field analysis tool is used as part of action research to identify user resistance to change and then identify strategies to deal with the resistance. This technique is designed to be done in an anonymous way and involve the users directly in identifying the solutions to their own resistance issues.

The chapter concluded with a list and discussion of the most often identified critical success factors for improving the probability of successful information systems implementation.

QUESTIONS

13.1 List and define the three phases of implementation.

13.2 What is conversion, and what are the two basic conversion strategies?

13.3 What are the three main factors to consider when choosing a conversion strategy?

13.4 What are some of the advantages and disadvantages of the different conversion strategies?

13.5 What is one of the important consequences that the development team almost always faces during the activation stage of implementation?

13.6 What are some of the reasons that new or changed information systems meet with user resistance?

13.7 Briefly discuss a common misperception about why a new information system fails.

13.8 How do the stages of organizational change relate to the stages of information system implementation?

13.9 What is force field analysis and how is it used to deal with user resistance to information systems?

13.10 What are some of the critical success factors for information system implementation?

REFERENCES

BEAN, A., R. NEAL, M. RADNOR, and D. TANSER, "Structural and Behavioral Correlates of Implementation in U.S. Business Organizations," *Implementation Operations Research/Management.* New York: American Elsewier Publishing, 1975, pp. 77–132.

CORBITT, G.F., and R.J. NORMAN, "CASE Implementation: Accounting for the Human Element," *Information and Software Technology* (November 1991).

CORBITT, G.F., R.J. NORMAN, and M.C. BUTLER, "Assessing Proximity to Fruition: A Case Study of Phases in CASE Technology Transfer," *International Journal of Software Engineering and Knowledge Engineering,* 1, no. 2 (June 1991), 189–201.

KEEN, PETER, "Information Systems and Organizational Change," *Communications of the ACM,* 24, no. 1 (January 1981), 24–33.

LEWIN, K., *Field Theory in Social Science.* New York: Harper & Row, 1958.

LEWIN, KURT, "Group Decision and Social Change," *Readings in Social Psychology.* New York: Henry Holt and Company, 1952, pps. 459–473.

MORRISON, D.E., "Making Change User-Friendly: Managing the Impact of Technology on People," *Information Executive* (Spring 1989).

NORMAN, R.J., and G.F. CORBITT, "The Operational Feasibility Perspective," *Journal of Systems Management* vol. 42, no. 10 (October 1991).

NORMAN, R.J., and G.F. CORBITT, "A Cure for Technology Implementation Headaches," *Chief Information Officer Journal,* 3, no. 4 (Spring 1991), 35–38.

NORMAN, R.J., G.F. CORBITT, M.C. BUTLER, and D.D. MCELROY, "CASE Technology Transfer: A Case Study of Unsuccessful Change," *Journal of Systems Management,* Association of Systems Management (May 1989).

ROBBINS, S.P., *Organizational Behavior* (5th ed.). Englewood Cliffs, NJ: Prentice-Hall, 1991.

ROGERS, EVERETT, *Diffusion of Innovations.* New York: The Free Press, 1972.

SCHULTZ, R., M. GINZBERG, and H. LUCAS, "A Structural Model of Implementation" in Randall Schultz and Dennis Slevin (eds.), *Implementing Operations Research/Management Science.* New York: American Elsewier Publishing, 1984, pps. 55–88.

SORENSON, R.E., and D.E. ZAND, "Improving Implementation of OR/MS Models by Applying the Lewin-Schein Theory of Change" in Randall Schultz and Dennis Slevin (eds.), *Implementing Operations Research/Management Science.* New York: American Elsewier Publishing, 1984, pps. 217–236.

SWANSON, E.B., *Information Systems Implementation.* Boston: Irwin, 1988.

WOOD-HARPER, TREVOR, "Research Methods in Information Systems: Using Action Research" in E. MUMFORD et al. (eds.), *Research Methods in Information Systems,* North Holland: Elsevier Science Publishers, 1985, pps. 169–191.

ZAND, D.E. and R.E. SORENSON, "Theory of Change and the Effective Use of Management Science," *Administrative Science Quarterly,* 20, no. 4 (December 1975), 532–545.

ZMUD, R.W., and J.F. COX, "The Implementation Process: A Change Approach," *MIS Quarterly,* 3, no. 2 (June 1979), 35–43.

Module A

Information Systems Planning

MODULE OBJECTIVES (YOU SHOULD BE ABLE TO)

1. Define information systems planning.
2. Describe a generic information systems planning methodology.
3. Describe the input to information systems planning.
4. Describe the output from information systems planning.
5. Describe the information architecture.
6. Discuss why it is important to perform information systems planning.
7. Discuss potential problems that could arise as a result of not engaging in information systems planning.
8. Name some information systems planning techniques and methodologies.

INTRODUCTION

Do you have a plan for what you will do today, tomorrow, this coming weekend, next week, next month, during semester break, during spring break, during the summer months, after graduation? Do you have a plan for completing your education? You probably do. That plan consists of a list of courses taken, in progress, and yet to be taken, along with a tentative semester or quarter when you would like to take or have to take the courses. Suffice it to say that most professional men and women and aspiring professional men and women have professional plans. That's why DayTimer, Franklin, and other calendaring systems are so popular.

Dating from the early days of business computing in the 1950s, information systems have been developed or acquired according to some plan. Quite often the plan was based on a manager's or group of managers' prioritization of information systems projects. Many times the prioritization was based on the political clout of the managers or the criticalness of the need for the proposed information system to the business operation. Politics and "fire fighting" are rarely based on solid business decision making aimed at making the greatest impact for the business as a whole rather than a small part of the business.

The purpose of **information systems planning** is to identify and prioritize the information systems applications whose development and implementation would most benefit the business as a whole. The scope of information systems planning could be the entire business, a division, a plant, or some other significant organizational unit. Its measurable objective is to look for opportunities to exploit information technology to support the objectives of the business unit being considered.

Organizations that have an ongoing, rigorous, and formalized information systems planning activity usually separate this activity from the *Analysis* activity as shown in Figure A.1. Organizations that focus their planning activities primarily on project (detail) level planning usually incorporate their planning activity as part of *Analysis* as shown in Figure 1.8 in Chapter 1.

Similar to systems analysis and design, information systems planning has inputs and outputs. As illustrated in Figure A.1, the inputs to information systems planning are existing information system details along with the business's current mission and goals. The output from information systems planning is the information systems plans

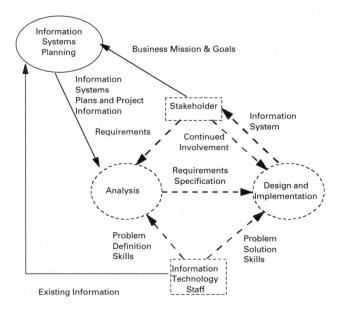

Figure A.1   Information Systems Planning Inputs and Outputs

and projects, which then become the input to the analysis phase of the systems development life cycle, as shown in Figure A.1.

Information systems planning is an ongoing activity which must be repeated frequently to ensure that information systems continue to be developed according to the information systems plan, and to update the process with any changes that are occurring due to management decision or other external business factors. The frequency of information systems planning varies from business to business and even from week to week, month to month, or quarter to quarter depending on how dynamic the business climate is. A reasonable guideline is to revisit information systems planning at least once every quarter of a year.

A GENERIC INFORMATION SYSTEMS PLANNING METHODOLOGY

There are many different information systems planning methodologies—perhaps as many as there are businesses doing information systems planning. Figure A.2 depicts a simple approach to information systems planning for your consideration and understanding of the concept. Businesses desiring you to be directly involved in information systems planning for them will train you in their specific methodology. Note that Figure A.2 is an exploded detail version of the information systems planning symbol shown in Figure A.1.

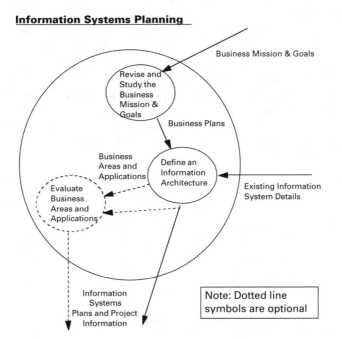

Figure A.2   A Generic Information Systems Planning Methodology (Adapted from J. L. Whitten, L. D. Bently, and V. M. Barlow, *Systems Analysis & Design Methods,* 3rd ed., Boston: Irwin, 1994, p. 103.)

Doing information systems planning correctly is difficult if a business or business unit does not have a well-articulated mission and goals statement because this is fundamental to doing information systems planning. If information systems are to truly add value to the business, they must address the mission and goals.

The first activity within the generic information systems planning methodology is to study and revise the business mission and goals. This sounds a bit misleading. The systems analysts, users, and managers who are participating in this activity do not actually change the business's mission and goals. They may need to modify the current version of the business mission and goals that they used the last time they performed information systems planning with the revised mission and goals that they have received from the business or business unit. A good guideline to remember is that corporatewide missions and goals for large businesses (e.g., *Fortune* 1000 companies) change very infrequently, while small businesses or small business units of large companies may change their mission and goals on a more frequent basis. For example, IBM as a corporation has had the same mission for many years, while its Personal Computer Division has changed some of its goals (and perhaps even modified its mission) over the last few years as it determines the best way to impact its market.

In order for this first activity to be successful, senior management within the business or business unit must be willing to participate in the identification and articulation of the mission and goals. Without their input and support, this activity might as well be skipped. Figure A.3 presents the video store mission statement and a partial list of its goals.

The key deliverable from the first activity is business plans. A business that is keen on doing business planning will already have such plans. If it does, the plans may or may not get refined in terms of information system needs as they are passed on to the second activity within information systems planning. If the business does not have business plans, then they are created during the first activity and passed along to the second activity.

Once business plans are known and available to information systems planning, the information systems planning team can create an information architecture for in-

---

MISSION STATEMENT

To be the video store of choice by successfully providing a generous selection of home video products for sale or rental at competitive prices.

GOALS

1. Increase market share and maintain profitability
2. Offer superior customer assistance and browsing environment
3. etc...

Figure A.3   Video Store Mission Statement and Goals (partial)

formation systems, the second activity within the generic information systems planning methodology presented here. An **information architecture** is a plan for selecting the appropriate information technology and information systems that will best support the business mission and goals. The first time that an information architecture is developed for a business and depending on the business size, it could take six months to a year to create. Another factor that seriously determines the length of time to complete this activity is the amount of dedicated time the planning team is allowed to work on it. If they work part-time on it due to other commitments, it will take much longer.

Along with the business plans, existing information systems details are input to this activity. As with all human activities within systems development, people's comments, desires, biases, and ideas are also considered input to each of these activities. The information architecture plan consists of many separate but related planning documents. For example, the following planning documents may be developed:

1. A people architecture, referred to as an organization chart.
2. A telecommunications network architecture.
3. A data architecture, called an enterprise data model, depicting all business data.
4. An applications architecture showing all information systems applications along with their interfaces to each other.
5. A technology architecture laying out plans for the information technology that should be used to develop future systems.

Looking at the preceding list of planning documents, it isn't hard to see why this activity takes so long. Many large businesses have abandoned the idea of creating a corporatewide information architecture because it takes so long and the business units it would represent are so dynamic that by the time the plan is complete, it would have to be revisited and revised.

The output from this activity is the information systems plans and project information which go one of two places in the generic information systems planning methodology. This output can go directly into systems analysis as depicted by the solid arrow leaving the information systems planning circle in Figure A.2. Or it can become input to an optional activity called "evaluate business units and applications" along with another input—business areas and applications. Refer to Figure A.2 again for the illustration of this.

Assuming the third activity is part of the information systems planning methodology in use by your business, business areas and applications are input to it along with the information systems plans and project information mentioned previously. Business areas are groups of logically related business functions and activities, independent of the business structure. For example, an order entry and order fulfillment process may be identified as a business area and application. The order entry and order fulfillment process runs across several organizational units, such as sales department, order process department, accounts receivable department, warehouse department, and shipping department. The information system that supports order entry and order fulfillment should integrate aspects of each of these departments in order for the process to flow smoothly.

In this third activity, the business areas and applications are evaluated and prioritized according to their importance to the mission and goals of the business. Once determined, they can be assigned project priority and when appropriate will be moved to the analysis activity of systems development as an information systems development project.

As mentioned earlier, information systems planning is not a static, one-time activity. It is dynamic and ongoing. The first time it is done, it will take much longer than subsequent planning meetings to revise and update the plans.

WHY ENGAGE IN INFORMATION SYSTEMS PLANNING?

As stated earlier, information systems planning is a process that attempts to employ information system resources in a manner that aligns itself with and supports the business mission and goals of the business or business unit. By doing so, the information systems plan can be directly linked to the overall business plan for the business or business unit. This can have a profound impact on the competitive strategy of the business or business unit.

Another reason to engage in information systems planning is to avoid a number of problems that can be suffered by the business if it does not engage in information systems planning. Figure A.4 lists some representative problems and each is briefly discussed here. For example:

1. Rushed work near the end of a project, causing the team to minimize critical implementation tasks.
2. Missed opportunities to take advantage of product differentiation, and productivity and management situations.
3. Loss or reduction of credibility with users because of missed deadlines.
4. High prices paid for hardware, software, and telecommunications equipment because of short deadlines to acquire.

- rushed work near the end of a project, causing
  the team to minimize critical implementation tasks
- missed opportunities to take advantage of product
  differentiation, and productivity and management
  situations
- loss or reduction of credibility with users because
  of missed deadlines
- high prices paid for hardware, software, and
  telecommunications equipment because of short
  deadlines to acquire
- disruption of users when system resources peak
  out causing slowdowns and response time delays

Figure A.4   Potential Problems When Not Doing Information Systems Planning

5. Disruption of users when system resources peak out causing slowdowns and response time delays.

Note also that information systems planning does not guarantee that a business will avoid each of the foregoing problems, but it does increase the probability of limiting or minimizing them due to the practice of proper planning activities.

INFORMATION SYSTEMS PLANNING TECHNIQUES AND METHODOLOGIES

There are many information systems planning techniques and methodologies as mentioned at the beginning of this module. There have been a few which have been well articulated in the information systems and business literature over the years. The most common techniques are Jack Rockart's critical success factors (CSF) and Michael Porter's value chain analysis. Both of these techniques are used to perform business planning and can also be used to perform information systems planning.

A few of the commercially available information systems planning methodologies include IBM's Business Systems Planning (BSP), one of the oldest methodologies being used today, Foundation by Andersen Consulting, and Navigator by Ernst & Young. Each of these methodologies is well documented and supported by software. Many commercially available CASE technology software packages contain information systems planning modules, some including the ones mentioned here.

SUMMARY

Information systems planning is becoming an important part of information systems development within larger businesses in order to more fully align information systems goals with the business goals, and to justify the funding of information systems projects. A generic information systems planning methodology was presented and illustrated along with problems that could arise should a business fail to do adequate information systems planning. Finally, a few information systems planning techniques and methodologies were presented as examples of commercially available approaches to do information systems planning.

QUESTIONS

A.1 What is the purpose of information systems planning?

A.2 What are the inputs and outputs to information systems planning?

A.3 What are each of the planning steps within a generic information systems planning methodology? Describe their purpose.

A.4 Describe some of the problems that could arise as a result of no information systems planning.

REFERENCES

BURCH, J.G., *Systems Analysis, Design, and Implementation.* Boston: Boyd & Fraser, 1992.

EMERY, J.C., *Management Information Systems: The Critical Strategic Resource.* New York: Oxford University Press, 1987.

GRAY, P., W.R. KING, E.R. MCLEAN, and H.J. WATSON, *The Management of Information Systems.* Chicago: The Dryden Press, 1989.

PORTER, M., "How Information Gives You Competitive Advantage," *Harvard Business Review,* 63, no. 4 (July–August 1985), 149–160.

PORTER, M., *Competitive Advantage: Creating and Sustaining Superior Performance.* New York: Free Press, 1985.

WHITTEN, J.L., L.D. BENTLEY, and V.M. BARLOW, *Systems Analysis & Design Methods* (3rd ed.). Boston: Irwin, 1994.

ZACHMAN, J., "A Framework for Information Systems Architecture," *IBM Systems Journal,* vol. 26, no. 3 (1987).

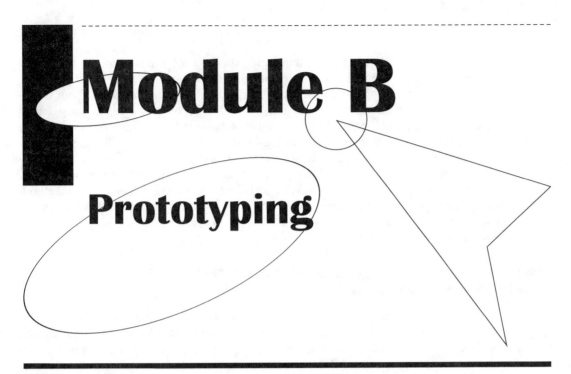

Module B

Prototyping

MODULE OBJECTIVES (YOU SHOULD BE ABLE TO)

1. Define prototyping.
2. Discuss prototyping as a technique and methodology.
3. Discuss the differences between product prototypes and information systems prototypes.
4. Discuss prototyping benefits.
5. Discuss the risk of prototyping.
6. Name some prototyping synonyms.
7. Describe the enabling technologies for prototyping.
8. Describe how prototyping can be initiated in a business.

Without a doubt, we all know that a **prototype** is the first or primary type of anything created. Prototypes abound in the real world for several reasons, such as (1) for proof of viability, (2) due to cost and time to build the real thing, and (3) improved understanding of the real thing before it is built. For example, television producers create one or more pilots (prototypes) of new shows to see if there is a viable audience for the show. Automobile manufacturers create concept cars (prototypes) of new model cars prior to incurring the huge cost and time commitment for retooling to build a production version of the car. The recently completed Biosphere II experiment (prototype) in Arizona increased researchers' understanding of humans living for an extended period of time in a fully contained ecological system.

In information systems, systems analysts think of a prototype as a working model or replica of the real system. **Prototyping** is the name given to the process systems analysts go through to create the prototype. There is much debate about prototyping being considered a full-fledged methodology or just a technique used to develop information systems. My perspective is that it is a technique because it can be incorporated into almost any methodology just as other kinds of techniques, such as joint application design (JAD), data flow diagrams, and structure charts can be included within a methodology. Nonetheless, I recognize that there are those who say that prototyping is a strategy for creating an information system; hence, it is a methodology. From either perspective, the point is that prototyping must have a certain amount of discipline and rigor within it in order to contribute to the success of information systems development.

PROTOTYPING'S PLACEMENT WITHIN A SYSTEMS DEVELOPMENT LIFE CYCLE

Figure B.1 illustrates a traditional systems development life cycle that could be supported by a variety of methodologies, techniques, and tools. Although prototyping can be used in a number of ways within the life cycle, its use is being depicted in a generally accepted way in Figure B.2. Note that prototyping has a collapsing effect on the formal boundaries between definition, design, and construction.

In 1984, Bernard Boar authored a book in which he presented his view of prototyping. His prototyping approach was mildly controversial because it came out during a period in software development when the "experts" were saying that the "right" way to develop information systems was to rigorously analyze the user requirements first and then plunge into the details of design and implementation. Boar championed prototyping as a legitimate way of discovering user requirements during the analysis phase of systems development. His view of the prototyping development life cycle is illustrated in Figure B.3.

Since the mid-1980s, prototyping has been widely praised as an effective way of exploring alternative human interfaces for an information system, especially in recent years with the introduction of the graphical user interface. Prototyping has also been very useful in assisting with the eliciting of requirements from a user who is unsure of the very nature of the information system that he or she wants created. Such is the case for many decision support systems and executive information systems.

PRODUCT VERSUS INFORMATION SYSTEMS PROTOTYPING DIFFERENCES

Although the concept of prototyping for information systems is the same as prototyping for other products, there are three general differences:

1. For product prototypes, development of the prototype usually occurs over a long period of time; for information systems prototypes, development of the prototype usually occurs over a short period of time.

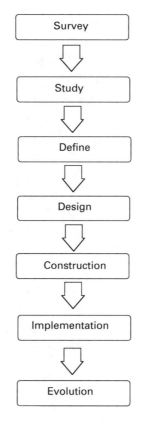

Figure B.1   Generic Systems Development Life Cycle

2. For product prototypes, the prototype usually costs more than the production models; for information systems prototypes, the prototype usually costs less than the production information system.

3. For product prototypes, multiple production models are usually manufactured after the prototype; for information systems, usually only one production version is created.

PROTOTYPING BENEFITS

The prototyping concept has been in existence forever it seems. Applying it successfully to information systems has occurred over the last two dozen years. The benefits of prototyping that have been justifiably touted by its advocates include (1) shorter development times, (2) improved information system quality, and (3) economics. Each is discussed here.

Shorter development times can often be achieved because prototyping encourages greater user involvement in the development process, thus allowing the real requirements to be determined in a shorter amount of time. The user can actually see a

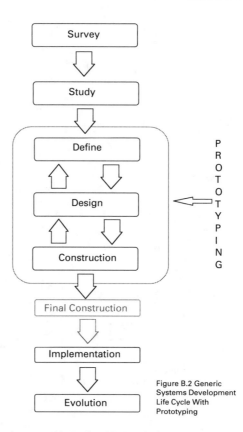

Figure B.2 Generic
Systems Development
Life Cycle With
Prototyping

Figure B.2   Generic Systems Development Life Cycle With Prototyping

working model of his or her system rather than just written words and diagrams of the system. Users are much better at telling systems analysts what they like and don't like in the system after they see it than before they see it.

The second benefit, improved information system quality, also comes from the user being more involved in the prototyping process and actually seeing the prototype. The iterative nature of prototyping gives opportunities to refine and purify the system.

Economic benefits are often obtained with prototyping due primarily to a shorter systems development time frame often achieved when prototyping is used. This translates into either lower development costs, or increased benefits in the form of increased revenue or improved services obtained sooner than would be without the use of prototyping, or both.

PROTOTYPING'S RISK

Although the benefits of prototyping are significant, it is not without risk. Systems analysts must guard against too little analysis and too much prototyping. The temptation is to

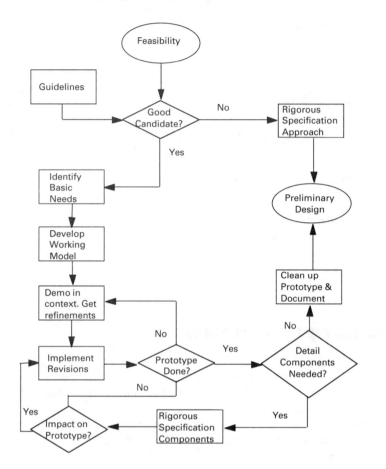

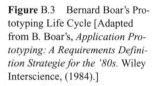

Figure B.3   Bernard Boar's Prototyping Life Cycle [Adapted from B. Boar's, *Application Prototyping: A Requirements Definition Strategie for the '80s.* Wiley Interscience, (1984).]

begin prototyping prematurely before the problem domain has been defined. If prototyping begins too soon in the systems development life cycle, the development time can actually take more time than traditional, nonprototyped development methods. When the problem domain is not well defined prior to prototyping, the user may believe that anything and everything should belong in the information system. The systems analyst, not having a well-defined problem domain either, may keep accepting additions to the system every time he or she reviews the prototype with the user. Neither the user nor the systems analyst has a specification document to follow so the project becomes laden with undocumented assumptions and expectations. These misunderstandings can place significant stress on any systems development project and even cause it to fail if not corrected.

PROTOTYPING SYNONYMS

Prototyping is often referred to in the literature by other names. There may be some subtle differences between each of the techniques listed in Figure B.4, but for the most part each is similar to prototyping.

- Trial-and-Error Approach

- Evolutionary Development

- Iterative Development

- Adaptive Design

- Hueristic Development

- Rapid Prototyping

- Rapid Application Development (RAD)

Figure B.4   Prototyping Synonyms

ENABLING TECHNOLOGIES FOR PROTOTYPING

There are several technologies that are either required or desirable when prototyping. The greater the number of these technologies in place, the greater the probability for successful prototyping, risks as discussed earlier notwithstanding. A caution flag should be raised for a business that implements several of these technologies at the same time it expects to begin a prototyping project because this can place an overwhelming burden on the team members to learn new technologies while they are attempting to utilize prototyping.

Figure B.5 lists each of the technologies. Prototyping is virtually impossible if the development team does not have personal workstations or on-line access via display terminals to the development personal or mainframe computer. Most modern businesses understand the benefits of on-line technology and, therefore, have it already installed and operational.

Relational database management systems (DBMS) and fourth generation languages (4GL) are also commonplace in larger businesses. An all COBOL and file-based software development environment would increase its prototyping success probability by implementing these technologies.

Introducing an industry standard query language such as Structured Query Language (SQL) or even a Query By Example (QBE) type language would also significantly facilitate the prototyping effort. Users and developers could use either of these technologies to quickly show results from user requests for information.

Computer-aided software engineering (CASE) technology can be a powerful prototyping technology. Users and developers can create display screen samples, database definitions, report definitions, and module logic using Structured English or

- On-Line Access for Developers

- DBMS/4GL - Relational or Object

- Query Language - SQL or QBE

- CASE With Reverse Engineering Support

- Simulation Capability

- Testing Facilities

Figure B.5   Enabling Technologies for Successful Prototyping

pseudocode. Some CASE products can even generate COBOL, C, or other programming language code from these definitions.

Simulation can also assist prototyping. Certain applications lend themselves nicely to simulation and having such a technology available during prototyping of these types of applications would greatly enhance the chances for success. Airline flight simulators are a popular example of a simulation capability used to train pilots.

A final enabling technology to assist prototyping is that of testing facilities. The kinds of testing facilities being referred to here are the systems analyst workbench types of testing facilities which lend computer-aided testing support to the prototyping environment. Automated test generators to create test data fit well here along with other computer-aided test support. In addition, some prototypes have to prove their viability in a variety of environments, such as different operating system environments—Windows, Mac, Unix, for example, and different hardware environments. Having physically good testing facilities including a variety of workstations, monitors, printers, networks, and so on can contribute significantly to improved prototyping in the situations that need these.

DOES PROTOTYPING WORK?

Selecting an appropriate software development strategy for each software development project is crucial to building an on-time, within budget, successful information system. Even though the waterfall (and its many variations) system development life cycle remains the dominant paradigm, evolutionary or prototyping methods are making a significant contribution to systems development.

A number of prototyping case studies have been reported over the years in the literature. A recent article evaluated a number of these case studies and summarized some of the results of these studies in a way that is helpful for this chapter's discussion of prototyping. It needs to be pointed out that the reported results represent the collective case study researchers' perceptions of the effects of prototyping in the case study each reported on. Figure B.6 summarizes six software product attributes and four human and staffing factors that are often affected by prototyping. Notice that seven of the ten items in the figure show a more positive prototyping effect while the remaining three show a more negative prototyping effect. The bottom line on prototyping is that each software development project should assess these and other factors when deciding to prototype or stick to the long-standing SDLC.

HOW TO INITIATE PROTOTYPING

A business that desires to incorporate prototyping into its development activities should consider the following items:

1. Get management and customer approval and support.
2. Put as many enabling technologies in place as possible and begin using them.
3. Pick a small pilot project that has a high probability of being successful.
4. Select and train progressive staff members to be on the project and be the prototyping "champions."

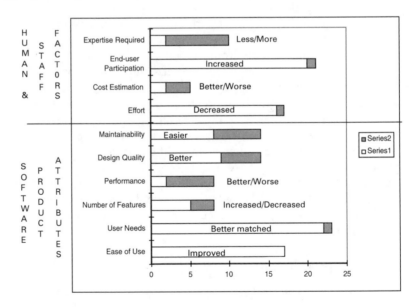

Notes:
1) A total of 39 case studies were evaluated; when a bar does not add up to 39 it means that one or more studies did not address that factor or attribute.
2) Series 2 (grey) items represent the opposite of the Series 1 (white) items. Example: white = Improved; then grey = not improved.

Figure B.6 Summary of Prototyping Effects from Case Studies [Based on Figures 1 and 2 in: V. Scott Gordon and James M. Bieman, "Rapid Prototyping: Lessons Learned," *IEEE Software*, 12, no. 1 (January 1995), 85–95.]

SUMMARY

Prototyping as a technique or a methodology is receiving more attention in systems development projects than ever before. Benefits include shorter development time, higher-quality systems, and improved economics. The risks are minimal but systems analysts must guard against doing too little analysis and too much prototyping. User involvement is enhanced and often the user feels less intimidated by the systems development process. Certain enabling technologies lend strong support for prototyping's success, and getting management's support and selecting appropriate projects and staff are essential for successful prototyping.

QUESTIONS

B.1 What are the main differences between prototyping for an information system and prototyping for other products?

B.2 List and briefly discuss some of the advantages of prototyping.

B.3 What are some of the other names with which prototyping is synonymous?

B.4 How is CASE technology used in the prototyping process?

B.5 What are some of the major concerns that need to be considered before incorporating prototyping into development activities?

REFERENCES

BOAR, B., *Application Prototyping: A Requirements Definition Strategy for the 80s.* NY: Wiley Interscience, 1984.

BOAR, B., *Application Prototyping—A Project Management Perspective.* New York: American Management Association Membership Publications Division, 1985.

DAVIS, ALAN M., *Software Requirements: Analysis & Specification.* Englewood Cliffs, NJ: Prentice Hall, 1990.

FLAATTEN, PER O., DONALD J. MCCUBBREY, P. DECLAN O'RIORDAN, KEITH BURGESS, *Foundations of Business Systems* (2nd ed.). New York: The Dryden Press, A Harcourt Brace Jovanovich College Publisher, 1992.

GORDON, V. SCOTT, AND JAMES M. BIEMAN, "Reported Effects of Rapid Prototyping on Industrial Software Quality," *Software Quality Journal* (June 1993), pp. 93–110.

GORDON, V. SCOTT, AND JAMES M. BIEMAN, "Rapid Prototyping: Lessons Learned," *IEEE Software,* 12, no. 1 (January 1995), 85–95. (Note: This article has a large list of additional prototyping articles included within it.)

Module C

Computer-Aided Software Engineering (CASE)

MODULE OBJECTIVES (YOU SHOULD BE ABLE TO)

1. Define CASE.
2. Define the goal of CASE.
3. Describe the objectives of CASE.
4. Describe the architecture of CASE.
5. Describe the stages of CASE.
6. Describe the benefits of CASE.
7. Describe the issues of CASE.

INTRODUCTION

Since the mid-1980s, workstation and personal computer-based information systems development environments have been significantly enhanced with the introduction, acquisition, and acceptance of computer-aided software engineering (CASE) technology. CASE is more than just software. It is an information system because it incorporates all five components—hardware, software, people, procedures, and data. Therefore, CASE is an information system that helps people develop information systems.

Although CASE has its roots in the late 1960s and early 1970s, the current workstation-based, graphical user interface CASE has its roots in the early 1980s, just after the introduction of the IBM personal computer. One of the first commercially available CASE drawing packages was DFD Draw.

The overriding goal of CASE technology is the automation of the entire information systems development life cycle process using a set of integrated software tools, techniques, and methodologies. Wow! What a lofty goal! In reality, the software engineering community has been pursuing this goal ever since the invention of the computer, and the pursuit may very well continue throughout the remainder of your lifetime. Why? Well, the information systems development process has been very similar to someone shooting at a moving target. The goal remains the same, but the hardware technology keeps changing, which causes or allows software engineering researchers to make changes and enhancements in the information systems development process activities.

As CASE pursues the long-term goal of automation of the entire information systems development life cycle, a number of shorter-term objectives are being addressed as shown in Figure C.1. Over the years numerous organizations have cited improvements in a number of the objectives listed in the figure, and they hope to see additional ones as they continue to rely on CASE or its successor for developing information systems.

CASE ARCHITECTURE

The architecture of generic CASE products is often depicted in a hierarchical manner as illustrated in Figure C.2. Historically, CASE vendors tended to develop either **upper or front-end CASE** products or **lower or back-end CASE** products primarily because of the cost and time to develop and support these products. Several years ago a prominent software engineering researcher projected that it would cost as much as $80 million for a CASE vendor to create a "complete" CASE product with over 100 development tools in it.

- Improve productivity of developers

- Improve software quality (zero defects)

- Speed up the systems development process

- Reduce the cost of systems development

- Automate design recovery and reverse engineering of systems

- Automate systems documentation

- Automate programming code generation

- Automate validation and verification (error checking)

- Automate project management tasks

- Promote improved control over the development process

- Integrate systems development steps and tools

- Promote software reusability and portability

Figure C.1 CASE Technology Objectives

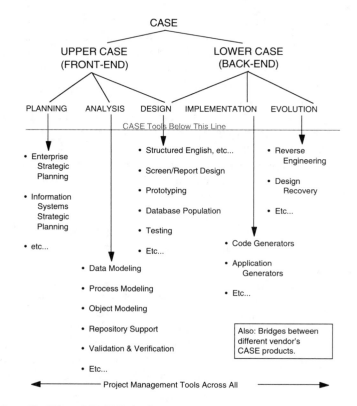

Figure C.2   Hierarchical View of CASE Technology

The upper CASE products include development tools that are to be used during the planning, analysis, and early design phases of information systems development. The lower CASE products include development tools that are to be used during the design, implementation, and evolution phases of systems development. This created an artificial gap for the outputs generated from using the upper CASE tools to be able to be utilized as the inputs to the lower CASE tools. Recognizing this, CASE vendors quickly developed "software bridges" between upper CASE and lower CASE, which would allow importing and exporting between the different vendors' CASE products. In recent years several upper CASE vendors have merged with lower CASE vendors in order to create a more seamless CASE product.

The use of CASE has proven valuable in a number of areas, but perhaps none more so than in the area of drawing and assisting with the verification and validation of user specifications. For example, data flow diagrams, entity-relationship diagrams, hipo charts, Warnier-Orr diagrams, and object-oriented diagrams are just a few of the diagrams that most upper CASE products can assist with. Without CASE, systems analysts/software engineers must draw these types of diagrams by hand or with a generic software drawing package, such as Microsoft's Powerpoint. In an information system of any consequence, this can mean dozens of diagrams of the user's detail requirements. Changes, which occur quite frequently, would necessitate the redrawing of many diagrams or portions of diagrams which would consume valuable time. With CASE, redrawing diagrams is quite easy and quick. Combine that with automatic diagram validation by the CASE drawing software and CASE becomes even more valuable to the software developers.

THE STAGES OF CASE USAGE

One research study revealed that many businesses go through five distinct stages as they integrate CASE into their development efforts. These stages were identified as **disenchantment, resignation, commitment, implementation, and maturity,** and each is briefly discussed here.

Usually after an initial period of time in which the systems analysts are excited about CASE, they begin to experience the limitations of CASE and become disenchanted with it. Nonetheless, management has made a significant investment to bring CASE into the business so the systems analysts are resigned to the fact that they need to live with CASE and accept its current limitations. Once over this mental hurdle, the systems analysts get serious about making it work and commit to using it as long as management continues to support and endorse it. CASE now moves into a much broader implementation, making it the status quo for systems development within the business. When this point is reached, CASE has reached a maturity level within the business. That does not mean that the ultimate CASE is in place, but that CASE is a major part of that organization's information systems development efforts. The organization will no doubt keep enhancing its use of CASE and upgrading to newer versions of the CASE package as they become available from the vendor or switch to a new automated support paradigm as it becomes available in the future.

THE BENEFITS OF CASE

As mentioned earlier, many CASE vendors have merged in order to create a more seamless integration across the entire systems development life cycle. Over the years businesses have collectively reported receiving, at least partially, the benefits outlined in Figure C.3. More progressive information systems development organizations are constantly pursuing these benefits more fully.

THE ISSUES OF CASE

Earlier the limitations of CASE were alluded to. CASE does have limitations and will continue to have limitations into the future. A constant debate among software engineering researchers and practitioners is that of **rigidity versus flexibility.** Rigidity brings with it a certain amount of rigor or discipline, whereas flexibility tends to invite creativity and variability. Information systems development has traditionally favored flexibility, whereas CASE is attempting to introduce a higher level of rigor or

- Makes the use of structured and OO techniques practical

- Enforces a more rigorous software engineering discipline

- Improves software quality through automated validation

- Makes prototyping more practical

- Simplifies software maintenance and evolution

- Speeds up the systems development process

- Frees the developers to focus on the creative part of

    problem solving for the system

- Encourages evolutionary/incremental development

- Enables the reuse of model and software components

- Provides improved project management tools

Figure C.3   The Benefits of CASE Technology

discipline into the practice of information systems development. The debate continues, and it is actually a very healthy and worthwhile debate for the software engineering community.

Not all systems analysts, researchers, and software engineering managers are "sold" on CASE, even though it has been available for over a dozen years. As discussed in an earlier chapter, organization culture and infrastructure are strong contributors to embracing or criticizing CASE technology.

- True model to model integration within CASE

- Stronger coupling between design models and code generation

- Stronger client (workstation) to server (host computer)

    integration for repository population and maintenance

    for systems development workgroups

- Stronger connection between CASE and other technologies

    such as DBMS and 4GLs

- Improved automated support for existing and future

    systems development methodologies

- Solidification of standards for CASE

- Improved maintenance and reverse engineering support in CASE

- Improved project management support

- Improved awareness of organizational change issues for CASE

- Stronger commitment from management for CASE

Figure C.4   CASE Technology Issues

Other less philosophical issues continue to be considered and improved upon with newer versions of CASE technology. Some of these issues are listed in Figure C.4. As one might expect, each one of these issues in its own right is a major issue deserving of significant thinking, planning, debate, and resolution.

SUMMARY

The creation of software continues to be a lengthy process. Therefore, its automation via CASE or its successor is absolutely essential in order to compete effectively in a global economy. CASE has proven itself over and over again as a viable approach to developing information systems, even though systems analysts are faced with major changes in the way that they develop systems. The benefits of CASE are now equaling or exceeding its cost so more businesses can justify its use. The more systems analysts use CASE, the higher their expectations of what it can do for them, hence, the continual identification of issues to be addressed by future generations of CASE. CASE is for real!

QUESTIONS

C.1 Briefly define the goal of computer-aided software engineering (CASE).

C.2 What are some of the main objectives of CASE?

C.3 What are some of the characteristics that differentiate upper from lower or front-end from back-end CASE products?

C.4 Briefly list and discuss some of the different stages organizations go through when trying to integrate CASE into their development efforts.

C.5 What are some of the benefits that can be gained from using CASE technology?

C.6 What characteristic of CASE technology is a major break from a more traditional information systems development?

REFERENCES

BROWN, A. W., D. J. CARNEY, E. J. MORRIS, D. B. SMITH, and P. F. ZARRELLA, *Principles of CASE Tool Integration.* NY: Oxford University Press, 1994.

CHEN, M., and R.J. NORMAN, "The Evolution Towards Integrated Computer-Aided Software Engineering Environments," *IEEE Software* (March 1992).

CHIKOFSKY, E., and R.J. NORMAN, "History of CASE Technology," in *Encyclopedia of Software Engineering,* John Marciniak (ed.). New York: John Wiley Publishing Co., 1994.

CHIKOFSKY, E. (ed.). *Computer-Aided Software Engineering (CASE)* (2nd ed.). Los Alamitos, CA: IEEE Computer Society Press, 1993.

COOKE, DANIEL E., *The Impact of CASE Technology on Software Processes.* Teaneck, NJ: World Scientific Publishing Company, 1994.

FORTE, G., and R.J. NORMAN, "A Self-Assessment of CASE Technology by the Software Engineering Community," *Communications of the ACM* (April 1992).

GANE, C., *Computer-Aided Software Engineering: The Methodologies, The Products, and The Future.* Englewood Cliffs, NJ: Prentice Hall, 1990.

HUGHES, C.T., and J.D. CLARK, "The Stages of CASE Usage," *Datamation,* February 1, 1990, pp. 41–44.

MARCINIAK, John J. (ed.), *Encyclopedia of Software Engineering,* vol. 1. New York: John Wiley & Sons, 1994.

McCLURE, C., *CASE is Software Automation.* Englewood Cliffs, NJ: Prentice Hall, 1989.

NORMAN, R.J., and M. CHEN, "Working Together to Integrate CASE," *IEEE Software* (March 1992).

Various editors, *Proceedings from CASE 'XX Workshops* (where XX starts in 1987 and continues near yearly through 1995), IWCASE, Inc., Burlington, MA.

Module D

Software Process Improvement

MODULE OBJECTIVES (YOU SHOULD BE ABLE TO)

1. Describe the difference between an immature and a mature systems development organization.
2. Define and describe each of the five maturity levels of the SEI Capability Maturity Model.
3. Describe a generic systems development process improvement model.
4. Briefly describe one other capability model.

INTRODUCTION

The information systems development life cycle (SDLC) consists of activities, products, and resources. **Activities** are actions performed in an SDLC and can be anything from high-level analysis to the compilation or testing of a program. **Products** are the documents and programs produced during the SDLC. **Resources** consist of things such as people, time, money, and equipment that are used during an SDLC. The SDLC is also referred to as a **software process** in some of the literature and research addressing software development.

Software engineering researchers around the world are committed to improving the software process. This chapter discusses software process improvement in the framework of presenting one software process improvement effort—the Software Engineering Institute's Capability Maturity Model. There are many other software process improvement efforts being utilized around the world. This one was chosen for this chapter due to the number of publications that present, discuss, and critique it along with its general acceptance in the United States and elsewhere.

In 1986, the Software Engineering Institute (SEI) at Carnegie Mellon University with assistance from the Mitre Corporation began developing a systems development process maturity framework that would help developers improve their development process. Initially called the Software Process Maturity Model, version 1 was documented in Watts Humphrey's 1989 book. Subsequent to that, version 1.1, renamed the Capability Maturity Model (CMM), has been described in the July 1993 issue of the *IEEE Software* journal. The new version has taken into account four years of experience with the model in addition to contributions from hundreds of expert reviewers.

IMMATURE AND MATURE SYSTEMS DEVELOPMENT ORGANIZATIONS

An organization desiring to set sensible goals for information systems development process improvements requires an understanding of the difference between an immature and a mature systems development organization. In an immature systems development organization, systems development processes are often improvised by the developers and their managers. Even if a particular development process has been specified, it is rarely adhered to or enforced.

Reactionary is the word that best describes the immature systems development organization. Its managers are usually focused on solving crises and "putting out fires." Their schedules and budgets are routinely exceeded because they are not based on realistic estimates. When artificial deadlines are imposed, software product functionality and quality are often compromised.

Finally, an immature systems development organization has no objective way to evaluate software product quality or solve product or process problems that arise. Activities intended to enhance software product quality, such as user involvement, design reviews, and testing, are often short-changed or eliminated in order to stay on or close to schedule.

The mature systems development organization has an organizationwide ability to manage systems development and maintenance. The managers can accurately communicate the systems development process to developers, and developer activities are carried out according to the planned process.

The mandated development processes are usable and consistent with the actual way that the work gets done, and are updated when necessary with improvements that have been developed through controlled pilot tests and cost-benefit analyses. Developer roles and responsibilities are understood within a project and across the organization.

The managers of a mature systems development organization continually monitor the quality of the system products and the process that creates them. They adhere to an objective, quantitative method for evaluating system product quality and analyzing problems with the product and the process. Schedules and budgets are based on historical performance and are realistic and are usually achieved.

Finally, a mature systems development organization uses a disciplined systems development process because all the participants understand and value doing so, and the business has an infrastructure to support them doing so.

The theory of what was just discussed contrasting the immature and the mature software development organization is easily presented in a book. The practical application and assessment of the theory is not so simple. In fact, it can be quite difficult as most software development organizations have elements of both an immature and a mature software development process. Most software development managers, if pressed to be completely honest, would be able to say whether their software development process is more immature than mature or vice versa.

Process improvement efforts, such as the SEI's CMM, are intended to assist software development managers, their developers, and the business they work for. They are not intended to lay blame or shame on them. Failure of a software development organization's manager to admit where the organization is in terms of process improvement is often based on fear of reprisal for admitting the truth about the software development process.

THE FIVE MATURITY LEVELS OF THE SEI CAPABILITY MATURITY MODEL

A systems development process is the set of activities, methods, practices, and transformations that developers use to develop and maintain information systems. As a software development organization matures, its development process becomes better defined and more consistently utilized throughout the organization. Institutionalization of its systems development process via policies, standards, and organization structures is a natural outgrowth of this maturity. Do not equate maturity with years of existence in this situation. For this discussion equate maturity with wisdom and wisdom being the application of knowledge. So a software development organization matures as it applies the generally accepted knowledge of the software development process to its own software development efforts.

In keeping with the **continuous process improvement** notion, the CMM provides a framework for organizing evolutionary steps into five maturity levels, as shown in Figure D.1, each of which lays successive foundations for continuous process improvement as the organization moves up the steps. Each level assumes that the organization has achieved all lower-level requirements set forth within the CMM guidelines. Figure D.2 summarizes the key process areas that are addressed by the CMM.

Level 1, **Initial,** is generally characterized by an organization which does not have a stable environment for developing and maintaining systems. Rarely can this organization make development commitments that can be met with an orderly development process. The results are a series of development crises in which development team members abandon planned procedures and revert to coding and testing.

Despite the chaotic nature of a level 1 organization, it can develop systems that work even if they are over budget and behind schedule. Success is dependent on having an exceptional project manager and on the heroics of individual developers on the project. Being able to repeat the performance on a subsequent project is almost totally

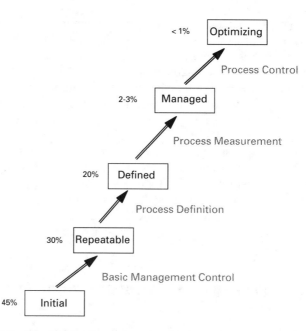

Figure D.1   SEI Capability Maturity Model

dependent on having the same competent people on the project. Thus, at level 1, capability is an individual characteristic rather than an organizational characteristic.

Level 2, **Repeatable,** is generally characterized by policies and procedures being in place and adhered to for systems development. The planning and commitments for new projects are based on experience with similar projects. Basic management controls are a part of each development project, and managers track costs and schedules and identify problems in meeting commitments when they arise.

The user requirements and the work products developed to satisfy them are baselined, and their integrity is controlled. Projects have standards that are followed; however, processes may differ among projects in level 2 organizations. The process capability of these organizations is disciplined because project planning and tracking are stable and earlier successes can be repeated.

Level 3, **Defined,** is characterized by an organizational standard and consistent systems development process due to its stability and repeatability on successive projects. This process is documented and includes both a developer process component as well as a management process component. Due to standardization, the organization as a whole should be able to exploit effective systems development practices.

At level 3, a group such as a systems development process group is responsible for the organization's system development process activities and ensures that an organizationwide training program is in place to provide developers and managers with the knowledge and skill to use the process effectively.

Individual project teams tailor the organization's standard systems development process to create their own defined process, which considers the project's unique char-

LEVEL	DESCRIPTION	KEY PROCESS AREAS

Key Process Areas in the Capability Maturity Model

LEVEL	DESCRIPTION	KEY PROCESS AREAS
5.	Continuous process improvement is enabled by quantitative feedback from the process and from piloting innovative ideas and technologies	Defect Prevention Technology change management Process change management
4.	Detailed measures of the software process and product quality are collected. Both the software process and products are quantitatively understood and controlled	Quantitative process management Software quality management
3.	The software process for both management and engineering activities is documented, standardized, and integrated into a standard software process for the organization. All projects use an approved, tailored version of the organization's standard software process for developing and maintaining software	Organization process focus Organization process definition Training program Integrated software management Software product engineering Intergroup coordination Peer reviews
2.	Basic project management processes are established to track cost, schedule, and functionality. The necessary process discipline is in place to repeat earlier successes on projects with similar applications	Requirements management Software project planning Software project tracking and oversight Software subcontract management Software quality assurance Software configuration management
1.	The software process is characterized as ad hoc, occasionally even chaotic. Few processes are defined, and success depends on individual effort and heroics	--------- (just doing it anyway possible)

Figure D.2   Key Process Areas in the Capability Maturity Model [Adapted from: M. C. Paulk, "How ISO 9001 Compares with the CMM," *IEEE Software*, 12, no. 1 (January 1995), 77.]

acteristics and needs. Thus, each project's systems development process is the same as for a subset of the organization's systems development process.

Because development organizations at this level have a well-defined process, management has good insight regarding the technical progress of each project. In addition to this, included in the organization's standardized systems development process are guidelines for:

1. Standards
2. Establishing readiness criteria
3. Developing inputs
4. Procedures for performing work
5. Work verification procedures such as peer reviews
6. Developing outputs
7. Determining completion criteria

Level 4, **Managed,** is best characterized by an organization that sets quantitative quality goals for both its systems development products and its systems development processes. Productivity and quality are measured for each of the important systems development processes across all projects as part of an overall organizational measurement program.

The organization maintains a process database to collect and analyze the data from all projects' defined processes. These organizationwide measurements establish a quantitative foundation for evaluating any projects' processes and products. Individual projects control their processes and products by reducing the variation in their performance so that it fits within the organization's acceptable quantitative boundaries.

Meaningful variations in a project's process performance can be distinguished from random variation, and when known limits of the process are exceeded, management can take action to correct the situation. Level 4 capability allows an organization the ability to predict trends in its systems development process and product quality within the established quantitative boundaries.

Level 5, **Optimizing,** is characterized by the entire systems development organization being focused on continuous process improvement. The organization has the means in place to identify its development process and product weaknesses and proactively strengthen them with the goal of zero defects. The process database (from level 4) is used to perform cost-benefit analyses for new technologies as well as changes to the systems development processes. Innovations that exploit the best systems development practices are identified and transferred throughout the organization.

Project teams analyze defects to determine their causes, evaluate the processes to prevent known types of defects from recurring, and disseminate lessons learned to other projects. The reduction of rework is an objective for each level, but at level 5 it becomes a primary focus because rework is a primary cause of inefficiency in systems development.

The level 5 organizations are continuously striving to improve their systems development process and products. Improvement occurs both in incremental advancements made to the existing systems development process and by innovations in technologies and methods. The technology and process improvements are both planned and managed as ordinary business activities within the level 5 organizations.

As shown in Figure D.1, estimates have been suggested for the percentage of worldwide software development organizations that mostly exhibit being at each one of the CMM levels. Note that the estimates show that 95 percent or all worldwide software development organizations have a software development process that is either at or below the CMM's Defined Level 3.

A GENERIC SYSTEMS DEVELOPMENT PROCESS IMPROVEMENT MODEL

A systems development organization desiring to advance up the levels of the CMM could do so by following this simplified systems development process improvement model. It lends itself to the acronym ICASE.

I = **Investigate.** Investigate the current status of the organization's systems development process. We need to know where we are before we can determine a plan to move to the next level.

C = **Create.** Create a vision within the organization of the desired process. Get the developers and managers to buy into the notion of improving the systems development process.

A = **Actions.** Establish a list of the required process improvement actions necessary in your organization.

S = **Select.** Select a plan to accomplish the required actions.

E = **Execute.** Commit the resources necessary to execute the plan.

Finally, repeat this process as you begin and continue your movement through the CMM levels.

THE ISO 9000 PROCESS IMPROVEMENT METHODOLOGY

One other model, **ISO 9000,** developed by the International Standards Organization (ISO), **refers to a series of five related quality management standards.** The most comprehensive of the standards encompasses 20 business functions, such as contract review, design control, document control, purchasing, inspection and testing, and training and maintenance. The standards are geared to two-party transactions, and they assess a supplier's ability to fulfill the customer's contractually specified requirements. ISO 9000 registered sites are audited every three years and receive less comprehensive "surveillance visits" every six months.

The ISO 9001 standard within the ISO 9000 series of standards pertains to software development and maintenance. It identifies the minimal requirements for a quality system along with continuous process improvement of the software development process. The Paulk article referenced at the end of the chapter compares the ISO 9001 standard with the CMM. He lists, compares, and describes 20 common areas between the two models.

SUMMARY

This chapter has presented a brief overview of software process improvement. The process by which software is created needs constant improvement due to many factors, and software process improvement is the activity that attempts to do this. Immature and mature software development environments were presented followed by a description of the SEI Capability Maturity Model's five levels of software development maturity. After this a generic systems development process improvement model was presented that followed an ICASE acronym. The chapter concluded with a section briefly discussing the ISO 9000 process improvement standard.

As mentioned before, the SEI CMM is not the only systems development measurement model. There is also the ISO 9000 series of standards along with several proprietary ones available through consulting organizations around the world. However, the CMM is perhaps the most publicly documented one of them all.

QUESTIONS

D.1 What distinguishes an immature from a mature systems development organization?

D.2 What are the five maturity levels of the SEI Capability Maturity Model? Briefly describe each.

D.3 Briefly describe the generic systems development process improvement model called ICASE.

D.4 What is the ISO 9000 model, and what is its role in software process improvement?

REFERENCES

BOLLINGER, T.B., and C. MCGOWAN, "A Critical Look at Software Capability Evaluations," *IEEE Software,* 8, no. 4 (July 1991), 25–41.

HUMPHREY, W.S., *Managing the Software Process.* Reading, MA: Addison-Wesley Publishing Company, 1989.

IEEE1074, *IEEE Standard for Developing Software Life Cycle Processes.* Los Alamitos, CA: IEEE Computer Society Press, 1991.

JENNY, B., "Raising the Standard with ISO 9000:1994," *Monash University FCIT Faculty Newsletter* (September 1994), pp. 3–4.

PAULK, M.C., "How ISO 9001 Compares with the CMM," *IEEE Software,* 12, no. 1 (January 1995), 74–83.

PAULK, M.C., B. CURTIS, M.B. CHRISSIS, and C.V. WEBER, "Capability Maturity Model, Version 1.1," *IEEE Software,* 10, no. 4 (July 1993), 18–27.

PUTNAM, L.H., and W. MYERS, *Measures for Excellence.* Englewood Cliffs, NJ: Prentice Hall, 1992.

Module E

The Systems Development Challenge

MODULE OBJECTIVES (YOU SHOULD BE ABLE TO)

1. Discuss the systems development challenge.
2. Describe systems development's quadruple constraint.
3. Describe the information technology management issues.
4. Describe the most common systems development risks.
5. Describe systems analysis and design versus software engineering.
6. Describe a generic systems development architecture.
7. Explain SDLC, *methodology, technique,* and *tool.*

INTRODUCTION

Headline: "Software Crisis! Software Crisis! Software Crisis!" Year after year, newspapers and trade journals alike continue to cry "software crisis." One only needs to look at the definition of *crisis,* "a short-term situation," to realize that the headlines should be reading something like "chronic software situation!" instead of "software crisis!" The software and systems development industry has been in and continues to be in a software situation that is chronic. Hence, the purpose of this chapter is to help you understand why the software situation is chronic with an eye toward improving it. Whereas most chronic medical conditions have no remedies, I believe software's chronic condition does.

The quality (zero defects) of the resulting products of systems development is a function of the productivity of the process of systems development as shown in Figure

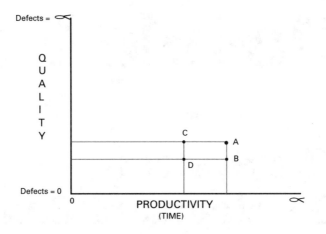

Figure E.1   Software Quality Versus Productivity

E.1. For discussion purposes, systems development process productivity includes both the time to complete the entire development process and resulting products as well as the quality of the development process. Therefore, **the systems development challenge is one of product quality versus process productivity.** Given an infinite amount of time, one can certainly "dot every i and cross every t" in the systems development process to achieve zero defects, but this is a utopian situation that very few, if any, projects ever have.

If Figure E.1's point *A* represents an example baseline for quality and productivity, the challenge is to change this point in at least one of three desirable ways:

1. Point *B* represents an improvement in the product quality given the same amount of productivity.
2. Point *C* represents no change in the quality of the product but an increase (reduced time) in the productivity to achieve this quality.
3. Point *D* represents an improvement in the product quality and an increase (reduced time) in the productivity to achieve it.

Software is generally recognized as being the pacing factor for automation around the world. As hardware technologies continue to evolve on a rapid time schedule, systems development continues to make improvements at a snail's pace. Hence, the resulting software products continue to lag behind the hardware. Combine this with the complexities of today's software requirements for functionality, interoperability across hardware platforms, reusability, and integration complexities, and software ends up even further behind.

Another factor in the systems development challenge that has been widely recognized is the **cost of software development,** as illustrated in Figure E.2. Software development and its associated deliverables, such as the software, documentation, training, and conversions, represent upward of 90 percent of the total systems cost today. The most significant component of this cost is the cost of people.

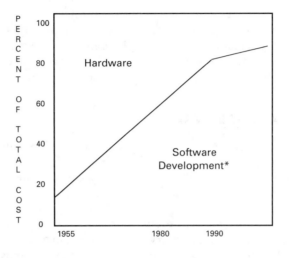

* Includes items such as software, documentation, training, conversion, etc.

Figure E.2   The Cost of Software/Systems Development

SOFTWARE DEVELOPMENT'S QUADRUPLE CONSTRAINT

Have you ever been on a teeter-totter at a playground? If you have or just know about them, you know that they are about balance. Have you ever seen a juggler spin plates on the tops of sticks? Just as the juggler finishes spinning the last plate, he or she must hurry back to the first plate to spin it again to avoid having it fall off the stick. Very impressive! Again, an example of a balancing act.

From a management and organizational perspective, software development is a balancing act of sorts also. Software development has what is known as a quadruple constraint. Its **quadruple constraint** is a balancing act between four tensions—**budgets, schedules, people, and capability.** Budgets deal with the financial aspects of a systems development project. Schedules address the time aspect of a systems development project. People represent the available resources to work on the systems development project, and capability relates to the readiness or training of the people who will be involved in the systems development project.

In a utopian environment, a systems development manager would have an infinite budget, with infinite amounts of time and people, and the people would be world-class participants. In the real world, however, systems development managers constantly deal with shrinking budgets and reduced time to develop coupled with reduction of work force (people) and reduced training funds and time. The systems development manager who can effectively manage (balance) these components will make a strong contribution to the overall success of each systems development project.

INFORMATION TECHNOLOGY MANAGEMENT ISSUES

The systems development challenge is further complicated by the dominant issues that information technology managers face. Some of these issues change over time, while others remain a constant for these managers. Some of the more pressing issues they are addressing as we head into the late 1990s are shown in Figure E.3. The issues are generally understood as listed; therefore, no discussion is presented here.

SYSTEMS DEVELOPMENT RISKS

There is risk associated with every systems development project. Capers Jones estimates that very few projects have more than 15 risk factors at any time, but many projects have half a dozen of them simultaneously. Here are the most frequently occurring ones:

1. **Inadequate measurement.** As much as 90 percent of all U.S. companies, government agencies, and military services have inadequate measurement programs to account for the costs associated with information systems development projects. The most overlooked portion of the true costs is the cost associated with the user being involved. An organizational culture oriented toward excellence and "management by fact" is the best way of reducing and preventing this risk. Management by fact includes a strong measurement component.

2. **Excessive schedule pressure.** Irrational and excessive schedule pressure may affect as much as 65 percent of systems development projects. Unfortunately, by the time excessive schedule pressure becomes visible, it is usually too late to control it.

3. **Management malpractice.** The root causes of management malpractice can be traced back to inadequate training in basic project management tasks such as sizing,

---

- Right-sizing the information technology organization. This could include issues such as reduction of personnel, outsourcing, and migration to client-server technology

- Project backlogs of several years. The queue of user requests for new and changed systems is not getting any shorter

- Projects being late and over budget. This relates to the maturity level of the organization's systems development processes as described in an earlier chapter

- User dissatisfaction with both the systems development process and the deliverable products. User expectations continue to increase

- Hardware and software integration issues. User expectations for interoperability across multiple vendor hardware and operating system platforms is becoming commonplace

- The merger and centralization of voice, data, and video technologies

- Concerns for effectively implementing notions such as continuous process improvements, total quality management, and applying for the Malcolm Baldridge Quality Award

Figure E.3   Information Technology Management Issues

estimating, planning, tracking, measurement, and assessment. There are several ways to address this risk, among them being appropriate project management training and mentorship.

4. Creeping user requirements. The rate of growth of creeping user requirements is about 1 percent per month. So, for a three-year project, about one-third of delivered functionality will have been added after the requirements phase. The use of prototypes and techniques like joint application development (JAD) can counteract creeping user requirements.

5. Canceled projects. The cancellation rate for systems development projects is directly proportional to the overall size of the system and is acute above 10,000 function points (a measurement technique) or 1 million source statements. Cancellation rates approach 50 percent for these large systems. Reduction of the other risks mentioned here are ways to counteract canceled projects.

The risks discussed here are just the tip of the iceberg as far as number of risks go. The good news is that new methods such as process assessments similar to the SEI Capability Maturity Model discussed in an earlier chapter and the use of functional metrics are proving to be effective in identifying risks before it is too late.

SYSTEMS ANALYSIS AND DESIGN VERSUS SOFTWARE ENGINEERING

Another aspect of the systems development challenge is the terminology chasm that exists between business information systems development environments and virtually all other systems development environments. For example, in business systems development circles systems development is referred to as systems analysis and design. In almost all other environments the same thing is referred to as software engineering. Also in business information systems circles the people who do this type of work are referred to generally as systems analysts and programmers, whereas in almost all other circles these same people are referred to as software engineers.

There are many other, albeit less significant, terminology differences too numerous to list here. One example that comes to mind, however, is the use of the word *persistence* within the software engineering community to communicate the notion of more permanent storage of data. The business information systems developers refer to *persistence* primarily as tables, files, and databases. The terminology gap is lessening for newly minted college graduates, but there are still tens of thousands of systems analysts and programmers not yet conversant in the cross-over terminology.

A few years ago I asked dozens of experienced systems analysts and programmers to respond to the following statement. "Give me a few words that you feel best describes your experiences with systems analysis and design." Those responding to the statement have consistently chosen the words shown in Figure E.4. With a few exceptions, I believe this to be a pretty depressing list. Who in their right mind would want to be a systems analyst with a list like this?

- Artistic/Creative

- Highly Cognitive

- Black Art

- Miracles

- Seat-of-the-Pants

- Coding = Productivity

- Let the maintenance team worry about that

- Sleepless Nights

- Long Hours

- Frustration

- Ulcers/Pills

- Loss of Hair

- Broken Homes/Marriages/Relationships

- Over Budget

- Late Projects

Figure E.4   Words and Phrases Describing Systems Analysis and Design

On the other hand, I then asked these same systems analysts and programmers to give me a "few words that come to your mind when you think of [software] engineering," and they responded with the list in Figure E.5. Quite a different looking list from Figure E.4. This list, even if it does not truly mirror reality, is a great psychological starting place. To sum up these two lists very loosely, I would say the first speaks of an "out of control" condition, whereas the second speaks of an "in control" condition. Even though these are only perceptions that were voiced to me, most times people's perceptions are reality to them. One of the challenges facing the software development industry is addressing the psychological issues surrounding software development.

A SYSTEMS DEVELOPMENT ARCHITECTURE FOR THE 1990S

In their 1982 *Advanced System Development/Feasibility Techniques* book, Cougar, Colter, and Knapp presented a historical systems development chart covering precomputer through a prediction of fifth-generation systems development techniques. Although their fifth-generation vision did not happen exactly as they suggested, they were correct in suggesting the idea of a complete systems development environment from which systems would be built.

- Scientific

- Standards

- Structure

- Details

- Rigor

- Quality

- Productivity

- On-time Projects

- Within Budget

Figure E.5   Words and Phrases Describing Software Engineering

In Figure E.6, we present a generic systems development architecture that has been taking hold within the information systems development community since the late 1980s. In the broadest sense, a systems development organization should adopt its own **systems development life cycle** (SDLC) framework as the overall strategic umbrella for this architecture. The SDLC represents the organization's strategic systems development philosophy and is supported with policies and general procedures for systems development. Figure E.7 shows the two components that are addressed by the strategic SDLC. The management strategy component addresses general systems development issues related to planning, scheduling, estimating, monitoring, and feedback. The development strategy component addresses general systems development issues related to:

1. A work breakdown structure (WBS) as the organization standard for systems development. Included here could be the traditional phases of systems development such as feasibility study, investigation, analysis, general design, detail design, and so on.

2. A WBS precedence structure as the organization standard for systems development. Included here could be the organization's philosophy regarding development precedence such as the waterfall, stairstep, circular, spiral, iterative, and so on.

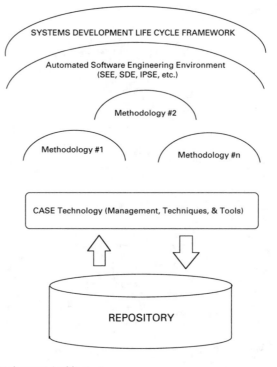

Figure E.6   Systems Development Architecture

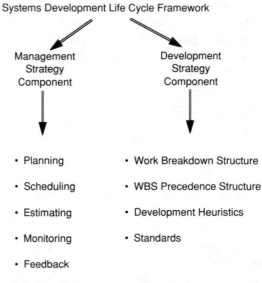

Figure E.7   SDLC Framework Components

3. Development heuristics that the organization desires to support. Included here could be the best practices used successfully by veteran systems developers.
4. Standards for systems development.

Referring again to Figure E.6, the next umbrella layer within the organization's systems development architecture is the **software engineering environment** (SEE). The SEE is a workstation-based workgroup information system used to develop information systems. Each project team member has this automated SEE operational on his or her workstation and each workstation is connected to a server-type host computer to facilitate and coordinate workgroup activities of the team.

In a sense, the SEE becomes the braintrust or the request broker of the organization's systems development strategies, tactics, and operations. Each software engineer would interact through his or her workstation with the SEE to make use of the project-specific **methodologies** and **CASE techniques and tools** that have been integrated with the SEE and catalogued or stored within the organization's systems development **repository.**

The repository would be populated with two kinds of information: **management and guidance,** and **project models** as shown in Figure E.8. The management and guidance component of the SEE's repository would assist both the project managers and the developers. The managers would be able to do their planning, scheduling, and controlling with the assistance from the SEE and could even get expert assistance and guidance from the SEE regarding these tasks. Likewise, the developers would follow the project's methodology and could ask for expert assistance with doing so from the SEE. In addition, all of the models that the developers create as a result of analyzing and designing a new or changed information system would be stored in the project's repository and expert assistance in using each of the modeling tools and techniques could be available on-line via the SEE.

SDLC, METHODOLOGY, TECHNIQUE, AND TOOL

The final systems development challenge topic discussed here deals with the general confusion or misunderstanding among practitioners regarding the terms *SDLC, methodology, technique,* and *tool.* Although the discussion here is an attempt to clarify these terms, realize that not all of the industry's professional colleagues agree with this explanation.

In the broadest sense, a systems development life cycle (SDLC) is a strategic or philosophical view of systems development from an organizational perspective. As described in the prior section, an organization's SDLC should have both a management and a developer component to it along with associated general guidelines discussed earlier. Why should an organization have a strategic SDLC? For at least five reasons:

1. The SDLC can promote **standardization** for the organization's systems development efforts.
2. The SDLC can facilitate consistent and more effective **communication** among project managers and developers.

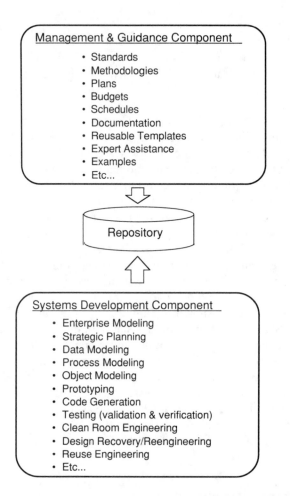

Figure E.8   Software Engineering Environment

3. The SDLC can assist with the **institutionalization** of the organization's systems development efforts.

4. The SDLC can support one or more **methodologies,** even ones customized specifically for a certain project.

5. The SDLC can support multiple **automated techniques and tools.**

A **methodology** is a specific, step-by-step strategy for completing all or parts of the SDLC. Some colleagues consider the SDLC and a methodology to essentially be the same thing. Methodologies usually impose their own detailed standards, techniques, and tools on the SDLC, which either complement or override SDLC established standards. There are literally thousands of methodologies in existence because every time an organization creates a variation of an existing methodology, it creates, in essence, a new one.

For example, there are at least six prevailing object-oriented analysis and design methodologies in industry today created by Booch, Coad/Yourdon, Jacobson, Martin/ Odell, Rumbaugh et al., and the Fusion team. In reality, these six have probably given birth to hundreds of variations as organizations create "home-brewed" versions of these. Other industry-recognized methodologies are structured programming, structured analysis, structured design, and so on, each of which has also given birth to hundreds of "home-brewed" variations.

Methodologies can generally be classified as belonging to one of three "schools of thought"—process oriented, data oriented, or object oriented. Certain methodologists are fanatical about their support and enforcement of only a process-, data-, or object-oriented methodology approach to developing information systems. The reason for this is mostly an economical and time-to-market decision on the part of the methodologist. In my view, that is a rather narrow and limited strategy as certain systems tend to "cry out" for development using a process-centered methodology, while others "cry out" for a data-centered or object-oriented methodology.

Techniques and tools usually provide support for one or more methodologies. Today it is often hard to distinguish between a technique and a tool, since most of them are automated. For example, an argument could be made that a structure chart is a technique used in structured design, but an equally valid argument could be made that a structure chart is a tool used in structured design. Software developers tend to refer to most of these items as tools and the way the tool is used is considered the technique. For example, the Coad/Yourdon object-oriented modeling symbol set and notation is the tool, and how a software engineer couples the symbols and notations together is the technique for using the tool.

Other techniques are more obvious, such as joint application design (JAD) and prototyping. Each of these techniques can support a number of methodologies and each of these techniques has one or more supporting tools to assist with performing the technique.

SUMMARY

The systems development challenge is real. This chapter has presented a number of the challenges facing software development managers and developers for your consideration and awareness. Someone once said, "Systems development would be easy if we did not have to deal with users." This might be true to a certain degree, but the systems development industry is still in its infancy or, at best, its "teen-age" years, still struggling to discover better, faster, and more economical ways to develop information systems.

In his *Harvard Business Review* article, Dr. Fred McFarlan is quoted as saying that one serious deficiency exists within the systems development community, and that deficiency is ". . . the lack of recognition that different projects require different managerial approaches." So the moral is, "It's not know-how that counts so much . . . it's know when!"

QUESTIONS

E.1 What is the main challenge of information systems development?

E.2 What is meant by the quadruple constraint in software systems development?

E.3 What are some of the more common issues that information technology managers face?

E.4 What are some of the risks or problems that may arise in a systems development project?

E.5 Discuss what distinguishes systems analysis and design from software engineering.

E.6 Why is it important for a systems development organization to have its own systems development life cycle (SDLC)?

E.7 What are the two components of a strategic SDLC and what does each address?

E.8 What is a systems development repository, and what purpose does the repository have in a systems development life cycle framework?

E.9 Why is it necessary for an organization to have a strategic SDLC?

E.10 What, if any, differences exist between techniques and tools?

REFERENCES

COUGAR, D., M. COLTER, and R. KNAPP, *Advanced System Development/Feasibility Techniques.* New York: John Wiley & Sons, 1982.

JONES, CAPERS, "Sick Software," *Computerworld,* December 13, 1993, pp. 115–116.

JONES, CAPERS, *Assessment and Control of Software Risks.* Englewood Cliffs, NJ: Prentice Hall, 1994.

MCFARLAN, F.W., "Portfolio Approach to Information Systems," *Harvard Business Review* (September–October 1981).

OLLE, T.W., H.G. SOL, and C.J. TULLY, (eds.), *Information Systems Design Methodologies: A Feature Analysis.* Amsterdam: North-Holland Publishing Co., 1983.

OLLE, T.W., H.G. SOL, and A.A. VERRIJN-STUART, (eds.), *Information Systems Design Methodologies: A Comparative Review.* Amsterdam: North-Holland Publishing Co., 1982.

OLLE, T.W., et al., *Information Systems Methodologies: A Framework for Understanding.* Wokingham, England: Addison-Wesley, 1988.

Module F

Project Management

MODULE OBJECTIVES (YOU SHOULD BE ABLE TO)

1. Describe the source of projects.
2. Describe several reasons for failed projects.
3. Describe several definitions of project failure.
4. Describe and demonstrate how to create and use a PERT network.
5. Discuss the PERT network's strengths and weaknesses.
6. Describe and demonstrate how to use a Gantt chart.
7. Discuss the Gantt chart's strengths and weaknesses.

INTRODUCTION

Projects are probably nothing new to you. At any given time you no doubt have a written or mental list of "little" projects that are pending, such as studying for an exam, making or buying a birthday present for your best friend, fixing the leaky kitchen sink faucet, and so on.

Information systems development is almost always done under the auspices of a project. Two of the differences between your "little" projects and a systems development project are that (1) you are probably the only resource involved in your "little" projects, while systems development projects usually involve many individuals, and (2) your "little" projects are truly "little," while systems development projects can last months, even years. Systems development projects come from one of three sources:

1. A directive or mandate from some person, such as a president, vice president, or senior manager, or an organization, such as a labor union or the U.S. Internal Revenue Service.

2. An opportunity to exploit. The opportunity usually results in increased revenues and/or profits, reduced costs, or increased or improved services.

3. A problem to solve. Something usually isn't working correctly and needs fixing.

Information systems development project management is a process of directing the development of an acceptable information system at a minimum of cost within a specified time period. A project manager is usually responsible for all of the normal managerial functions of planning, staffing, organizing, directing, and controlling the project. Thus, a project manager needs a different skill set from that of a systems analyst or programmer in order to effectively perform his or her job.

There are numerous commercially available project management software packages to assist a project manager with his or her normal managerial functions. In addition, some of the commercially available CASE software products also include project management aids. Having these automated support tools is usually necessary with today's complex systems development projects, but it is not sufficient for success. It takes a competent project manager along with a competent development team to achieve success. Someone once said, "a fool with a tool is still a fool!" How true this is in systems development.

Systems development project failures still continue to occur with serious regularity around the world. Most are never documented in industry trade press due to the professional stigma of having a failed project. Some of the more common reasons for systems development project failures are:

1. Use of undisciplined development methodologies or approaches.
2. Inadequate or not understood or appreciated systems development tools.
3. Project scope was not clearly defined in the beginning.
4. Use of no or poor estimating techniques.
5. Schedule delays.
6. Belief in the mythical person-month syndrome. For example, a manager with this mindset thinks that if two developers can do or finish up a project in six weeks, then simply adding a third developer to the project will allow it to be completed in four weeks. The reality is much different than this.

There are different ideas about what constitutes a systems development project failure. Among them are the following:

1. The organization completely abandons the project at some point prior to its implementation.
2. The organization must rework a significant amount of the project, so much so that they deem it a failure but do go ahead and initiate the rework.

3. The delivered information system is okay, but the project was way over time and budget, therefore, it is deemed a failure.
4. The delivered information system does not meet the user requirements or expectations; therefore, it is deemed a failure.

TWO TOOLS: PERT NETWORK AND GANTT CHART

There are two commonly used project management tools, a Program Evaluation and Review Technique (PERT) network and a Gantt chart. A PERT network is a graphical representation of project tasks laid out in the form of a critical path network, as illustrated in Figure F.1, and a Gantt chart shows project tasks and their durations in a bar chart format, as illustrated in Figure F.2. Both of these graphs are useful in the planning and estimating of a project prior to its inception. Once the project is underway, the actual results can replace the estimates (PERT network) or appear along with the estimates (Gantt chart) in order to reflect project actuals as they occur. This gives the project manager the opportunity to make any necessary adjustments based on what actually is happening during the project.

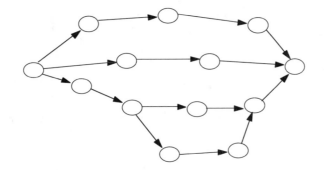

Figure F.1   Program Evaluation and Review Technique (PERT) Chart

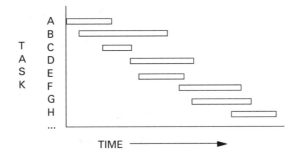

Figure F.2   Gantt Chart

As the PERT network and Gantt chart are discussed, you should know that there are several different ways to draw each of these diagrams. In addition, many of the high-quality project management software packages include additional information with each of the diagrams. Only a subset of all this information has been selected to demonstrate each of these diagrams here.

The PERT Network

The PERT network is a visual precedence diagram of the project tasks. Each project determines its own tasks and associated times and dependencies. The tasks are usually selected from a standardized list of possible tasks established by the systems development organization. The tasks are identical for constructing the Gantt chart for the same project. In fact, commercially available project management software can automatically draw the PERT network and the associated Gantt chart once you have input the required project data.

Referring to Figure F.3, the notation used in this example of a PERT network is as follows. The circle represents either a starting or ending node for a task. There will only be one starting node and one ending node for a PERT network. With the exception of the starting and ending nodes, each node in the network will have at least one task start-

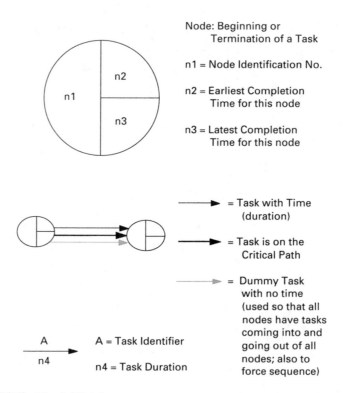

Figure F.3   PERT Chart Symbol Notation

ing with the node and at least one task ending with the node. Each node has three numbers within it. The number on the left side of the node, n1 in the figure, is the node identification number. Nodes are numbered starting with one and incrementing by one for each additional node in the network. The number in the upper right quadrant of the node, n2 in the figure, is the earliest completion time for the node. The number in the lower right quadrant of the node, n3 in the figure, is the latest completion time for the node. Both of these numbers, n2 and n3 in the figure, will be explained in more detail a little later when an example of a PERT network is discussed and illustrated.

The thin-lined arrow represents a task that is not on the network's critical path. The critical path is the longest time path from the project's starting node to the project's ending node. The thick-lined arrow represents all tasks that lie on the network's critical path. The letter above or to the left of each task arrow, letter A in the figure, is the task identifier. Each task will have a unique letter identifier, and you can use the task identifier to look up the actual task name on a task list report. The task identifier is used in order to avoid cluttering the diagram with the actual task names. The number that is below or to the right of each task arrow, n4 in the figure, is the expected duration time for completing that task. All duration times must be in the same time denomination, such as hours, days, weeks, or months.

There is one special task called a **dummy task,** and it always has a duration time of zero. The purpose of the dummy task is to make sure that all nodes, except the first and the last, have at least one task ending at the node and at least one task starting at the node. In other words, a network is not allowed to have any *dangling* nodes within it. A dangling node is one that either has no task ending at it or no task starting from it. A dummy task does create a task dependence; therefore, you should give some thought to which starting or ending node you will connect the dummy task to.

A PERT Network Example

As a simple PERT network example is discussed, keep in mind that some projects have hundreds of tasks making the project's network very intricate. Figure F.4 lists the steps necessary to create a PERT network. With project management software support, steps 1 through 4 could be supported via a generic template of project tasks for the type of project you are working on. You then edit the generic template until it accurately reflects your project's task needs. Steps 5 through 9 can be done automatically by most project management software.

Starting with step 1, seven essential tasks have been identified for the project. To avoid getting hung up on realism issues here, I have used task names that are meaningless. Figure F.5a shows the seven tasks. According to step 2 each of the seven tasks is assigned a task identification letter (A–G) as shown in Figure F.5b.

Step 3 requires that the duration time be determined for each task. There are several ways of doing this, among them being a weighted average method which will be used here. The duration is an estimate of the time to complete the task based on three estimates—optimistic (O), most likely (M), and pessimistic (P). Estimates are just that; however, quality estimates should be based on experience and project task simi-

1. Make a list of the project tasks

2. Assign a task identification letter to each task

3. Determine the duration time for each task

4. Determine task dependencies

5. Draw the PERT network, number each node,

   label each task with its task identification letter,

   connect each node from start to finish,

   and put each task's duration on the network.

6. Determine the need for any dummy tasks

7. Determine the earliest completion time for each task node

8. Determine the latest completion time for each task node

9. Verify the PERT network for correctness

Figure F.4   Steps to Create a PERT Chart

larities with other projects. The weighted average formula weighs the most likely estimate four times more than the optimistic and pessimistic. The weighted most likely
estimate is added to the optimistic and pessimistic estimates and the resulting number
is divided by six to arrive at the estimated duration for the task as shown in Figure F.6.

Step 4 requires that task dependencies be determined. In other words, we need to
know enough about the content of each task as well as the combination of the tasks in
order to know which tasks must precede one or more other tasks due to some dependency
between them. In the example shown in Figure F.7, it is determined that task A has no
precedent (it is the first task), and that tasks B through G have one precedent task each. It
is okay for more than one task to have the same precedent task as another task, as tasks B
and C in the example both have task A as a precedent. It is also okay for a task to have
more than one precedent task, even though this example does not have such a condition.
For example, if task G was really dependent on tasks A, C, and D, then under the Precedents column in the figure, task G would list "A, C, D." In so doing, the resulting PERT
network would have tasks A, C, and D all ending with the node that starts task G.

Step 5 requires that (1) the network nodes are numbered, (2) each task is labeled with its task identification letter, (3) each task's duration is put on the network,
(4) the nodes are connected from start to finish, and finally (5) the PERT network is
drawn. Figure F.8 is my drawing of the PERT network according to this step. In addition, step 6 has been accomplished, which says to determine the need for any dummy

a) Task Names (abstract)

aaaaaaaaaaaaaaaaa

bbbbbbbbbbbbbbbb

cccccccccccccccccc

dddddddddddddddd

eeeeeeeeeeeeeeeee

ffffffffffffffffffffffffffffff

gggggggggggggggg

b) Task Identifiers and Names (abstract)

Task ID	Task Name
A	aaaaaaaaaaaaaaaaa
B	bbbbbbbbbbbbbbbb
C	cccccccccccccccccc
D	dddddddddddddddd
E	eeeeeeeeeeeeeeeee
F	ffffffffffffffffffffffffffffff
G	gggggggggggggggg

Figure F.5 PERT Chart Tasks

tasks. Task ID H was created as a dummy task because node 6 was left dangling and needs to be connected into the network. I arbitrarily determined that it should connect with node 7; however, in a real project care should be given to the best connection for dummy tasks.

Step 7 requires that the earliest completion time (ECT) be computed for each node in the network. This is done by working from node 1 to node 7 in sequential node order. ECT for node number 1 is zero since node number 1 represents the beginning of the project and no tasks have been performed yet. ECT for all remaining nodes is determined by the following formula:

ECT for node X = ECT of preceding node + duration for node X

Note: If there is more than one preceding node (e.g., in Figure F.8, node 7 has nodes 4, 5, and 6 preceding it), then make a list of candidate ECTs by using the formula with each of the preceding nodes. Select the **largest** of the candidate ECTs as the ECT for node X. Why do we select the largest candidate ECT? Because that is the earliest that node X can possibly be completed with all preceding tasks. Figure F.9 shows the PERT network with ECTs indicated along with the calculations of a couple of the ECTs.

Task ID	Task Name	O	M	P	Duration
A	aaaaaaaaa.....	3	4	5	4
B	bbbbbbbbb.....	2	3	4	3
C	ccccccccc.....	1	2	3	2
D	ddddddddd.....	4	6	8	6
E	eeeeeeeee.....	1	2	3	2
F	fffffffff.....	1	2	3	2
G	ggggggggg.....	4	6	8	6

Note 1: O = Optimistic time estimate

M = Most likely time estimate

P = Pessimistic time estimate

Note 2: weighted average formula for duration:

$$D = (O+4M+P)/6$$

(example for Task ID A: $D = (3+4(4)+5)/6 = 24/6 = 4$

Note 3: it is strictly coincidental and based on the

simple numbers chosen for O,M and P that the duration

time is identical to the most like (M) time

Figure F.6   PERT Chart Tasks with Durations

Task ID	Task Name	O	M	P	Duration	Precedents
A	aaaaaaaaa.....	3	4	5	4	none
B	bbbbbbbbb.....	2	3	4	3	A
C	ccccccccc.....	1	2	3	2	A
D	ddddddddd.....	4	6	8	6	B
E	eeeeeeeee.....	1	2	3	2	C
F	fffffffff.....	1	2	3	2	D
G	ggggggggg.....	4	6	8	6	D

Figure F.7   PERT Chart Tasks with Durations and Precedents

Step 8 requires that the latest completion time (LCT) be computed for each node in the network. We do this by working from node 7 backward to node 1 in descending sequential node order. LCT for node 7 is the same as node 7's ECT, since node 7 represents the end of the project and all tasks have been performed. LCT for all remaining nodes is determined by the following formula:

LCT for node X = LCT of following node – duration for node X

Note: if there is more than one following node (e.g., in Figure F.9, node 2 has nodes 3 and 4 following it), then make a list of candidate LCTs by using the formula with each

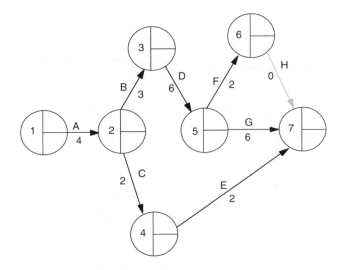

Figure F.8   PERT Chart with Task IDs, Node Numbers, and Durations

of the following nodes. Select the **smallest** of the candidate LCTs as the LCT for node X. Why do we select the smallest candidate LCT? Because that is the latest that node X can possibly be completed and still keep the project on time. Figure F.10 shows the PERT network with LCTs indicated along with the calculations of a couple of the LCTs.

   Slack time is calculated for each node by subtracting ECT for a node from its LCT. Any node that has zero slack time is on the **critical path.** Figure F.10's critical path is highlighted by the thick-lined task arrows. There will always be at least one critical path from node 1 through to the last node in the network. All tasks that are on the critical path must be cumulatively completed on schedule in order for the entire project to finish on schedule. If a task on the critical path exceeds its estimated duration to complete, then one or more other tasks on the critical path must be completed ahead of its estimated duration time in order to get back on the original project completion schedule. Nodes that have slack time can actually be delayed as much as the slack time without negatively affecting the estimated completion time for the entire project. If a node with slack time is delayed longer than the slack time, then this could have a negative effect on the project's completion date; therefore, project managers need to pay attention to all tasks, not just the ones on the critical path.

   As actual completion times are input to the PERT network after the project begins, they replace the estimates and in so doing could possibly change the critical path.

PERT Network Strengths and Weaknesses

The PERT network can be a valuable tool for project managers, but the tool is only as good as the data that are input to it. Its strengths include:

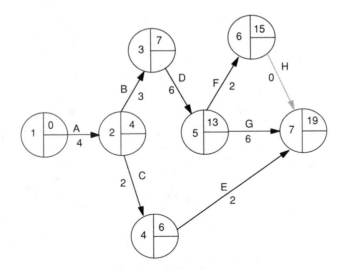

Node 5 ECT = 7 (node 3 ECT) + 6 (D's duration) = 13

Node 7 ECT = 15+0 OR 13+6 OR 6+2 = 19 (largest one)

Figure F.9 PERT Chart with Task IDs, Node Numbers, Durations, and Earliest Completion Times

1. The PERT network is continuously useful to project managers prior to and during a project.
2. The PERT network is straightforward in its concept and is supported by software.
3. The PERT network's graphical representation of the project's tasks help to show the task interrelationships.
4. The PERT network's ability to highlight the project's critical path and task slack time allows the project manager to focus more attention on the critical aspects of the project—time, costs, and people.
5. The project management software that creates the PERT network usually provides excellent project tracking documentation.
6. The use of the PERT network is applicable in a wide variety of projects.

The PERT network is not without weaknesses which should be pointed out:

1. In order for the PERT network to be useful, project tasks have to be clearly defined as well as their relationships to each other.
2. The PERT network does not deal very well with task overlap. PERT assumes that following tasks begin after their preceding tasks end.
3. The PERT network is only as good as the time estimates that are entered by the project manager.

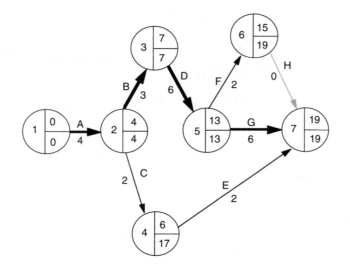

Node 3 LCT = 13 (node 5 LCT) - 6 (D's duration) = 7

Node 2 LCT = 17-2 OR 7-3 = 4 (smallest one)

Figure F.10   PERT Chart with Task IDs, Node Numbers, Durations, Earliest Completion Times, Latest Completion Times, and Critical Path

4. By design, the project manager will normally focus more attention on the critical path tasks than other tasks, which could be problematic for near-critical path tasks if overlooked.

The Gantt Chart

The Gantt chart is based on a two-dimensional graph scale. Referring to Figure F.11, each of the significant project tasks is listed along the vertical axis of the graph, and the estimated elapsed calendar time to complete the entire project is listed along the horizontal axis. An appropriate calendar time interval, such as days, weeks, or months, is selected for the horizontal axis.

The Gantt chart is at its best for visually showing each of the project's task status at any moment in time simply by drawing a vertical bar from top to bottom on the chart at the calendar time you are interested in. Once drawn, a visual inspection of the shading within each of the bars on the chart gives you an indication of project task status for each task. The Gantt chart is also useful for showing any overlapping or parallel tasks.

The Gantt chart does not clearly show task dependence, even though it does show task start and stop times, and you can clearly see that certain tasks start after others have already begun or are already finished.

Some Gantt chart strengths include:

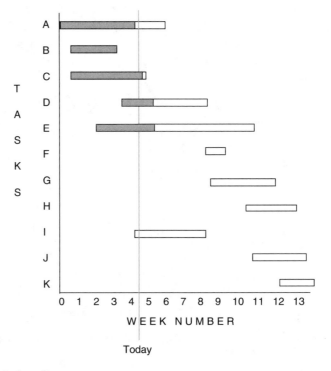

Figure F.11   Sample Gantt Chart

1. Being able to see the status of each project task at any point in time.
2. Being able to see overlapping or parallel tasks.

   A few of its weaknesses include:

1. Not being able to definitely tell from the Gantt chart whether the entire project is on time, behind time, or ahead of schedule.

2. Not showing task dependencies.

A Gantt Chart Example

Figure F.12 shows the Gantt chart for the data from the earlier PERT network examples. As you can see, tasks B and C overlap, tasks D and E overlap, and tasks F and G overlap.

SUMMARY

Project management is a very important part of information systems development. In fact, the literature suggests that ineffective project management contributes to most

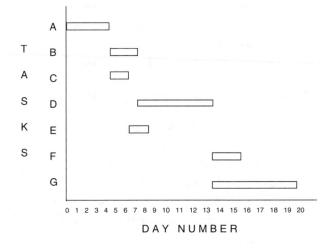

Figure F.12   Gantt Chart Using PERT Chart Sample Data

systems development delays and failures. In this chapter project sources—directives, opportunities, and problems—were discussed. Several reasons for systems development project failures were presented.

The last part of the chapter introduced the PERT network and the Gantt chart. A description of each was presented along with the mechanics for constructing the PERT network. Both strengths and weaknesses were cited for PERT networks as well as Gantt charts.

QUESTIONS

F.1 What are some of the sources or reasons that lead to the development of information systems?

F.2 What are the two main guidelines that underlie the management of an information systems development project?

F.3 Distinguish a project manager from a systems analyst or programmer.

F.4 What are some of the reasons for systems development project failures?

F.5 What are some of the results that would cause a systems development project to be considered a failure?

F.6 Briefly describe what is represented in a PERT network.

F.7 What does each number inside a node (in a PERT chart) represent?

F.8 What is a critical path in a PERT chart?

F.9 What is the purpose of a dummy connection in a PERT chart?

F.10 What is the process involved in determining the duration time for different tasks in an information systems development project? (Include the formula used to make the final computation.)

F.11 Discuss the importance of dependents and precedents in creating a PERT chart.

F.12 What is the formula for computing the earliest completion time (ECT) of a given node in a PERT chart?

F.13 What is the rule for selecting an ECT when a node has more than one preceding node?

F.14 What is the formula for computing the latest completion time (LCT) of a given node in a PERT chart?

F.15 What is the rule for selecting an LCT when a node has more than one preceding node?

F.16 What is the relationship between slack time and the critical path?

F.17 What is the significance of the tasks along the critical path?

F.18 What is the significance of paths that have slack time?

F.19 Briefly discuss some of the strengths and weaknesses of PERT charts.

F.20 What is the main function of a Gantt chart?

REFERENCES

MARCINIAK, JOHN J. (ed.), *Encyclopedia of Software Engineering,* vol. 1. New York: John Wiley & Sons, 1994.

WHITTEN, J.L., L.D. BENTLEY, and V.M. BARLOW, *Systems Analysis & Design Methods* (3rd ed.). Boston: Irwin, 1994.

Module G

Communication and Electronic Meetings

MODULE OBJECTIVES (YOU SHOULD BE ABLE TO)

1. Describe the three groups that make up an information systems development partnership.
2. Describe the communication interactions of a systems analyst.
3. Name each of the systems development project communication opportunities.
4. Discuss informal communication opportunities.
5. Discuss technical reviews.
6. Discuss oral presentations and reports.
7. Discuss the two types of problem-solving sessions.
8. Discuss a problem-solving session strategy.
9. Discuss electronic meetings.

Of all the topics in this book, communication is the one topic that every one of us already has years of experience with. Yet, business colleges usually require one or more communication courses for undergraduates as well as a course for graduate and executive MBA students. You may be thinking, "Why is there so much formal education on the topic of communication when we already have so much experience with it by the time we reach college?" That's a good question. Certainly there would be several aspects of these courses that would be new information to most people, since it deals with oral and written communication in business. In fact, in a recent executive MBA course dealing with communication, the students, all of whom are mid-level or higher managers, rated the communication course as one of the most valuable in the curriculum.

As you think about it, communication plays a role in almost everything you do in life. Your friends, your jobs, and your course grades have all been affected and influenced—good or bad—by your communication. Information systems managers year afteryear list good communication skills as one of the major skills they are looking for when hiring graduating students. With this in mind, the next section briefly discusses communication in the context of information systems development.

COMMUNICATION WITHIN AN INFORMATION SYSTEMS DEVELOPMENT PROJECT

Today's information systems development projects tend to have at least three distinct groups of individuals within the organization that come together in a partnership for the duration of the project. Figure G.1 illustrates these three groups, and the intersection of the three represents those members from each group who participate in a specific project partnership. The user group is most familiar with the problem domain. Management, which includes the steering committee, is most concerned with the goals and objectives of its business unit and its return on investment. The information systems development staff is most familiar with technology and the systems development process. Working in cooperation to exploit the strengths of each group can lead to a successful information systems development effort as well as a successful information system.

A software engineer must interact and communicate with diverse individuals and audiences during a systems development project, as illustrated in Figure G.2. Each of these individuals and audiences has a stake in the outcome of the project and is often referred to as a stakeholder. Often it is largely up to the software engineer to work with the stakeholders to see that their needs and business interests are included in the system. When conflict or differences arise, the software engineer attempts to facilitate resolution among the stakeholders. If necessary, management may even be required to resolve certain differences.

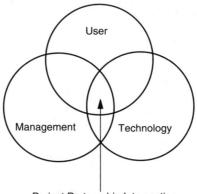

Figure G.1   Information Systems Project Partnership

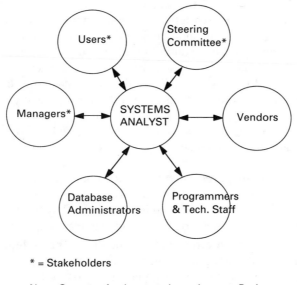

* = Stakeholders

Note: Systems Analyst may be acting as a Project
Manager for some of these interactions

Figure G.2   Software Engineer Interactions with Individuals During Systems Development

SYSTEMS DEVELOPMENT PROJECT COMMUNICATION OPPORTUNITIES

A software engineer once said, "Systems development projects would be easy if we did not have to interact with the users." Although unrealistic, there is a certain amount of truth to that statement. Systems development projects are packed with communication opportunities. For discussion purposes, these communication opportunities are classified as shown in Figure G.3.

The first group, mostly informal in nature, is the most prolific and frequently occurring of the four classifications listed in the figure. **Informal meetings and discussions** occur both at planned and unplanned times as well as at planned and un-

- Informal meetings, discussions, phone calls, and e-mail

- Technical Reviews (Walkthroughs)

- Presentations and Reports (Oral and Written)

- Problem Solving Sessions (Meetings)

Figure G.3   Project Communication Opportunities

planned locations. For example, a couple of users could just show up at your office one morning and want to briefly discuss something with you. Or standing in the lunch line in the cafeteria, you and a user chat about a project situation. Or while playing golf with a few managers, you informally discuss the project. **Phone calls and e-mail** are two other channels for informally discussing project aspects with team members.

Technical reviews, also referred to as **walk-throughs,** are a quality assurance check on one or more products developed during the systems development project. Technical reviews are usually formal in that they are scheduled and conducted anytime one or more products are ready to be reviewed. There are four groups of participants in the technical review:

1. The systems development staff author(s) of the product(s) being reviewed. As each product is reviewed for conformity and agreement with the user specifications, questions come up that need to be addressed by the author of the product, so this person needs to be present to answer and help clarify questions that arise.

2. One or more administrators to facilitate the review. This person(s) should be as neutral as possible regarding the project in order to facilitate without bias. This person is needed to keep the review on task and moving forward.

3. One or more secretaries to record the discrepancies and questions that surface during the review. It is helpful to have this person be someone familiar with technical jargon in order to accurately capture the discrepancies and questions.

4. The reviewers, usually users, who compare the product(s) with the specifications looking for problems, omissions, and possible errors.

On the surface, a technical review is considered a valuable part of the overall systems development project. However, there is always a concern that it not turn into a "we versus them" battle. Some of the common pitfalls that must be avoided are (1) attempting to review too large a product, (2) not providing enough time to prepare and review the product, and (3) criticism that deteriorates into personal attacks on individuals. An incident of any of these three pitfalls can lead to problems for the project.

Oral presentations and reports are the third type of systems development communication opportunities. There is such a significant variation in the types of reports that could be assembled during a project that they will not be covered here. Most systems development methodologies include "report type" guidelines within them which give structure to the project team members who are preparing reports.

Oral presentations have a few general guidelines that include:

1. Knowing your audience. On the one hand, it isn't necessary to personally know each member of your audience, but it is important that the presentation content and delivery be tailored to fit the type of audience you are speaking to. In the opinion of most communication experts, this is the number-one rule for all oral communication. Delivering an introductory talk to experienced individuals will tend to bore them and put them to sleep. Delivering an advanced talk to beginners will tend to overwhelm and demoralize them. Finding the right balance of content and delivery takes some skill, but without it we would not be effective communicators.

2. Use of a presentation outline similar to one containing (1) an introduction to the topic, (2) your main points that need to be covered, (3) questions to involve your audience (optional depending on a number of factors), and (4) a conclusion and/or summary.

3. Use of visual aids to enhance your presentation. People tend to retain more information if they both hear it and see it.

4. Anticipate audience questions during and after the presentation.

5. Ask questions of the audience at appropriate times during or after the presentation in order to engage them proactively. Doing so tends to increase their understanding of the topic and enhances your communication with the audience.

The last type of systems development communication opportunities is the **problem-solving session,** more commonly referred to as a **meeting.** This type of meeting is more formal than those discussed earlier that take place at random times throughout a day. These meetings are planned, scheduled, and conducted with certain objectives in mind. Figure G.4 illustrates the variations of problem-solving sessions. The tele-

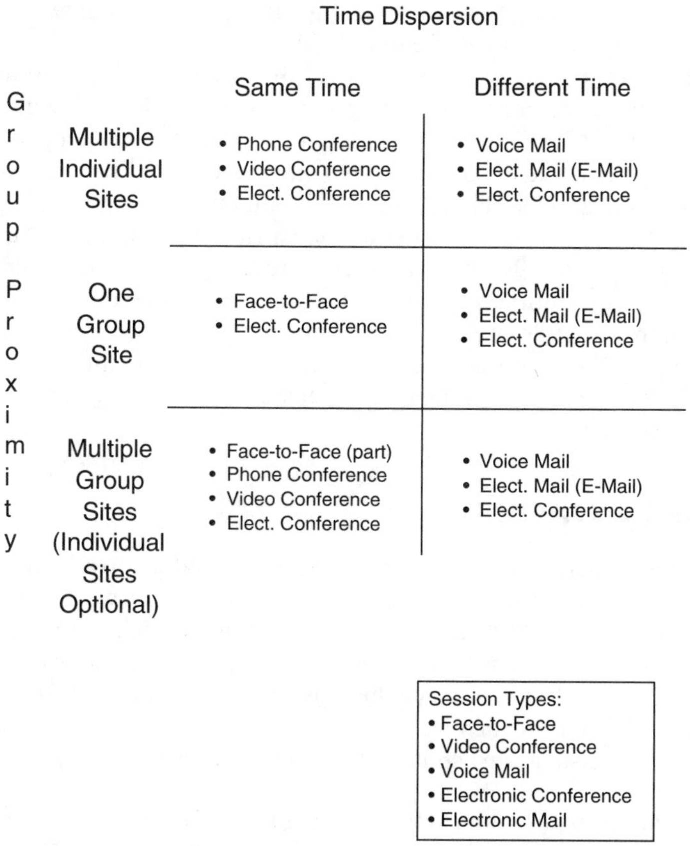

Time Dispersion

		Same Time	Different Time
G r o u p	Multiple Individual Sites	• Phone Conference • Video Conference • Elect. Conference	• Voice Mail • Elect. Mail (E-Mail) • Elect. Conference
P r o x i m i t y	One Group Site	• Face-to-Face • Elect. Conference	• Voice Mail • Elect. Mail (E-Mail) • Elect. Conference
	Multiple Group Sites (Individual Sites Optional)	• Face-to-Face (part) • Phone Conference • Video Conference • Elect. Conference	• Voice Mail • Elect. Mail (E-Mail) • Elect. Conference

Session Types:
• Face-to-Face
• Video Conference
• Voice Mail
• Electronic Conference
• Electronic Mail

Figure G.4 Problem-Solving Session (Meeting) Situations

phone is a common media for conducting both multiple individual and multiple group site meetings at the same time. Face-to-face meetings are common for same time, one group site as well as part of a same time, multiple group site meeting. Electronic conferencing technology is generally required for all different time dispersion meetings and can also be used to assist with all same time meetings.

Even though there are several variations for meetings, they can be classified into one of two types: face-to-face and conference style. **Face-to-face meetings** are just that—face-to-face. The attendees of the meeting are physically together in the same location at the same time. As few as two people can hold a face-to-face meeting, and the upper limit may only be limited to seating capacity of some large stadium, such as the Rose Bowl which holds over 100,000 people. Face-to-face meetings for problem solving certainly have an attendee practical limitation of one to two dozen depending on the objectives of the meeting. Face-to-face meetings have been the standard in business, education, and government for centuries. Although they are still very effective for problem solving, face-to-face meetings inherently suffer the problems of (1) schedule (e.g., people having to physically come together in the same place at the same time), and (2) cost (e.g., the cost associated with bringing the people together, such as out-of-pocket expenses for travel, accommodations, meals, and so on, and reduced, limited, or lost direct productivity while traveling). In a global economy, face-to-face meetings are becoming ever more difficult to hold.

Conference-style meetings cover all other types of formal problem-solving sessions. For the most part, conference-style meetings have only been practical since the proliferation of the telephone around the world. Conference telephone meetings are still a viable alternative to having limited attendance, brief face-to-face meetings. Today's technology and associated costs make this type of meeting very attractive. Technologies such as satellite communications, fiber optics, the picture-phone, which was reintroduced in 1993 after a failed introduction two dozen years earlier, interactive and high-definition television, and high-speed teleconferencing are all shaping the conference-style meeting of the not too distant future. As the twenty-first century approaches, conference-style meetings are not intended to completely replace face-to-face meetings, but are expected to be utilized as a cost-effective and natural alternative.

PROBLEM-SOLVING SESSION STRATEGY

Meetings should be planned, scheduled, organized, and conducted as efficiently as possible. No one has a shortage of meetings to attend in a given day or week, so we would all really like to give and/or get value from every meeting we attend. I worked for a sales manager once who used to have an 8:00 a.m. meeting every Monday morning whether it was needed or not. It was probably his way of making sure the staff was at work bright and early at the start of each week!

The following outline is a suggested strategy for an effective meeting:

1. Determine a need and purpose for each meeting ahead of time.

2. **Create an agenda** with suggested times for each item to be covered during the meeting and distribute it several days ahead so potential attendees can think about it.

3. **Schedule and make the necessary arrangements** for the meeting, such as place, size of space to accommodate the group, type of seating needed for this type of meeting, audio-visual aids, and so on.

4. **Conduct the meeting** being sure to keep it on task and on schedule.

5. **Follow up the meeting with written minutes** of the results and distribute them to each of the attendees.

ELECTRONIC MEETINGS TO SUPPORT GROUP WORK

A new generation environment to support collaborative group work, termed *electronic meeting systems (EMS)* by one research group, has emerged since the mid-1980s. The EMS environment consists of networked PCs (either local or distributed), a facilitator workstation, a software set of EMS workgroup tools, and a facilitator person and a large-screen projection system (for same place and time meetings).

The electronic meeting is a technique used to enhance group meetings of any type. It is not intended to completely replace verbal dialog between group members, but to complement dialog to its best advantage. To illustrate, you have called a face-to-face meeting with several users to discuss their initial requirements for a new information system. You choose a technique known as brainstorming to identify their requirements. As you begin, you say something like, "As you think of them, tell me your requirements for the information system, and I will write them on the board." Users, one after another in random order, begin to tell you their requirements, and you write each of them on the board making a consolidated list. Over time, you notice that some users are speaking much more than others, and eventually even they exhaust their ideas. In one hour's time you have amassed a list of about 40 system requirements. Using an EMS to complement this same meeting, each user sits at a PC and enters his or her system requirements into the computer and the network collects each response and displays the consolidated list on the large screen at the front of the room. Typing tends to stop after about 30 minutes, and you end up with a list of about 60 requirements.

A few observations regarding the preceding two scenarios are in order. First, you notice the time was cut in half using the EMS. This is not guaranteed but many studies have reported this type of time savings on average. Second, you notice that more ideas, systems requirements in this example, were generated by the group using the EMS. This too has been reported consistently in studies. Many EMS users indicate that the meeting itself was more "satisfying" to them compared to traditional meetings. Finally, although difficult to prove without directly asking the participants, it appears that more users were actively engaged in the EMS brainstorming session than the verbal one. Studies of EMS have suggested that use of the EMS with its inherent anonymity removes or reduces some of the participants' fears of peer intimidation, supervisor-subordinate concerns, and shyness or introversion.

There are at least four characteristics of problem-solving sessions that have an effect on the outcome or results of using an EMS, as illustrated in Figure G.5. One study suggested that meeting outcomes—effectiveness, efficiency, participant satisfaction—depend upon the interaction within the meeting process of these group, task, and context factors with the EMS components the group uses.

The **group** characteristic relates to issues such as group size, proximity to each other, composition of the group (e.g., peers, subordinates), cohesiveness of the group, and so on. The **task** characteristic refers to the activities required of the group to accomplish the task (e.g., brainstorming, voting, and so on.). The **context** characteristic refers to issues such as the group's organization culture, the time pressure the group is under, the evaluative tone for the group working together, the reward structure for the group, and so on. Finally, the **technology** characteristic refers to all of the issues surrounding the actual view of and use of the EMS to assist with the group work (e.g., ease of understanding and use).

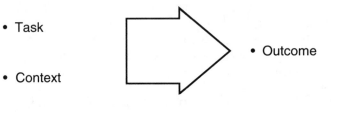

- Group

- Task

- Context   • Outcome

- Technology

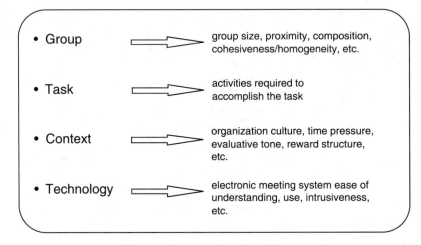

• Group	⇒	group size, proximity, composition, cohesiveness/homogeneity, etc.
• Task	⇒	activities required to accomplish the task
• Context	⇒	organization culture, time pressure, evaluative tone, reward structure, etc.
• Technology	⇒	electronic meeting system ease of understanding, use, intrusiveness, etc.

Figure G.5   Characteristics of Problem Solving Sessions that Have an Effect on the Outcome in an Electronic Meeting

SUMMARY

Communication skills continue to be much sought after skills when hiring decisions are made for software engineers. Two candidates with equal technical skills and education may boil down to which of the two has communicated the best during the interviewing process.

Information systems development has its own opportunities to communicate—informal ways, technical reviews, presentations and reports, and problem-solving sessions. The systems analyst has many different communication channels to work with during a systems development project, each important to the success of the project. Electronic meetings have been shown to be an effective complement to work done by groups.

QUESTIONS

G.1 What are the three elements that make up a project partnership?

G.2 What does each element contribute to a project partnership?

G.3 How do the systems analyst and a project's stakeholders interact in a project partnership?

G.4 What are some of the most basic and commonly used methods of communication between systems analysts and users?

G.5 What is a technical review and who are its participants?

G.6 Briefly describe some of the problems that can occur with a technical review.

G.7 What are some of the important concepts that should be utilized when communicating through an oral presentation?

G.8 List and briefly describe a few specific types of problem-solving sessions.

G.9 What are some of the problems associated with a face-to-face meeting?

G.10 Briefly discuss a few strategies that can help make meetings more effective.

G.11 What are the different parts of an electronic meeting system (EMS)?

G.12 What are some of the advantages of using an EMS as compared to a more traditional, face-to-face type of meeting?

G.13 List and briefly describe the characteristics of problem-solving sessions that have an effect on the outcome in an electronic meeting.

REFERENCES

NUNAMAKER, J.F., A.R. DENNIS, J.S. VALACICH, D.R. VOGEL, and J.F. GEORGE, "Electronic Meeting Systems to Support Group Work," *Communications of the ACM,* 34, no. 7 (July 1991), 40–61.

Module H

Business Process Reengineering

MODULE OBJECTIVES (YOU SHOULD BE ABLE TO)

1. Define and discuss business process reengineering.
2. Discuss two principal reasons for doing business process reengineering.
3. Discuss the major lessons learned from those who have done business process reengineering.
4. Describe the core of business process reengineering.
5. Describe one business process reengineering strategy.

INTRODUCTION

Business process reengineering. Just another fad? Peter F. Drucker asserts, "Reengineering is new, and it has to be done." I don't think it's a fad either, as businesses have moved beyond their pioneering experiences learning key lessons about what works and what doesn't, about the most common mistakes companies make, and about what executives can do to put their efforts in the win column. Business process reengineering is also known by a number of other synonyms, such as business process redesign, business transformation, process innovation, business reinvention, and change integration.

Business process reengineering is the search for, and implementation of, radical change in business processes to achieve breakthrough results. Notice the word *radical* in the definition. This is strong medicine, not always successful, and almost always accompanied by some organizational pain.

396

While most traditional change efforts start with what exists and fix it up, many reengineering advocates say that a clean sheet of paper is its chief tool. Business process reengineering is not a bottom-up continuous improvement program either. The reengineers start from the future and work backward, as if unconstrained by existing methods, people, or organizational structures. They often begin the process by asking the visionary question, "If we were a new company, how would we run this place?" The end result? "Throw everything else away and do it that way!"

Business process reengineering is serious work. Here is a partial list of what one large organization had to do over a two-year period to implement part of its reengineering effort: rewrite job descriptions for hundreds of employees, create new recognition and reward systems, revamp the computer system, conduct massive retraining, make extensive changes in financial reporting, writing proposals and contracts, and deal with suppliers, manufacturing, shipping, installation, and billing. That's radical and serious change!

Business process reengineering isn't easy and it isn't free. There is a cost associated with the reengineering process—both financial and cultural—which must yield a good return to the organization, otherwise it should not reengineer. Reengineering advocates suggest that the best corporate candidates for reengineering are companies facing big shifts in the nature of competition, such as banking and financial services, as well as the telecommunications industry, all of which have been significantly affected due to deregulation.

Some say fear and greed are the two principal reasons to reengineer. If you can imagine a startup rival, unburdened with high overhead, competing more effectively than your organization, then you may be a candidate for business process reengineering. Likewise, if you think your reengineering effort would afford you the opportunity to "eat your competition for lunch," you may also be interested in it.

LESSONS LEARNED FROM ORGANIZATIONS THAT HAVE DONE IT

All change is a struggle. Dramatic, across-the-organization change, such as business process reengineering, is war. As with war, strategy and tactics are critical for operational success. This is a new type of war, dynamic by its very nature, and there are several lessons that have been learned by the organizations that have undertaken this war:

1. Get the strategy straight first. Determine ahead of time what business you want to be in and how you intend to make money in it. This decision will dictate the company's aim in reengineering, since reengineering is about operations and only strategy can tell you what operations really matter.

2. Lead from the top. Reengineering is cross functional; therefore, someone high up in the organization who has clout needs to be the reengineering champion. Often it is not enough to have a department head be this person, but rather it should be the

CEO or COO along with a core team of first-rate people from all relevant organizational units, including human resources and information systems.

3. Create a sense of urgency. Business process reengineering will break apart under political pressure or die out after a few easy gains unless the case for doing it is compelling, urgent, and constantly refreshed. Fear, resistance, and cynicism are inevitable as the reengineering team begins to uncover problems and throw around radical ideas for solutions.

4. Design from the outside in. Focus should start from the customer's perspective. Do not spend too much time studying existing work flows.

5. Combine top-down and bottom-up initiatives. On the surface, business process reengineering with its emphasis on strong leadership, technology, and radical change seems to be the opposite of total quality management and continuous process improvement environments, which are essentially participative programs. Quite the contrary, advocates claim that reengineering works best in these types of environments, since it's these very people who have to live with the results of the reengineering effort.

6. Communication and expectations. Communication surfaces once again as an important lesson to be learned with yet again another major information systems concept. Business process reengineering (BPR) goals, objectives, and fundamentals must be communicated well to all involved in order to improve the probability of success for the BPR process. Expectations also need to be kept in check, as with all other information systems projects. People's expectations, if distorted, can cause severe problems with a BPR process because they will be expecting more from the process than can actually be delivered.

In the final analysis, business process reengineering is an approach used to change the way people work; hence, organization culture counts big. The change will not occur merely because management wills it.

THE CORE OF BUSINESS PROCESS REENGINEERING

For over a hundred years, segmentation of tasks has been at the core of industrial processes. Its reversal or reaggregation of those tasks is the core of business process reengineering, reversing the primary set of governing principles that has shaped industry around the world. Organizations are beginning to view processes as the web of the organization rather than the traditional functional orientation. In many instances, enabling technology is facilitating the reaggregation of labor tasks.

Business process reengineering consultancies have developed a variety of concrete methodologies, techniques, and training tools that make the reengineering effort more of a discipline and less of a crap shoot. Those pursuing reengineering efforts can avail themselves of many tools and techniques, such as brainstorming workshops, electronic meetings, best-practices databases, benchmarking analyses, and core business process identification activities.

BUSINESS PROCESS REENGINEERING IS ORGANIZATIONAL CHANGE

In another section of this book organizational change was discussed in some detail. Business process reengineering is, without a doubt, classic organizational change. Senior managers who undertake a reengineering project must have an appetite for change. In other words, they understand what would happen over time if they do not change. They must also be tolerant of ambiguity and chaos as the reengineering effort begins since many reengineering projects are able to identify point A (where they are) and point B (where they want or need to be), but are not real certain about the way to get from point A to point B.

A BUSINESS PROCESS REENGINEERING STRATEGY

One business process reengineering strategy used by A.D. Little Inc. uses the following approach:

1. Assess how well the client is meeting the expectations of its stakeholders, including customers, shareholders, and employees.
2. Figure out which of the stakeholders' expectations must be met to stay even with competitors, and which expectations should be met to garner a competitive advantage.
3. Figure out how to redesign processes to meet those expectations.
4. Map out information technology solutions to fit the redesigned processes.
5. Develop and implement the new processes along with all supporting details such as infrastructure, training, technology, and so on.

There are many other strategies available for doing business process reengineering. The vote is not in yet regarding its overall success or lack thereof in the United States. Articles praising and criticizing it continue to appear in the trade press.

SUMMARY

Business process reengineering defined as an all-or-nothing proposition is doomed to fail. Only a few organizations will be able to withstand the stresses of a radical, enterprisewide "all-or-nothing" redesign. A new definition of the concept is needed that does not have to start from scratch and produce dramatic results.

Business process reengineering must be viewed as a continuum—process improvement on one end, process transformation on the other. It must be both a top-down and bottom-up approach. In the same vein, business process reengineering need not be enterprisewide. Processes can be viewed as infinitely divisible, and considerable value can be gained by reengineering the inventory management process of a large warehouse, for example. The term itself may not survive, but the concept will, even as it is refined and expanded along with improved methodologies, techniques, and tools.

QUESTIONS

H.1 Give a single definition for what may be called business process redesign, business transformation, process innovation, business reinvention, and change integration.

H.2 What are some of the specific tasks that are often involved in business process reengineering?

H.3 What are some of the most important considerations when undertaking business process reengineering?

H.4 Briefly discuss one example of a business process reengineering strategy.

REFERENCES

BENNETT, KEITH, "Legacy Systems: Coping with Success," *IEEE Software,* 12, no. 1 (January 1995), 19–23.

DAVENPORT, Thomas H., *Process Innovation: Reengineering Work through Information Technology.* Boston, MA: Harvard Business School Press, 1993.

DEDENE, GUIDO, and JEANE-PIERRE DEVREESE, "Realities of Off-Shore Reengineering," *IEEE Software,* 12, no. 1 (January 1995), 35–45.

HAMMER, M., and J. CHAMPY, *Reengineering The Corporation: A Manifesto For Business Revolution.* New York: HarperBusiness, 1993.

MERLO, E., P. GAGNE, J. GIRARD, K. KONTOGIANNIS, L. HENDREN, P. RANANGADEN, and R. DEMORI, "Reengineering User Interfaces," *IEEE Software,* 12, no. 1 (January 1995), 64–73.

STEWART, T.A., "Reengineering, The Hot New Managing Tool," *Fortune,* 128, no. 4, August 23, 1993, 40–48.

SAGER, IRA (ed.), "Beyond Reengineering," *Information Week* (a supplement to), May 10, 1993.

SNEED, HARRY M., "Planning the Reengineering of Legacy Systems," *IEEE Software,* 12, no. 1 (January 1995), 24–34.

WONG, KENNY, S.R. TILLEY, H.A. MULLER, and M.D. STOREY, "Structural Redocumentation: A Case Study," *IEEE Software,* 12, no. 1 (January 1995), 46–54.

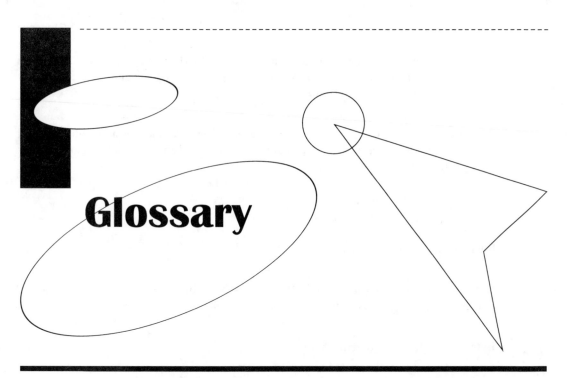

Glossary

Note: Terms with (OMG) following their definition are taken from the official OMG glossary contained in the *Object Management Architecture Guide,* Object Management Group, Inc., Framingham, MA, 1992.

Abstraction—The principle of ignoring those aspects of a problem domain that are not relevant to the current purpose in order to concentrate more fully on those that are.

Attribute—(1) one or more bytes combined to create a meaningful piece of data; (2) a data-valued characteristic defined for a class; attributes are used to maintain the state of the instances of a class (OMG).

Attribute data dictionary—See *Data dictionary, attribute.*

Basic service—A service that is implied and available to every class; six basic services are create object, get and set attribute values, add and remove object connections, and delete.

Behavior—The observable effects of a request (OMG).

Black box testing—The portion of the system that is being tested from an outside perspective.

Bottom-up (analysis, design, programming, testing)—Starts with the details of the system and proceeds to higher levels by a progressive aggregation of details until they collectively fit the requirements for the system.

Business model—A document that identifies specific goal and mission statements for levels within a business such as corporate level, division level, region level, department level, section level, and so on.

Business policy—See *Policy.*

Business procedures—See *Procedures.*

Business process reengineering (BPR)—The search for and implementation of radical change in business processes to achieve breakthrough results.

Byte—A single alphanumeric character.

Cardinality—See *Object connection constraint.*

Class—See *Class-with-objects;* a class has no objects associated with it.

Class-with-objects—A set or collection of abstracted objects that share common characteristics.

Cohesion—A measure of the strength of the interrelatedness of statements within a module (service).

Coupling—A measure of the strength of the connection between modules (services).

Database—A collection of one or more related folders or files.

Data dictionary, attribute—An alphabetized list of attribute names, which class or class-with-objects it belongs to, and details regarding its definition and editing rules.

Data element—See *Attribute.*

Data management component (DM)—A grouping of classes of objects that provide an interface between problem domain objects and a database or file management system. In an object model, such objects most often correspond to specific problem domain objects that need support for persistence and searching.

Decision table—A tool for graphically documenting a set of complex decision situations.

Decision tree—A tool for graphically documenting a set of complex decision situations.

Directory—See *Folder.*

Economic feasibility—A measure of the benefits of a new or changed information system compared to the costs to create and operate it. This is often called cost-benefit analysis.

Encapsulation (information hiding)—The notion that a software component (module, subroutine, method, and so on) should hide a single design decision.

Enterprise model—See *Business model.*

Event—Something that happens at a point in time; an action which occurred.

Feasibility—See *Economic, Operational,* and *Technical feasibility.*

Feasibility study—An activity done to determine the viability of a proposed systems development project from economic, operational, and/or technical perspectives. See *Economic, Operational,* and *Technical feasibility* for further discussion.

Field—See *Attribute.*

File—One or more related records.

Folder—A collection of one or more related files.

Heuristic—Rule of thumb or guideline that can usually be applied to a situation.

Human interaction component (HI)—A grouping of classes of objects that provides an interface between the problem domain objects and people. In an object model, such objects most often correspond to specific windows and reports.

Information architecture—A plan for selecting the appropriate information technology and information systems that will best support the business mission and goals.

Information hiding—See *Encapsulation.*

Information systems development life cycle—See *Systems development life cycle.*

Information systems planning—Strategy or technique used to identify and prioritize the information systems applications whose development and implementation would most benefit the business as a whole.

Information system's responsibility—The portion of the problem domain for which the information system is responsible.

Inheritance—A principle for managing complexity. All lower-level or children nodes in an inheritance hierarchy inherit the characteristics of the parent node.

Instance connection—A relationship mapping one object's needs with other objects in the problem domain in order to fulfill its responsibilities.

Joint application development (JAD)—A technique used during requirements determination which makes use of facilitated groups of users that collaborate in concentrated work sessions to define needed system functions, screens, reports, expectations, and data elements.

Message—A request for a service to be performed sent from a sender to a receiver. The sender sends a command and optionally some arguments; the receiver performs the service; the receiver returns control to the sender, along with some optional results.

Method—See *Service.*

Methodology—The packaging of methods and techniques together. The way something gets done. The purpose of a methodology is to promote a certain problem-solving strategy by preselecting the methods and techniques to be used.

Model component—See *Object model component.*

Multiple inheritance—All lower-level or children nodes in an inheritance hierarchy inherit the characteristics of two or more parent nodes.

Multivalue attribute—An attribute which is the opposite of the single-value attribute because it can have multiple data values at any moment in time.

Mutually exclusive value attribute—An attribute in which the presence or absence of its data value is dependent upon the presence or absence of one or more other attribute data values.

Normalization—A set of seven process steps used to simplify and reduce redundancy in data structures. Data structures are classified as being in either no normal form, first normal form, second normal form, third normal form, fourth normal form, Boyce-Codd normal form, fifth normal form, and domain-key normal form.

Notation—A set of symbols used to communicate or represent something.

Object (instance)—An abstraction of a person, place, or thing within the problem domain of which the information system must be aware.

Object connection—Represents that one object has a need to know another object; a relationship exists between them.

Object connection constraint—An expression of "who an object knows." It indicates how many other objects an object is aware of in the object connection. Each number or number pair expresses a lower and upper numeric boundary range for the relationship.

Object model component—A grouping of classes of objects. The groupings are problem domain (PD), human interaction (HI), data management (DM), and system interaction (SI). These components enable the separation of concerns concept by grouping objects according to domain (PD) or technology (HI, DM, SI).

Object-oriented pattern—See *Pattern.*

Operational feasibility—A measure of whether or not a new or changed information system will work in a certain business's culture.

Organizational change—A systematic effort to improve the functioning of some human system.

Pattern—A template of objects with stereotypical responsibilities and interactions; the template may be applied again and again, by analogy. Pattern instances are building blocks used to assemble effective object models.

Persistence—The ability to store objects longer than the time of program execution.

Persistent data—Data that are stored and held over time. Similar to a file or database.

Planned change—See *Organizational change.*

Policy—A set of rules that governs some activity within a business; policies often are the basis for decision making within the business.

Polymorphism—(1) The ability to take on different forms. For example, H_2O can take on three forms—water, steam, or ice; (2) giving the same name to services in different classes. Those services may do the work differently; yet those services produce the same kind of results.

Problem domain—The business problem or function being planned, analyzed, designed, and ultimately implemented as the information system.

Problem domain component (PD)—A grouping of classes of objects that directly corresponds to the problem being modeled. Objects in this component are technology-neutral and have little or no knowledge about objects in the human interaction, data management, and system interaction components.

Problem domain service—A service that is specific to a problem domain and must be explicitly documented in a class or class-with-objects.

Procedures—Step-by-step instructions for carrying out a policy or accomplishing some task(s).

Prototype—A working model of an information system.

Prototyping—An activity used to demonstrate the feasibility or some other aspects of the proposed information system in order to more fully understand the user's real requirements or to improve the chances of user acceptance of the proposed information system.

Pseudocode—See *Structured English.*

Record—One or more related attributes.

Requirements—The wants and/or needs of the user within a problem domain.

Response time—The average delay between a transaction or user request and the response to that transaction or user request.

Scenario—A specific time-ordered sequence of object interactions, one that exists to fulfill a specific need within the information system.

Service—The actions that the information system must perform in order to fulfill its purpose and meet the information system needs of the user.

Single-value attribute—An attribute that has only one value or state for itself at any moment in time.

Stakeholder—An organizational unit, individual, or group of individuals that affects or is affected by an information system.

Steering committee—Cross-functional, senior managers within a business, such as vice presidents or directors and the senior information systems manager or a designatee. The main role of this group is to conduct high-level reviews and evaluations of proposed information systems development projects and make recommendations for prioritization and resources for the projects.

State—The condition of an attribute at any moment in time for a specific object instance.

State-transition diagram—A diagram that shows (1) the states (values) of an object instance, and (2) the operations and exceptions that cause the transitions (changes) between the states.

Structured English—A concise and abbreviated version of our spoken language.

Subject matter expert (SME)—One who knows a great deal about a problem domain.

Subsystem—The next lower or detailed level system of the system of interest.

Suprasystem—The next higher or macro level system of the system of interest.

System—A set of interrelated components working together for a common purpose.

System interaction component (SI)—A grouping of classes of objects that provides an interface between problem domain objects and other systems or devices. A systems interaction object encapsulates communication protocol, keeping its companion problem domain object free of such low-level, implementation-specific detail.

Systems analyst—An individual who studies the problems and needs of a business in order to ascertain how hardware, software, people, procedures, and data can best accomplish improvements for the business.

Systems development life cycle (SDLC)—The process by which an information system maintains its usefulness to a business as it moves from inception to replacement.

Systems model—Six interrelated components that work together for a common purpose. The six components are boundary, inputs, processing, controls, outputs, and feedback.

Systems planning—See *Information systems planning.*

System's responsibility—See *Information system's responsibility.*

Technical feasibility—A measure of whether or not a new or changed information system is technically practical given the availability of technical resources and schedule limitations.

Throughput—The amount of work performed over some period of time.

Top-down (analysis, design, programming, testing)—Follows the functional decomposition approach to problem solving. It starts by defining an overview or high-level view of the system, then works its way into the details of the system, layer by layer, until reaching the bottommost details. The bottommost parts are dependent on the higher-level parts of the system.

Tuple—See *Record*. Tuples are the equivalent to records in relational database terminology.

Vendors—Those businesses which support the information systems development effort, such as consultants, hardware and software companies, training companies, telecommunications companies, documentation companies, and so on.

White box testing—The portion of the system that we are testing from an inside perspective.

OO Recommended Reading to Get Started—August 1995

BOOCH, G., *Object-Oriented Analysis and Design with Applications,* 2nd ed. Redwood City, CA: Benjamin/Cummings Publishing Co., 1994.

COAD, P., and J. NICOLA, *Object-Oriented Programming.* Englewood Cliffs, NJ: Prentice Hall, 1993.

COAD, P., D. NORTH, and M. MAYFIELD, *Object Models: Strategies, Patterns, and Applications.* Englewood Cliffs, NJ: Prentice Hall, 1995.

COAD, P., and E. YOURDON, *Object-Oriented Design.* Englewood Cliffs, NJ: Prentice Hall, 1991.

COAD, P., and E. YOURDON, *Object-Oriented Analysis,* 2nd ed. Englewood Cliffs, NJ: Prentice Hall, 1991.

HENDERSON-SELLERS, B., *A Book of Object-Oriented Knowledge.* Englewood Cliffs, NJ: Prentice Hall, 1992.

JACOBSON, I., M. CHRISTERSON, P. JONSSON, and G. OVERGAARD, *Object-Oriented Software Engineering: A Use Case Driven Approach.* New York: Addison-Wesley, 1992.

KHOSHAFIAN, S., *Object-Oriented Databases.* New York: John Wiley & Sons, 1993.

RUMBAUGH, J., M. BLAHA, W. PREMERLANI, F. EDDY, and W. LORENSEN, *Object-Oriented Modeling and Design.* Englewood Cliffs, NJ: Prentice Hall, 1991.

TAYLOR, DAVID A., *Object-Oriented Information Systems.* New York: John Wiley & Sons, 1992.

WIRFS-BROCK, R.I., B. WILKERSON, and L. WIENER, *Designing Object-Oriented Software.* Englewood Cliffs, NJ: Prentice Hall, 1990.

YOURDON, E., *Object-Oriented Systems Design: An Integrated Approach.* Englewood Cliffs, NJ: Prentice Hall, 1994.

Bibliography—Object-Oriented Technology*

REFERENCES†

ABBOTT, R.J., 1983, Program design by informal English descriptions, *Comms. ACM,* 26(11), 882–894.

ADAMS, M., and LENKOV, D., 1990, Object-oriented COBOL: the next generation, *Hotline on Obj-Oriented Technology,* 2(2), 12–15.

ALABISO, B., 1988, Transformation of data flow analysis models to object oriented design, *OOPSLA '88 Proceedings,* ACM, 335–353.

ALLABY, A. and ALLABY, M. (eds.), 1990, *The Concise Offord Dictionary of Earth Sciences,* Oxford Univ. Press, Oxford, 4l0 pp.

ANTEBI, M., 1990, Issues in teaching C++, *J. Obj.-Oriented Programming,* 3(4), 11–21.

AUER, K., 1989, Which object-oriented language should we choose? *Hotline on Obj.-Oriented Technology,* 1(1), 1, 3–6.

BAILIN, S.C., 1989, An object-oriented requirements specification method, *Comms. ACM,* 32(5), 608–623.

BALDA, D.M., and GUSTAFSON, D.A., 1990, Cost estimation models for the reuse and prototype software development life-cycles, *ACM SIGSOFT Software Engineering Notes,* 15(3), 42–50.

BANCILHON, F., and DELOBEL, C., 1991, Recent advances in O-O DBMS, TOOLS' 91 Tutorial Notes, 4th International Conference and Exhibition, March 4–8, Paris.

BARBER, G.R., 1991, The Object Management Group, *HOTLINE ON OBJ.-ORIENTEDTECHNOLOGY,* 2(5), 17–19.

---

* Portions of this bibliography are from: Henderson-Sellers, Brian, *A Book of Object-Oriented Knowledge,* Prentice Hall, New York, 1992.

† As of August 1995.

BECK, K., and CUNNINGHAM, W., 1989, A laboratory for teaching object-oriented thinking, *SIGPLAN Notices,* 24(10).

BERARD, E.V., 1993, *Essays on Object-Oriented Software Engineering,* Prentice Hall, New York, NY, 352 pp.

BERARD, E.V., l990a, Life-cycle approaches, *Hotline on Obj.-Oriented Technology,* 1(6), 1, 3–4.

BERARD, E.V., 1990b, Understanding the recursive/parallel life-cycle, *Hotline on Obj.-Oriented Technology,* 1(7), 10–13.

BERARD, E.V., 1990c, Object-oriented requirements analysis, *Hotline on Obj.-Oriented Technology,* 1(8), 9–11.

BERMAN, C., and GUR, R., 1988, NAPS – a C++ project case study, *Procs.USEINX C++ Conference,* 137–149.

BIELAK, R., 1991, The guessing game. A first Eiffel program, *Eiffel Outlook,* 1(1), 16–18.

BIRD, R., and WADLER, P., 1988, *Introduction to Functional Programming,* Prentice Hall, New York, 293 pp.

BIRTWISTLE, G., DAHL, 0.-J., MYRHAUG, B., and NYGAARD, K., 1973, *Simula Begin,* Studentliteratur (Lund) and Auerbach Pub. (New York).

BLAIR, G.S., GALLAGHER, J.J., and MALIK, J., 1989, Genericiry vs inheritance vs delegation vs conformance vs . . ., *J. Obj.-Oriented Programming,* 2(3), 11–17.

BLAIR, G.S., MALIK, J., NICOL, J.R., and WALPOLE, L., 1990, A synthesis of object-oriented and functional ideas in the design of a distributed software engineering environment, *Software Engineering Journal,* 5(3), 194–204.

BOBROW, D.G., 1989, The object of desire, *Datamation,* May 1, 1989, 37–41.

BOLLINGER, T.B., and PFLEEGER, S.L., 1990, Economics of reuse: issues and alternatives, *Inf. Software Technol.,* 32, 643–652.

BOOCH, G., 1983, *Software Engineering with Ada,* Benjamin/Cummings, Menlo Park, CA.

BOOCH, G., 1987, *Software Engineering with Ada,* 2nd ed., Benjamin/Cummings, Menlo Park, CA, 580 pp.

BOOCH, G., 1994, *Object Oriented Analysis and Design with Applications,* Benjamin/Cummings, Menlo Park, CA.

BOOCH, G., and VILOT, M., 1990a, Object-oriented design. Evolving an object-oriented design, *The C++ Report,* 2(8), 11–13.

BOOCH, G., and VILOT, M., 1990b, Object-oriented design. Inheritance relationships, *The C++ Report,* 2(9), 8–11.

BRACHA, G., and COOK, W., 1990, Mixin-based inheritance, *ECOOP/OOPSLA '91Procs.,* ACM, 303–311.

BURNHAM, W.D., and HALL, A.R., 1985, *Prolog Programming and Applications,* Macmillan, Basingstoke, UK, 114 pp.

CALDIERA, G., and BASILI, V., 1991, Identifying and qualifying reusable software components, *IEEE Computing,* 24(2), 61–70.

CARD, D.N., COTNOIR, D.V., and GOOREVICH, C.E., 1987, Managing software maintenance cost and quality, *IEEE,* 145–152.

COAD, P., and NICOLA, J., 1993, *Object-Oriented Programming,* Yourdon Press/Prentice Hall, New York, 582 pp.

COAD, P., "Object-Oriented Patterns," *Communications of the ACM,* 35, no. 9, (September 1992), 152–159.

COAD, P. and YOURDON, E., *Object-Oriented Analysis,* 2nd ed., Prentice Hall, Englewood Cliffs, NJ, 1991, 233 pp.

COAD, P. and YOURDON, E., *Object-Oriented Design,* Prentice Hall, Englewood Cliffs, NJ, 1991, 197 pp.

COAD, P., NORTH, D., and MAYFIELD M., *Object Models: Strategies, Patterns, and Applications,* Prentice Hall, Englewood Cliffs, NJ, 1995, 505 pp.

COLEMAN, D., ARNOLD, P., BODOFF, S., DOLLIN, C., GILCHRIST, H., HAYES, F., JEREMAES, P., *Object-Oriented Development: The Fusion Method,* Prentice Hall, Englewood Cliffs, NJ, 1994.

COLEMAN, D., and HAYES, F., 1991, Lessons from Hewlett-Packard's experiences of using object-oriented technology, in TOOLS4 (Procs. 4th Int. Conf. TOOLS Paris, 1991) (eds. J. Bezivin and B. Meyer), Prentice Hall, New York, 327–333.

COLLINS, R., 1990, Object orientation changes the face of computing, *Professional Computing,* March 1990, 10–11.

CONSTANTINE, L.L., 1989a, Object-oriented and structured methods. Toward integration, *American Programmer,* 2(7-8), 34–40.

CONSTANTINE, L.L., 1989b, Object Oriented and Structured Design Seminar, Digital Consulting Pacific, Sydney, 1989.

CONSTANTINE, L.L., 1990a, Objects, functions, and extensibility, *Computer Language,* 7, 34–36.

CONSTANTINE, L.L., 1990b, "The Object-Oriented Systems Symposium," Digital Consulting Pacific, Sydney, June 1990.

CONSTANTINE, L.L., 1990c, Objects by Teamwork, *Hotline on Obj.-Oriented Technology,* 2(1), 1, 3–6.

COX, B.J., 1986, *Object Oriented Programming: An Evolutionary Approach,* Addison-Wesley, Reading, MA, 274 pp.

COX, B.J., 1990a, There *is* a silver bullet, *Byte,* October 1990, 209–218.

COX, B.J., 1990b, Planning the software industrial revolution, *IEEE Software,* November 1990, 25–33.

DAVIS, A.M., 1988, A taxonomy for the early stages of the software development life cycle, *J. Systems and Software,* 8, 297–311.

DEMARCO, T., 1978, *Structured Analysis and System Specification,* Yourdon Press, New York.

DEUTSCH, L.P., 1989, Comment made during a panel session at OOPSLA '89, October, New Orleans.

DEWHURST, S.C., and STARK, K.T., 1987, Out of the C world comes C++, *Computer Language,* February 1987, 29–36.

DITTRICH, K.R. (ed.), 1988, *Advances in Object-Oriented Database Systems, Lecture Notes in Computer Science,* 334, Springer-Verlag, Berlin, 373 pp.

DOBBIE, G., 1991, Object oriented database systems: a survey, Procs. 14th Aust. Comp. Sci. Conf, UNSW, February 6–8, 1991, *Australian Computer Science Communications,* 13(1), 10–11.

DONALDSON, C.M., 1990, Dynamic binding and inheritance in an object-oriented Ada design, *J. Pascal, Ada and Modula-2,* 9(4), 13–18.

ECKEL, B., 1989, *Using C++,* McGraw-Hill, Berkeley, CA, 617 pp.

ECKEL, B., 1990, C++ notes, *The C Gazette,* Spring 1990, 70–72.

EDWARDS, J.M, and HENDERSON-SELLERS, B., 1991, A coherent notation for object-oriented software engineering, *Technology of Object-Oriented Languages and Systems: TOOLS5* (eds. T. Korson, V. Vaishnavi, and B. Meyer), Prentice-Hall, New York, 405–426.

ELMASRI, R., and NAVATHE, S.B., 1989, *Fundamentals of Database Systems,* Benjamin/Cummings, 802 pp.

FELDMAN, P., and MILLER, D., 1986, Entity model clustering: structuring a data model by abstraction, *The Computer Journal,* 29, 348–360.

FERGUSON, T., 1991, Operator overloading in C++, *J. Obj.-Oriented Programming,* 4(1), 42–48.

FIRESMITH, D.G., 1993, *Object-Oriented Requirements Analysis and Logical Design,* John Wiley & Sons, New York, NY, 575 pp.

GAMMA, E., HELM R., JOHNSON, R., and VLISSIDES, J., *Design Patterns: Elements of Reusable Object-Oriented Software,* Addison-Wesley Publishing, Reading, MA, 1995.

GHEZZI, C., and JAZAYERI, M., 1987, *Programming Language Concepts,* 2nd ed., John Wiley and Sons, New York, 428 pp.

GIBBS, S., TSICHRITZIS, D., CASAIS, E., NIERSTRASZ, O., and PINTADO, X., 1990, Class management for software communities, *Comms. ACM,* 33(9), 90–103.

GIBSON, E., 1991, Flattening the learning curve: educating object-oriented developers, *J. Obj.-Oriented Programming,* 3(6), 24–29.

GOLDBERG, A., 1985, *Smalltalk-80: The Interactive Programming Environment,* Addison-Wesley, Reading, MA.

GOLDBERG, A., 1991, Object-oriented project management, TOOLS '91 Tutorial Notes, Paris, March 1991.

GOLDBERG, A., and ROBSON, D., 1983, *Smalltalk-80: The Language and Its Implementation,* Addison-Wesley, Reading, MA.

GOLDBERG, A., and RUBIN, K., 1990, Talking to project managers: organizing for reuse, *Hotline on Object-Oriented Technology,* 1(10), 7–11.

GOLDSTEIN, S.C., 1990, Introducing OOPS through design—it's all in the words, *Hotline on Object-Oriented Technology,* 1(11), 1, 4–7.

GOLDSTEIN, T., 1989, The object-oriented programmer, *The C++ Report,* 1(5), May 1989.

HARMON, P., 1990, Object-oriented systems, *Intelligent Software Strategies,* 6(9), 1–16.

HECHT, A., 1990, Cute object-oriented acronyms considered fOOlish, *ACM SIGSOFT, Software Engineering Notes,* 15(1), 48.

HEINTZ, T.J., 1991, Object-oriented databases and their impact on future business database applications, *Inf. Management,* 20, 95–103.

HENDERSON-SELLERS, B., 1992, *A Book of Object-Oriented Knowledge,* Prentice-Hall, New York, NY, 297 pp.

HENDERSON-SELLERS, B., 1990, Three methodological frameworks for object-oriented systems development, Procs. 3rd Intl. Conf, *TOOLS3,* Sydney 1990 (eds. J. Bezivin, B. Meyer, J. Potter, and M. Tokoro), 118–131.

HENDERSON-SELLERS, B., 1991a, Metrics necessary for object-oriented development, submitted to *ACM SIGMETRICS.*

HENDERSON-SELLERS, B., 1991b, Hybrid object-oriented/functional decomposition methodologies, *Hotline on Obj.-Oriented Technology.*

HENDERSON-SELLERS, B., 1991c, Parallels between object-oriented software development and total quality management, *Journal of Information Technology,* 6(3), 15–19.

HENDERSON-SELLERS, B., and CONSTANTINE, L.L., 1991, Object-oriented development and functional decomposition, *Journal of Obj.-Oriented Programming,* 3(5), 11–17.

HENDERSON-SELLERS, B., and EDWARDS, J.M., 1990a, The object-oriented systems life cycle, *Comms. ACM,* 33(9), 142–159.

HENDERSON-SELLERS, B., and EDWARDS, J.M., 1990b, Object-oriented graphics: old wine in new bottles? *Procs. Graphics in Object-Oriented Software Engineering (GOOSE) Workshop,* OOPSLA '90, Ottawa, Canada, October 1990.

HENDERSON-SELLERS, B., and FREEMAN, C., 1991, Cataloguing object libraries, *ACM SIGSOFT Software Engineering Notes.*

HOPKINS, D., 1990a, An Eiffel experience, *ACS Bulletin (Victoria Branch),* June 1990, 5–8.

HOPKINS, J.W., 1990b, Object.oriented programming: the next step up, *J. Obj. Oriented programming,* 3(1), 66–68.

HOPKINS, J.W., 1991, What management needs to know, *TOOLS '91* Tutorial Notes, Paris, March 1991.

HOPKINS, T., and Warboys, B., 1990, Asset management and object-oriented technology, *Hotline on Obj.-Oriented Technology,* 1(11), 12–13.

HOWARD, G.S., 1988, Object oriented programming explained, *J. Systems Management,* 39(7), 13–19.

JACOBSON, I., CHRISTERSON, M., JONSSON, P., and OVERGAARD, G., 1992, *Object-Oriented Software Engineering,* Addison-Wesley, New York, NY, 524 pp.

JACKSON, I.F., 1986, *Corporate Information Management,* Prentice Hall, London, 338 pp.

JACKSON, M.A., 1983, *System Development,* Prentice Hall, London, 418 pp.

JALOTE, P., 1989, Functional refinement and nested objects for object-oriented design, *IEEE Trans. Software Eng.,* 15, 264–770.

JOHNSON, R.E., and FOOTE, B., 1988, Designing reusable classes, *J. Obj.-Oriented Programming,* 1(2), 22–35.

KERR, R., 1991, Object disorientation, *Hotline on Obj.-Oriented Technology* 2(5), 12–13.

KHOSHAFIAN, S., *Object-Oriented Databases,* John Wiley & Sons, New York, NY, 1993.

KHOSHAFIAN, S., 1990, Insight into object-oriented databases, *Inf. Software Technol.,* 32(4), 274–289.

KIM, W., and LOCHOVSKY, F.H. (eds.), 1989. *Object-Oriented Concepts, Databases, and Applications,* ACM Press New York/Addison-Wesley, Reading, MA, 602 pp.

KOENIG, A., and STROUSTRUP, B., 1990, Exception handling for C++. *J. Obj.-Oriented Programming,* 3(2), 16–33.

KORIENEK, G., and WRENSCH, T., *A Quick Trip To Objectland,* Prentice Hall Inc., Englewood Cliffs, NJ, 1993.

KORSON, T., and MCGREGOR, J.D., 1990, Object-oriented software design: a tutorial, *Comms. ACM,* 33(9), 40–60.

KORSON, T., and MCGREGOR, J.D., 1991, Technical criteria for the specification and evaluation of object-oriented libraries, Clemson University Technical Report #91-112.

KUHN, T., 1962. *The Structure of Scientific Revolutions,* Univ. Chicago Press.

LADDEN, R.M., 1989, A survey of issues to be considered in the development of an object-oriented development methodology for Ada, *Ada Letters,* 9 (2), 78–88.

LAHIRE, P., and BRISSI, P., 1991, An integrated query language for handling persistent objects in Eiffel, in *Technology of Object-Oriented Languages and Systems: TOOLS 4* (eds. I. Bezivin and B. Meyer), Prentice Hall, New York, 101–114.

LaLonde, W., and Pugh, J., 1990, Smalltalk as the first programming language: the Carleton experience, *J. Obj.-Oriented Programming,* 3(4), 60–65.

LaLonde, W., and Pugh, J., 1991, Subclassing <>subtyping<> is-a, *J. Obj.-Oriented Programming,* 3(5), 57–62.

Laranjeira, L.A., 1990, Software size estimation of object-oriented systems, *IEEE Trans. Software Eng.,* 16(5), 510–522.

Leathers, B., 1990a, Cognos and Eiffel: a cautionary tale, *Hotline on Obj.-Oriented Technology,* 1(9), 1, 3, 6–8.

Leathers, B., 1990b, OOPSLA panel: OOP in the real world, *ECOOP/OOPSLA '90 Proceedings,* ACM Press, New York, 299–302.

Linowes, J.S., 1988, It's an attitude, *Byte,* August 1988, 219–224.

Lippman, S.B., 1989, *C++ Primer,* Addison-Wesley, Reading, MA, 464 pp.

Loomis, M.E.S., 1990a, OODBMS. The basics, *J. Obj.-Oriented Programming,* 3(1), 77, 79–81.

Loomis, M.E.S., 1990b, OODBMS vs. relational, *J. Obj.-Oriented Programming,* 3(2), 79–82.

Loomis, M.E.S., Shah, A.V., and Rumbaugh, I.E., 1987, An object modeling technique for conceptual design, *Procs. ECOOP '87,* Springer, New York, 192–202.

Lorenz, M., 1993, *Object-Oriented Software Development,* Prentice Hall, New York, NY, 227 pp.

Love, T. *Object Lessons: Lessons Learned in Object-Oriented Development Projects,* SIGS Books, Inc., New York, NY, 1993.

Loy, P.H., 1990, A comparison of object-oriented and structured development methods, *ACM SIGSOFT, Software Engineering Notes,* 15(1), 44–48.

McCullough, P., and Deshler, N., 1990, WyCASH+: an application built within an OOP environment, *Hotline on Obj.-Oriented Technology,* 1(10), 1, 3–4.

Malhotra, A., Thomas, J.C., Carroll, J.M., and Miller, L., 1980, Cognitive processes in design, *J. Man-Machine Studies,* 12, 119–140.

Martin, J., 1993, *Principles of Object-Oriented Analysis & Design,* Prentice Hall, New York, NY, 412 pp.

Martin, J. and Odell, J.J., 1992, *Object-Oriented Analysis & Design,* Prentice Hall, New York, NY, 513 pp.

Meyer, B., 1988, *Object-Oriented Software Construction,* Prentice Hall, Hemel Hempstead, UK, 534 pp.

Meyer, B., 1989a, From structured programming to object-oriented design: the road to Eiffel, *Structured Programming,* 1, 19–39.

Meyer, B., 1989b, Writing correct software, *Dr Dobbs Journal,* December 1989, 48–63.

Meyer, B., 1989c, The new culture of software development: reflections on the practice of object-oriented design, *Procs. TOOLS '89* (Paris, November 13–15, 13–23).

Meyer, B., 1989d, Course notes for two day seminar, Object-oriented design and programming: a software engineering perspective, Sydney. November 1989, Interactive Software Engineering, Inc., 235 pp.

Meyer, B., 1990a, Lessons from the design of the Eiffel libraries, *Comms. ACM,* 33(9), 68–88.

Meyer, B., 1990b. Eiffel and C++: a comparison, unpublished technical note, *Interactive Software Engineering,* March 1990, 12 pp.

MEYER, B., 1991, *Eiffel: The Language,* Prentice Hall, New York.

MILLER, G., 1956, The magical number seven, plus or minus two: some limits on our capacity for processing information, *The Psychological Review,* 63(2), 81–97.

MILLIKIN, M., 1989, Object-orientation: what it can do for you, *Computerworld,* March 13, 1989.

MOREAU, D.R., and DOMINICK, W.D., 1989, Object-oriented graphical information systems: research plan and evaluation metrics, *J. Systems and Software,* 10, 23–28.

MULLIN, M., 1989, *Object Oriented Program Design With Examples in C++,* Addison-Wesley, Reading, MA, 303 pp.

NORDEN, P.V., 1958, Curve fitting for a model of applied research and development scheduling, *IBM Journal,* July 1958, 232–248.

NURICK, A., 1990, An OOP developer's success story, *Programmer's Update,* June 1990, 41–51.

PAGE-JONES, M., 1980, *The Practical Guide to Structured System Design,* Yourdon Press, New York.

PAGE-JONES, M., 1991a, Object-orientation: stop, look, and listen! *Hotline on Obj.-Oriented Technology,* 2(3), 1, 3–7.

PAGE-JONES, M., 1991b, TOOLS '91 Tutorial Notes, 4th International Conference and Exhibition. March 4–8, Paris.

PAGE-JONES, M., and WEISS, S., 1989, Synthesis: an object-oriented analysis and design method, *American Programmer,* 2(7-8), 64677, Summer 1989.

PAGE-JONES, M., CONSTANTINE, L.L., and WEISS, S., 1990, Modeling object-oriented systems: the Uniform Object Notation, *Computer Language,* 7 (10), October 1990.

PAPAZOGLOU, M.P., GEORGIADIS, P.I., and MARITSAS, D.G., 1984, An outline of the programming language Simula, *Comput. Lang.,* 9, 107–131.

PARNAS, D., 1972, On the criteria to be used in decomposing systems into modules, *Comms. ACM,* 15(2), 1053–1058.

PECKHAM, J., and MARYANSKI, F., 1988, Semantic data models, *ACM Computing Surveys,* 20, 153–189.

PFLEEGER, S.L., 1991, Model of software effort and productivity, *Inf. Software Technol.,* 33(3), 224–231.

POKKUNURI, B.P., 1989, Object oriented programming, *SIGPLAN Notices,* 24(11), 96–101.

POTTER, J., 1990, Eiffel 2.2, *J. Obj.-Oriented Programming,* 3(3), 84–88.

PRICE, R.T., and Girardi, R., 1990, A class retrieval tool for an object oriented environment, in *Procs. 3rd Intl. Conf, TOOLS3,* Sydney 1990 (eds. J. Bezivin, B. Meyer, J. Potter, and M. Tokoro), 26–36.

PRIETO-DIAZ, R., and FREEMAN, P., 1987, Classifying software for reusability, *IEEE Software,* 4(1), 6–16.

PUN, W., and WINDER, R., 1990, A design method for object-oriented programming, Department of Computer Science, University College London, Research Note RN/90/51, 17 pp.

PURCHASE, J.A., and WINDER, R.L., 1991, Debugging tools for object-oriented programming, *J. Obj.-Oriented Programming,* 4(3), 10–27.

RAJLICH, V., 1985, Paradigms for design and implementation in Ada, *Comms. ACM,* 28, 718–727.

RHOADES, C.E. Jr., 1990, Scientific programming, concurrency and object-oriented languages: a view from the battlements, *Procs. 3rd Intl. Conf., TOOLS3,* Sydney 1990 (eds. J. Bezivin, B. Meyer, J. Potter, and M. Tokoro), 109–116.

ROSENQUIST, C.J., 1982, Entity life cycle models and the applicability to information systems development lifecycles, *The Computer Journal,* 25(3), 307–315.

RUMBAUGH, J., 1987, Relations as semantic constructs in an object-oriented language, *OOPSLA '87 Proceedings,* ACM, 466–481.

RUMBAUGH, J., BLAHA, M., PREMERLANI, W., EDDY, F., and LORENSEN, W., 1991, *Object-Oriented Modeling and Design,* Prentice Hall, New York, 528 pp.

SAUNDERS, J.H., 1989, A survey of object-oriented programming languages, *J. Obj.-Oriented Programming,* 1(6), 5–11.

SCHMUCKER, K.J., 1986a, MacApp: an application framework, *Byte,* August 1986, 189–193.

SCHMUCKER, K.J., 1986b, *Object-Oriented Programming for the Macintosh,* Hayden Book Company.

SEIDEWITZ, E., 1989, General object-oriented software development: background and experience, *J. Systems and Software,* 9, 95–108.

SEIDEWITZ, E., and STARK, M., 1987, Towards a general object-oriented software development methodology, *Ada Letters,* 7, July/August 1987, 54–67.

SHAW, R.H., 1991, C++ without objects, *Borland Language Express,* 1(1), 10–13.

SHLAER, S., and Mellor, S.J., 1988, *Object-Oriented Systems Analysis: Modeling the World in Data,* Yourdon Press/Prentice Hall, 144 pp.

SIMSION, G.C., 1989, A structured approach to data modelling, *Aust. Comp. J.,* 21, 108–117.

SNYDER, A., 1986, Encapsulation and inheritance in object-oriented programming languages, *OOPSLA '86 Proceedings,* ACM Press, 38–45.

SOMMERVILLE, I., 1989, *Software Engineering,* 3rd ed., Addison-Wesley, Wokingham, UK, 653 pp.

STEIN, J., 1988. Object-oriented programming and databases, *Dr Dobbs Journal,* March 1988, 18–34.

STEWART, M.K., 1991, Object projects: what can go wrong, *Hotline on Obj-Oriented Technology,* 2(6), 15–17.

STROUSTRUP, B., 1986, *The C++ Programming Language,* Addison-Wesley, Reading, MA, 328 pp.

STROUSTRUP, B., 1988. What is object-oriented programming? *IEEE Software,* May, 1–19.

TAYLOR, DAVID A., 1992, *Object-Oriented Information Systems,* John Wiley & Sons, New York, NY, 357 pp.

TAYLOR, DAVID A., 1995, *Business Engineering with Object Technology,* John Wiley & Sons, New York, NY.

THOMAS, D., 1989a. In search of an object-oriented development process, *J. Obj.-Oriented Programming,* 2(1), 60–63.

THOMAS, D., 1989b, What's in an object? *Byte,* March 1989, 231–240.

THOMSETT, R., 1990, Management implications of object-oriented development, *ACS Newsletter,* October 1990, 5–7, 10–12.

TOPPER, ANDREW, *Object-Oriented Development in COBOL,* McGraw-Hill, Inc., New York, NY, 1995.

TROWBRIDGE, D., 1990, OOP needs changed programmer thinking, *Computer Technol. Rev.,* August 1990.

TURNER, J.A., 1987, Understanding the elements of system design, Chapter 4 in *Critical Issues in Information Systems Research* (eds. R.I. Boland, Jr. and R.A. Hirschheim), John Wiley and Sons, Chichester, UK, 97–111.

URLOCKER, Z., 1989, Teaching object-oriented programming, *J. Obj.-Oriented Programming,* 2(2), 45–47.

WALDO, J., 1990, O-O benefits of Pascal to C++ conversion, *The C++ Report,* (8), 1, 5–7.

WAMPLER, K.D., 1990, The object-oriented programming paradigm (OOP) and FORTRAN programs, *Computers in Physics,* 4(4), 385–394.

WAND, Y., and WEBER, R., 1989, An ontological evaluation of systems analysis and design methods, in *Information Systems Concepts: An In-depth Analysis* (eds. E.D. Falkenberg and P. Lindgren), Elsevier Science Publishers (North Holland), Amsterdam, 79–107.

WARD, P., 1989, How to integrate object orientation with structured analysis and design, *IEEE Software,* March, 74–82.

WARD, P.T., and MELLOR, S.J., 1985, *Structured Development for Real-Time Systems,* Yourdon Press, Englewood Cliffs, NJ, 156 pp.

WASSERMAN, A.I., PIRCHER, P.A., and MULLER, R.J., 1989. An object-oriented structured design method for code generation. *ACM SIGSOFT, Software Engineering Notes,* 14(1), 32–55.

WASSERMAN, A.I., PIRCHER, P.A., and MULLER, R.J., 1990. The object-oriented structured design notation for software design representation, *Computer,* March 1990, 50–63.

WEGNER, P., 1989, Learning the language, *Byte,* March 1989, 245–253.

WEGNER, P., 1990, Concepts and paradigms of object-oriented programming, *OOPS Messenger,* 1(1), 7–87.

WILSON, D.A., 1990, Class diagrams: a tool for design, documentation, and teaching, *J. Obj.-Oriented Programming,* 2(5), 38–44.

WINBLAD, A.L., EDWARDS, S.D., and KING, D.R., 1990. *Object-Oriented Software,* Addison-Wesley, Reading, MA, 291 pp.

WINSTON, A., 1990, Objective reality, *Unixworld,* April 1990, 72–75.

WIRFS-BROCK, A., and WILKERSON, B., 1989a, Variables limit reusability, *J. Obj.-Oriented Programming,* 2(1), 34–40.

WIRFS-BROCK, R.I., and WILKERSON, B., l989b, Object-oriented design: a responsibility-driven approach, *OOPSLA '89 Proceedings,* 71–75.

WIRFS-BROCK, R.I., and JOHNSON, R.E., 1990, A survey of current research in object-oriented design, *Comms. ACM,* 33(9), 104–124.

WIRFS-BROCK, R.I., WILKERSON, B., and WIENER, L., 1990, *Designing Object-Oriented Software,* Prentice Hall, New York, 341 pp.

WOODFIELD, S.N., 1990, Object-oriented software development, pp. 715–725 in *Procs. Int. Symp. on Water Quality Modeling of Agricultural Non-Point Sources, Utah, 1988,* U.S. Department of Agriculture, ARS-81, 881 pp.

WYBOLT, N., 1990, Experiences with C++ and object-oriented software development, *Procs. USENIX C++ Conference, 1–9.*

WYBOLT, N., 1991, Bootstrapping object-oriented CASE, *Hotline on Obj.-Oriented Technology,* 2(3), 13–15.

YAMAZAKI, S., KAJIHARA, K., HORI, M., and YASUHARA, R., 1990, Real time OOD object-oriented design techniques for communication control systems, *Procs. 3rd Intl. Conf., TOOLS3,* Sydney 1990 (eds. J. Bezivin, B. Meyer, J. Potter, and M. Tokoro), 199–206.

YOURDON, E., and CONSTANTINE, L.L., 1979, *Structured Design: Fundamentals of a Discipline of Computer Program and Systems Design,* Yourdon Press/ Prentice Hall, New York, 473 pp.

YOURDON, E., *Object-Oriented Systems Design: An Integrated Approach,* Prentice Hall, New York, 1994.

JOURNALS/MAGAZINES‡

Note: Individual OO articles appear monthly in almost every MIS, Software Engineering, and Computer Science scholarly and trade publication.

Communications of the ACM, Special issue on "Object-Oriented Experiences and Future Trends," Vol. 38, No. 10, October 1995, ACM, NY.

Journal of Object-Oriented Programming. Several issues per year. Both technical papers and regular columns are featured, together with other industry news. Regular columns on C++, OODBMS, Eiffel, etc. Occasional topic-focused special issues such as the July 1991 issue on OO Analysis & Design.

Hotline on Object-Oriented Technology. A monthly newsy publication aimed at management discussing managerial and technical issues within the framework of asking "how do we do it?" Includes lots of useful firsthand experiences with OO in commercial environments. Columns and occasional articles by many well-known names in OO.

Computer, Special issue on "Object-Oriented Technology," Vol. 28, No. 10, October 1995, IEEE Computer Society, NY.

Object-Oriented Strategies. A monthly newsletter for managers and developers of object-oriented systems. Published by Cutter Information Corporation, Editor: Paul Harmon, 617.648.8702 or 800.964.8702.

Object Magazine. Publication commenced May 1991. Published every month. Directed at software developers and endusers using, or intending to use, object technology on a daily basis. Technical and nontechnical issues are covered at a very readable level.

OOPS Messenger. Publication commenced October 1990. An ACM publication under the SIGPLAN committee. Aimed at rapid dissemination of OO ideas to the community.

The C++ Report, subtitled *The International Newsletter for C++ Programmers.* Feature articles, industry and product news, tutorials, etc.

Eiffel Outlook. Publication commenced April 1991. An independent magazine of interest to Eiffel users.

In addition, there have been special issues on OO in regular journals: for example, *Communications of the ACM,* September 1990, on OO design, September 1992 on Analysis Modeling; *IEEE Software,* May 1988; *Byte,* August 1981, on Smalltalk, also August 1986, March 1989, and October 1990, plus occasional papers in other issues of *Byte* such as, April 1990.

---

‡ As of October 1995

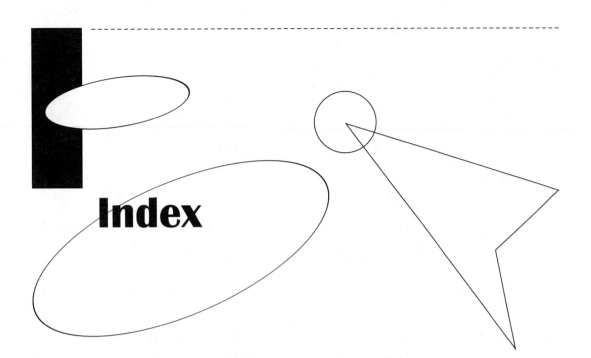

Index